Digital Cellular Radio

The Artech House Telecommunication Library

Preparing and Delivering Effective Technical Presentations by David L. Adamy

The Executive Guide to Video Teleconferencing by Ronald J. Bohm and Lee B. Templeton

The Telecommunications Deregulation Sourcebook, Stuart N. Brotman, ed.

Digital Cellular Radio by George Calhoun

E-Mail Stephen A. Caswell

The ITU in a Changing World by George A. Codding, Jr. and Anthony M. Rutkowski

Design and Prospects for the ISDN by G. DICENET

Communications by Manus Egan

Introduction to Satellite Communication by Bruce R. Elbert

Television Programming across National Boundaries: The EBU and OIRT Experience by Ernest Eugster

The Competition for Markets in International Telecommunications by Ronald S. Eward

A Bibliography of Telecommunications and Socio-Economic Development by Heather E. Hudson

New Directions in Satellite Communications: Challenges for North and South, Heather E. Hudson, ed.

Communication Satellites in the Geostationary Orbit by Donald M. Jansky and Michel C. Jeruchim

World Atlas of Satellites, Donald M. Jansky, ed.

Handbook of Satellite Telecommunications and Broadcasting, L. Ya. Kantor, ed.

World-Traded Services: The Challenge for the Eighties by Raymond J. Krommenacker

Telecommunications: An Interdisciplinary Text, Leonard Lewin, ed.

Telecommunications in the U.S.: Trends and Policies, Leonard Lewin, ed.

Introduction to Telecommunication Electronics by A.Michael Noll

Teleconferencing Technology and Applications by Christine H. Olgren and Lorne A. Parker

The ISDN Workshop: INTUG Proceedings, G. Russell Pipe, ed.

Integrated Services Digital Networks by Anthony M. Rutkowski

Writing and Managing Winning Technical Proposals by Timothy Whalen

The Law and Regulation of International Space Communication by Harold M. White, Jr. and Rita Lauria White

Digital Cellular Radio

GEORGE CALHOUN

Artech House

Library of Congress Cataloging-in-Publication Data

Calhoun, George, 1952-

Cellular mobile radio telephony/George Calhoun.

p. cm.

Includes bibliographies and index.

ISBN 0-89006-266-8

1. Cellular radio. I. Title.

TK6570.M6C35 1988

621.3845 dc19

88-11619

ARTECH HOUSE, INC.
685 Canton Street
Norwood, MA 02062

International Standard Book Number: 0-89006-266-8
Library of Congress Catalog Card Number: 88-11619

10 9 8

Contents

Preface

This book is divided into six parts. Part I is a brief essay on the current state of affairs in the mobile telephone industry, and on the confusion and excitement surrounding the "digital revolution."

Part II is a survey of the development of mobile-telephone systems in the United States, focusing on the difficulties currently being experienced by analog cellular radio — the problems which have convinced many that a new generation of technology must come in.

Part III is a presentation of some of the basic ideas and vocabulary of digital communication. Emphasis is given to the generic concepts and advantages of digital *transmission,* and to the emerging metaconcepts of the *integrated digital network.* Part III should be viewed as preparatory for the discussion of digital *radio* in Part V.

Part IV poses the basic design problems which confront the mobile communication architect. Particular attention is given to the challenge of designing a radio system to provide high-quality telephone circuits for users who are moving at high speeds through a radio-hostile urban environment. A second major consideration: how to design so as to manage interference between mobiles on the same channel and, thus, allow frequency reuse. Other design considerations include: designing for low cost, designing for compatibility with changing network characteristics, and designing for the competitive, nonmonopoly, marketplace.

Part V contains what some may consider the meat of this book: the analysis of the broad technological alternatives available to systems designers for the next generation of mobile systems.

Part VI addresses the other crucial question: How do we get there from here? How will the transition take place? Particular emphasis will be placed on the question of standards and the subsidiary issues of compatibility (backwards, forwards, and sideways) and openness to continuing technological progress.

I have not approached these questions with a narrow, particularistic solution or criterion in mind. My chief acknowledged bias is that I do not believe the current cellular technology is capable of righting itself. A few years ago this opinion would have been more controversial than it is today. Today the controversy begins when solutions are proposed.

I am virtually certain that the next generation will be digital. Again, this would have been more open to question a few years ago than it appears today. There are still, however, at least two nondigital proposals for the next generation — single sideband and narrowband FM — which have received some support. I have tried to review these fairly in Chapter 12, but I confess that I cannot actually regard them as viable alternatives for the long term.

With respect to other facets of the technological analysis, I admit to holding strong personal opinions on many of the issues addressed, and this may perhaps become evident in a few places. I believe above all else, however, that the choices for the next generation are still open; there is no definitive solution at hand.

I offer a few opinions on the regulator's dilemma — to standardize or not to standardize — in Part VI. It should be clear enough where I am stating my own opinions on this question, since they are not yet widely shared.

My focus is primarily on developments in the United States, particularly in the historical sections. The materials for a comparable historical and policy perspective on cellular developments in Japan or Europe are simply not readily available to me. I have tried to provide a brief summary of European developments in particular, as viewed from afar; I am sure that it will be found inadequate by any readers there. This deficiency will be felt most keenly in Part II and, to a lesser degree, in Part VI. It should be less of a problem in Parts III, IV, and V.

ACKNOWLEDGMENTS

I wish to acknowledge the support of many friends who helped in various ways with the development of this book. Ridge Bolgiano exercised a very patient tutorial in a wide range of technical matters. Of course, I bear sole responsibility for any technical errors. Bill Hilsman created the environment that permitted this work to germinate and to grow. Sherwin Seligsohn was, as ever, the "nourisher" of my creative efforts.

Among the many others who lent their assistance in different ways, I want to mention Steve Abramson, Hudson Barton, Chris Chacona, Paul Clements, Al Dayton, Charlie DiSanza, Jerry Fat, Maureen Ferrara, Mike Floyd, John Goetz, Gwen Kelly, Jud Kenney, Brian Kiernan, Gil Lavean, Richard Levine, Paul Levitsky, Dan Lin, Kathy Massey, Jim Mullen, Tim Plummer, Sid Rosenblatt, Dick Saunders, John Schaeffer, Jerry Schumacher, Kris Shelton, Herschel Shosteck, Craig Startt, and Jack Taylor. Of course, I must also thank Maria Menocal, who gracefully allowed this work to dominate a good part of her life as well as my own for many months.

Part I
INTRODUCTION

Chapter 1
THE UNCERTAIN FUTURE OF MOBILE TELEPHONY

1.1 MARCONI *versus* BELL

It is often said that we are on the verge of a "revolution" in mobile communications, a revolution that will ultimately liberate us as communication users from being tied down to a particular fixed location in the telephone network, and will provide us with an advanced voice and data communication capability in a highly portable package (one that is at least "briefcase-friendly," and preferably "body-friendly," in the current jargon), at a reasonable price. Today, for some, this mobile revolution is a grand vision about to become reality: like Gulliver on the beach, the whole network will soon awaken and brush away the wires that hold it fast, and rise to astound the Lilliputians (ourselves) with its far-reaching powers.

Despite a series of disappointing false starts, the communication world in the late 1980s *is* rapidly becoming more mobile for a much broader segment of communication users than ever before. Thirty years ago only a handful of people had ever used a mobile communication device of any kind. Today, the growth of Citizens Band Radio, one-way paging systems, "cordless" telephones, and especially the widespread use of radio systems in industry (at construction sites and transportation facilities, by security guards, in dispatch fleets, *et cetera*) have exposed millions of people to the technology of wireless communication and have influenced patterns of work and leisure on a wide scale.

The ability of this rapidly growing community of wireless communication users to interface to the existing *wireline* network, however, is still very limited. The "flagship" of the mobile services — interconnected mobile telephony — has experienced a rocky technical and regulatory history. For decades it remained a stunted niche business, for reasons that encompassed both technical and economic constraints, but which were rooted in the briar-patch of telephony politics and spectrum management (or mismanagement). Even the application of the newer "cellular" architecture, far from unsnarling these difficulties, has only amplified them. While equipment costs have come down, the economics of mobile telephone *service* remain out of reach for the vast majority of potential users. The technical difficulties of scaling up the mobile architecture for large urban networks continue to frustrate system engineers. Mobile telephony is hampered by obsolete technology standards that make it very difficult, perhaps impossible, to play into the emerging digital network services on the wireline side or to handle modern data-communication requirements. The lack of privacy in today's analog radio systems is not easily remedied (cosmetic legislation aside), and creates an imbalance between the service standards of the two networks. Finally, mobile telephony in the late 1980s has inherited a flawed regulatory process that has been designed and redesigned by a hundred interest groups, tweaked and critiqued by economists, lawyers, and engineers, scrapped and salvaged, challenged, upheld, and compromised in the courts (judicial and administrative) for more than twenty years, spanning the last years of the monolithic communication monopoly into the era of wide-open competitive markets. Regulatory goals have been put on the defensive, and regulatory mechanisms — in particular the licensing process — have devolved into a patchwork of partial solutions, from which it often seems that any semblance of policy rationale has been lost.

Today "cellular radio" stands in the spotlight, with its hopes illuminated and its deficiencies exposed, still pretending to a degree of technological permanence that once seemed more valid than it does today. We shall dwell at some length (in Part II) upon its shortcomings. The "mobile revolution" is, however, much larger than the current generation of cellular radio, and its immediate problems, although severe in some respects, should be recognized as developmental rather than fundamental. Consider a parallel case. Thirty years ago computers were bulky monstrosities: expensive, power-hungry, slow, finicky. Viewed from today's perspective, the basic technology was inadequate. Computers were too slow for many tasks, and too fragile for most environments. It would have been inconceivable to put such computers in a car, aboard an orbiting satellite, or on someone's desk. Yet, the breakthroughs came, and today observers would agree that computer applications are no longer restricted by technology — hardware capability and availability — as much as by the economics, architecture, and ingenuity of the software implementations. At least as far as conventional data-processing (nonreal-time) applications are concerned, the hardware is fast enough, cheap enough, and durable enough to go anywhere and to do almost anything that we are willing to pay programmers to develop and debug.

Mobile telephony today is in the same situation as the computer of the 1950s. Our goals for mobile communication — in terms of performance, cost, capacity, spectrum efficiency, portability — seem far beyond the reach of today's hardware. This, I believe, will prove to be a very temporary state of affairs. Imminent technical breakthroughs — some are already unfolding — will completely change our thinking about mobile telephony and transform our sense of the possible.

This book evaluates the problems of the particular technology of today's cellular mobile telephone systems in the light of a larger question: How and when will the two hemispheres of communication — the wireline network, with its hundreds of millions of users, and the emerging wireless systems of the future — grow together, *interconnect,* and become, as it were, *transparent* to one another?

Interconnection will spur the development of both systems, but especially that of the mobile hemisphere. Mobile communication has gone about as far as it can without interconnection. The greatest technical challenges today lie in adapting radio to the service standards and economic parameters of conventional telephony in a much more demanding transmission environment. At the same time, in mobile telephony advanced radio technologies are emerging that will cross-fertilize the conventional wireline telephone network to invigorate and expand basic (nonmobile) telephony beyond its current geographic and economic limits. The use of these technologies to effect *wireless access* to the existing network will in turn pull the evolution of mobile radio technology forward even faster. The problem of designing a better car phone will become subsidiary to the larger issue of wireline-wireless interconnection.

The greatest regulatory challenge lies not in the mechanics of licensing or spectrum allocation, but in making the conceptual leap from "mobile radio" (which has always connoted inherent limitations) to "mobile telephony" as a coequal arm of the telephone network. The radio-telephone stands at the intersection of two dissimilar and, until now, separate paths in the development of communication media. Our thinking about mobile communication often reflects the conflict between these two historical trajectories. This conflict permeates every tough issue in cellular radio regulation today: competition, service standards, privacy, cost, pricing, spectrum management.

To take one example, the radio tradition has inherited a red-in-tooth-and-claw competitive instinct, traceable back to De Forest and Marconi, Armstrong and Sarnoff, which regulatory policies reflect in their relatively unflinching attitude toward market forces. This perspective reaches full flower in the regulation of broadcast franchises, where much of the non-technical regulation is aimed at *curbing* the growth of the monopoly powers inherent in large networks. Local competition is unbridled, and completely free of public utility thinking. Pricing and most other economic decisions are completely within the prerogative of the franchisee, without review by state or federal authorities. Telephony, on the other hand, has always viewed competition with a jaundiced eye. For more than seventy years the telephone industry has been administered as a public utility, with state-regulated pricing and monopoly franchises (this has so far changed very little, even under "deregulation"). Most telephone managers have sincere difficulty in even conceiving of unrestricted competition and generally view the procompetitive steps of the past ten years as misguided. These divergent traditions clash head-on in the arena of mobile telephony. No policy has been more at war with itself than the attempt by the federal authorities simultaneously to foster and to contain competition in the field of mobile communication.

Another example: telephony has committed itself over many decades to a very high service standard, indeed, to the elusive and expensive ideal of *universal* service, based on average pricing and cost-sharing. Access to low-cost telephone service is virtually viewed as a citizen's right. This has produced a technological and operational conservatism based on the "philosophy of the 99-th percentile." Consider the immense resources that must be held in reserve to ensure that a telephone subscriber receives immediate access to a circuit ("dial tone") 99 out of every 100 times that he or she picks up a phone during the busiest hour of the day (the operational definition of a 1% blocking standard). For most subscribers it means (among other things) that there must be a completely dedicated local wire circuit connecting with the nearest switch — a costly facility that sits idle most of the time. The world of radio, on the other hand, has always been a world of catch-as-catch-can. Most channels are shared; blocking is often extremely common. Interference is normal. Even in the broadcasting business, there is no guarantee of service. If you want to live in the country, far from the transmitter towers, you should not expect the quality of TV reception your urban counterparts enjoy. It is up to you to spend the extra money for a larger antenna, or, perhaps, a satellite dish. Service standards in the radio world are dictated by market forces — "you get what you pay for." CB radio, cordless phones, and mobile telephony have all experienced severe degradations in service as the regulators in effect allowed demand to outstrip the allocated spectrum. "Poor reception" was, and often is, accepted as a fact of life by many users of radio. The idea that a customer has a "right" to a uniformly high quality of radio service is new and difficult to interpret.

Shall we view mobile telephony as the child of Guglielmo Marconi, or of Alexander Graham Bell? When "old radio assumptions" clash with "old telephone assumptions" it is very difficult to define a coherent regulatory program. This book delineates the conflicts between the Way of Radio and the Way of Telephony, and points toward the emerging synthesis of the two.

1.2 THE UNIQUE FAILURE OF MOBILE TELEPHONY

The telephone was introduced to the public in 1876 at the Centennial Exposition of the United States in Philadelphia. Alexander Graham Bell was able to transmit speech electrically, in one direction only, over a copper wire circuit of several hundred feet in length. Although not everyone present that day could immediately perceive its commercial value, the "speaking telegraph" was quickly perfected for adequate two-way communication and was offered for business and residential service the following year. Within a short time there were thousands, then tens of thousands, and soon hundreds of thousands of paying customers.

Forty years later, on the brink of America's entry into the first World War, the penetration rate in the United States had reached about 20 telephones for every 100 people, more than 15 million subscribers. Even such a modest achievement (two-thirds of American homes were still without telephones, and the vast majority of telephone subscribers were on party-line service) constituted an enormous engineering and financial undertaking. The deployment of Alexander Graham Bell's relatively simple invention was paced, and constrained, by the massive programs to construct and maintain a nationwide network of transmission facilities, millions of miles of poles, wires, cables — the "outside plant," as it came to be called — to link each user physically with the switchboards in the nearby "central office." Because of this expense, telephony was at first an overwhelmingly urban phenomenon; where population densities were high, many customers could be loaded onto each expensive mile of cable. Yet soon rural areas began to catch up, mainly through the ebullient efforts of the so-called independent telephone companies which sprang up after the Bell patents expired. By World War I, the number of telephones per 100 people was higher in Iowa and Kansas than in New York. Growth in rural areas and in the suburbs of the older cities meant longer distances to cover, more wire and more poles to reach each new subscriber. Beginning in the 1920s, the number of miles of new construction per subscriber began to rise, almost tripling in that prosperous decade. Costs began to climb; the network's appetite for copper became gargantuan. Scientists began to study ways of utilizing the copper plant more efficiently. They developed methods for transmitting two conversations over the same wire circuit, then four conversations, then dozens. A new impetus was given to the study of basic electrical phenomena, which led to further advances in physics, materials science, mathematics, and, ultimately, a clutch of Nobel prizes and the creation of the largest privately funded research effort in history. The story of the evolution of the wire network is in large part the story of a long struggle against the burden and expense of the physical wire plant. The scientists of the Bell system were led deeper and deeper into the study of electrical transmission, to understand in detail the characteristics of wire media, and to search for ways to utilize this costly transmission facility more and more efficiently.

Toward the end of the nineteenth century, while this struggle was only beginning, a young German scientist named Heinrich Rudolf Hertz discovered a strange and wonderful phenomenon: from an electric spark of sufficient intensity there seemed to emanate invisible waves of force which could be captured at a distant location by a suitably constructed receiving device. It seemed to be a realization of the ancient concept of "action at a distance." Classical physicists found this philosophically disturbing and postulated the existence of some impalpable intervening medium — the Ether — which actually transmitted these strange waves. Hertz's own experiments extended only over a few yards. A few years later, Guglielmo Marconi transmitted these waves over several kilometers, and began to call it Radio.

The early telephone engineers, caught up in the heroic and unrelenting struggle with copper wire physics and economics, looked upon the new phenomenon with awe. "It has been shown," wrote one John J. Carty in 1891, "that longer waves may be generated which are capable of electrical action, and which can be propagated through the densest fog and even through a stone wall with just as much ease as through the clearest atmosphere [1]." (Carty at the time was a young technician; he later became the head of engineering for AT&T and is generally regarded as the founder of the scientific tradition that led to the creation of the Bell Laboratories.)

Communication devices which exploited these "Hertzian" waves experienced a much slower technological gestation but a more rapid market development than the telephone. Radio broadcasting was introduced commercially in the United States in 1921. Within ten years, more than 50% of all American households boasted a radio set. Within twenty years, it was over 90%. The growth of television — another technology based on radio transmission — was even more rapid. Once it became readily available to the American public in 1946 it took only nine years to reach the

50% level and only fourteen years to reach 90%. Such inventions, inherently far more complex and costly than the telephone instrument, were able to spread so much more rapidly because they did not require a massive investment in wireline infrastructure.

John J. Carty, however, writing in 1891, had not foreseen either radio or TV broadcasting as we know it today. For him, the promise of Hertz's discoveries lay in another direction (see [1], p. 51): "A system of telephony without wires seems one of the interesting possibilities . . . The ether will transmit speech."

It was a bold guess. At first, only gross pulses of energy could be transmitted, and communication was limited to Morse code. As radio techniques were refined, it became possible to transmit speech and music (first accomplished in 1905 by Reginald Fessenden). Amplifiers boosted weak voice signals to traverse great distances. By the end of World War I, only the seemingly impenetrable radio patent tangle stood in the way of commercial development of the industry. The federal government, perceiving the military value of radio communications, cut through the Gordian knot by fostering the Radio Corporation of America, which was partly owned at the beginning by the telephone interests (AT&T). Radio came to America.

The possibility for utilizing radio devices for communicating with moving vehicles was quickly appreciated. The earliest commercial application of radio had been for communication with ships at sea. As early as 1921, the Detroit Police Department was conducting experiments with "mobile" radio — actually at first only a one-way broadcast which instructed police cars to respond by stopping to place a call back to the dispatcher from a pay phone. A few years later, two-way mobile systems were created. Throughout the 1930s, experience with mobile communication accumulated. It was World War II, however, and the sudden and pressing need for two-way mobile communication on a large scale, that gave the real impetus to mobile radio technology. Wireless communication revolutionized the battlefield, permitting the deployment of large armies, moving rapidly over great areas, with coordinated armor, infantry, and air support. It is impossible to imagine any of the characteristic tactical operations of that war functioning effectively without radio.

At the end of the war, the first licenses were granted for the provision of true mobile telephone service — in other words, to allow a user calling by radio from a moving vehicle to be interconnected into the public telephone network. The car phone was conceived. The war had brought to prominence a new type of radio technology — frequency modulated radio or FM — which permitted superior mobile voice communication. Interfaces to telephone switches had been established. Americans were buying automobiles in record numbers. Prosperity had returned. Along with television, mobile radio seemed poised for a postwar boom.

And then something went wrong. From the promise of those early commercial systems in the late 1940s, the actual deployment of mobile telephone systems proved painfully slow. More than forty years later, even after the deployment of "modern" cellular radio systems, the actual development of the market for mobile telephony was abysmal. In metropolitan Los Angeles — one of the densest traffic centers with more automobiles per capita and more daily "driving minutes" than any other large city, ideally suited for vehicular telephony — the penetration in the mid-1980s is considerably less than 1%! Moreover, based on current technology the available spectrum in Los Angeles is loaded to near capacity. Even to double this penetration will apparently involve very substantial technical and economic challenges.

Mobile telephony has undoubtedly set the record for the slowest penetration by any new technology to the mass marketplace. (See Figure 1.1.)

Cost is not the whole explanation. The cost of ordinary wireline telephony in the early years of the twentieth century was, in relative terms, *much* higher than the cost of mobile service today. (In 1900 a year's telephone service in New York City cost about $150.00, about 25% of the average annual household income; the full monthly cost of a cellular telephone today, about $155.00 [2] or about $1,860.00 per year, constitutes around 8% of the average household income.) Moreover, wireline telephony was burdened with the expense of construction of the physical wire network, as well as the ongoing maintenance of that network in the face of constant environmental assault.

Regulation is not the whole explanation. Mobile telephony has been highly regulated, undoubtedly overregulated. But the regulatory obstacles and disputes over technical standards that faced television in its early days would make frightening reading even for the most battle-scarred veterans of the cellular licensing wars. And conventional wireline telephony for the past seventy years has been *much* more regulated than mobile telephony, especially on rates and other economic matters.

Spectrum shortage is no explanation at all, but a symptom. It is true that for many years the allocations were inadequate in a few large cities. (There was never any shortage in smaller cities, however, and the available spectrum was virtually unused in rural areas.) In the 1980s, the spectrum available for mobile telephony was vastly increased. Architectural innovations under the heading of "cellular" radio were to have further increased the carrying capacity of mobile systems. Yet *within one year* after

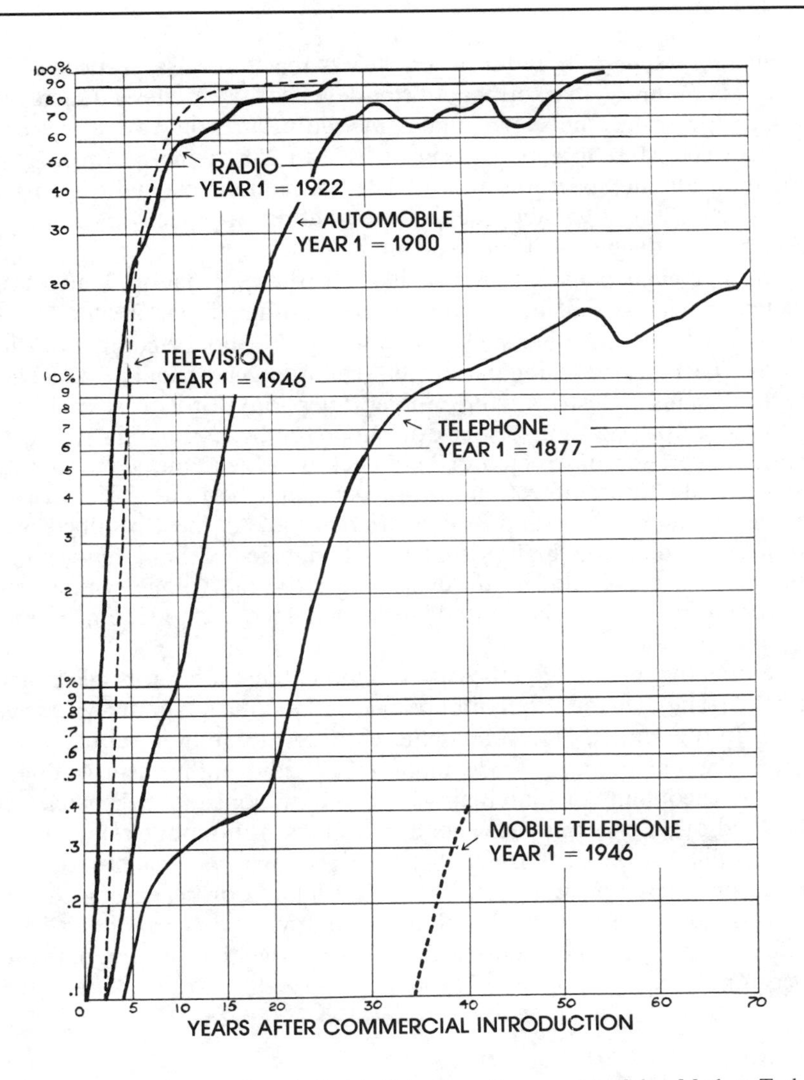

Figure 1.1 Market Penetration of Mobile Telephony Compared to Other Modern Technologies.

the start of commercial service, cellular operators in Chicago and other large cities were already projecting renewed spectrum shortages. The Federal Communications Commission (FCC) released additional spectrum, bringing the total allocation to more than 50 MHz (compared to approximately 2 MHz prior to the cellular allocations). Capacity projections from large metropolitan cellular operators in the late 1980s forecast the need for still more spectrum within a few years, reaching a saturation of the current spectrum at market penetration levels of 2% or less. At current levels of spectrum efficiency, cellular mobile radio would require something like 600 MHz to support, say, a 20% to 25% penetration level — more than all the spectrum allocated for all commercial television and radio broadcasting!*To label such a state of affairs a "spectrum shortage" is clearly a misdiagnosis of the problem.

Insufficient demand is most decidedly *not* the explanation. If anything, the indications have all been strongly in the other direction, tending to show a very large "pent-up" demand for affordable mobile communication services. Technical and regulatory uncertainties have clouded the early growth of many industries: conventional telephony, broadcast radio and TV, one-way paging, not to speak of phonographs, computers, video recorders, have been obstructed or held back by regulation, standards disputes, manufacturers' uncertainties, or consumer hesitancy. Yet in each case the market mechanism prevailed after a time. All these products were introduced at relatively high prices, but all managed within a few years to "ride the cost curves down" to levels where the marvelous symbiosis of high volume manufacturing and low pricing helped create true mass markets.

Somehow, uniquely in the case of mobile telephony, this mechanism has failed. The demand for noninterconnected mobile communications has surged. In the United States today there are several million subscribers of paging services, perhaps 20–30 million CB radio enthusiasts willing to brave the cacophony of shared airwaves, and millions more dispatch fleet radios and other two-way radio net applications. But mobile telephony — the essential "corpus callosum" between the two hemispheres of communication — has not kept pace. The realization is growing among industry observers that interconnected mobile telephony, *in its current configuration,* cannot become the mainstream mobile service that its designers once hoped for.

1.3 THE CELLULAR "DISASTER"

That current configuration — analog cellular radio — was conceived in the 1940s, planned in the 1960s, and launched in the 1980s. It was developed in response to the chronic problem of congestion and poor service in the mobile telephone business. Its architecture was designed

*Recognizing the realities of cell-splitting (see Chapter 4) and assuming a telephone grade of service and telephone traffic standards, e.g., 0.1 Erlangs per customer and no worse than a 2% blocking.

primarily by the experienced telephone engineers of the Bell system based on a powerful system concept that promised to solve all future spectrum shortages. At the core of the design is a kind of "breeder reactor" for the radio spectrum capable (in theory) of generating an almost limitless number of channels from a finite initial frequency allocation. At last (it was thought) mobile telephony would have the electromagnetic real estate to develop into a mainstream communication service. Systems were planned, factories readied, market forecasts analyzed, investors enticed. As it gathered momentum through the 1970s and reached the market in the 1980s, cellular radio was widely hailed as the mobile millenium. At the very least (it was thought), it would be the technological standard for decades to come.

In 1983, the first American commercial cellular system was turned on in Chicago.

By 1984, the Chicago system was already saturated in some of its cells [3]. A petition was filed with the FCC for additional spectrum relief. Most major urban cellular operators subsequently joined in the request. An additional 10 MHz of spectrum was released by the FCC. The head of one of the largest wireline cellular operators applauded the move as "a positive step," but added: "We still need more Megahertz [4]."

In 1985, the former head of Cellular Development for Motorola — widely acknowledged as one of the "fathers of cellular radio" — predicted that cellular "will go down in telecommunications history as one of the costliest engineering and marketing disasters ever [5]."

In 1986, AT&T terminated its efforts to market mobile cellular units [6]. Motorola's estimated market share for cellular mobile radios fell to about 8% (at least, according to some surveys). It was projected that of the then-current cellular mobile equipment suppliers "approximately ½ . . . likely are not economically viable at current rates of sales [7]."

By 1987, many of the largest and best-run cellular operators — precisely those who had led the initial charge into the cellular market — had abandoned the field entirely. Among the major companies selling out their substantial cellular holdings were Metromedia, MCI, and Graphic Scanning. Cellular franchises for many markets were being awarded to individuals — housewives, dentists, truck drivers, disabled senior citizens, and others — who were patently unprepared to build or operate mobile telephone systems. The FCC was inundated with mass-produced, prefabricated applications; charges of fraud clouded the licensing process.

By 1987, the rate of growth in the subscriber base was slowing down; revenue per customer was shrinking. Large numbers of customers were "disappearing" from the system. In short, the mobile millenium did not arrive. American cellular markets were in deep disarray by the late 1980s.

For the consumer, the equipment prices had moderated somewhat due to inventory sell-offs and other "loss leader" tactics on the part of sometimes desperate manufacturers and operators. Huge "bounties" paid by operators to sign up new customers allowed some distributors to retail cellular phones well below cost. Yet prices generally remained well above the $500.00 level regarded as crucial by mass merchandisers. Service prices remained high. Average monthly usage charges ran over $100.00 (not including toll charges, roaming charges, or the cost of the mobile equipment, installation, or insurance) — for traffic levels corresponding to less than 10 minutes of talktime per working day for the average user. At such high levels, the market penetration remained quite limited.

For cellular operators, the slow market development and persistently high costs began to stretch their business plans to the breaking point. Projected paybacks were from three to five years in major markets, stretching to seven years, to ten years, or longer in smaller markets and markets with intense competition. Customer service costs were reported by many operators to be on the rise. The marketing and selling costs of attracting and signing up each new subscriber were estimated at from one thousand to two thousand dollars — a very substantial and largely unanticipated start-up cost. Billing and collection problems have plagued many operators. According to one major cellular operator, more than 25% of an operator's subscriber base can be expected to cancel service every year. Another 10% or so will remain on the rolls, but cease using their cellular phones. Subscriber numbers must apparently be deeply discounted. Even more disturbing was the trend in the late 1980s for revenue per subscriber to *decline* — as much as 1% to 2% per subscriber per month in some cases [8]! As of this writing, it is doubtful that more than a handful of cellular operators are profitable. For equipment vendors, the reduced market was severely overcrowded with competing suppliers, leading to razor-thin profit margins on the mobile units. There were patent infringement suits and acrimonious charges of dumping against Japanese suppliers. A number of system suppliers withdrew from the business. The development costs of most suppliers have probably not been recouped.

For regulators, the unlucky attempt to "create" competition through the unlikely vehicle of a duopoly — only two licensed operators are permitted for each cellular market, one from the ranks of the wireline telephone companies and one from outside — has boomeranged. Many of the strongest potential operators have been driven out, reducing the likelihood of viable competition in many markets. At the start, cellular operators were to be scrutinized to ensure their ability to build and operate the proposed systems effectively. Awards would be decided on merit. By 1987, the Commission had abandoned any effort at rational assessment of the capabilities of prospective operators or their construction plans. The field has been openly ceded to speculators whose practices the FCC has felt a repeated need to caution against (it can do little else within the current framework).

The "disaster" is still shrouded in contradictory optimism. Industry spokesmen acknowledge the most alarming business or technical trends and in the same breath reaffirm the "tremendous potential" of the cellular business. Investors still flock to stock offerings of cellular companies, speculators still bid up the price of cellular franchise "lottery tickets," and the newspapers still tout the marvel of technology — "a phone in your car — imagine!" Like El Cid, propped in the saddle, festooned with battlegear, accompanied by trumpets and thunder, cellular radio can still inspire its adherents mightily. Meanwhile, the technologists are beginning to ponder how best to dispose of the corpse.

Mobile telephony *will* break out of today's constraints, but *not* with today's technology. The story of analog cellular radio will be written in vivid hindsight as one of the classic technological miscues of modern history, on a par with, say, the Zeppelin airship. The trend it represents is real, but the instrumentality is fatally flawed.

Mobile telephony will also be remembered as a regulatory fiasco. Indeed, the tendency of those in the business world will be to blame regulatory obstacles almost exclusively. But the regulators did not create the problem singlehandedly. Radio, television, indeed almost all communication services, and especially those involving the use of the radio spectrum, have been hampered to some degree by contradictory regulatory policies, and yet they have prospered.

What went wrong? What will it take to set it right? The questions are not academic. A new generation of mobile telephone systems is on the drawing boards. Many of the same issues will be confronted again. Yet what *are* the lessons of the analog cellular experience?

Analysis is difficult; many of the participants are still emotionally (and in some cases financially) committed to technical positions staked out over the past ten or fifteen years. There is no historical distance to provide for cool objectivity. As long as there are investments to be recovered, and careers staked on their recovery, there will be strong attachments to the current architectures, the current technologies, the current business arrangements, the current solutions. No consensus has emerged as to the positive lessons to be extracted from the experience. Some negative conclusions, however, can be stated.

Almost every industry proposal still begins with an appeal for "more spectrum." Yet additional spectrum alone cannot solve the problem. The current analog systems have too large an appetite for bandwidth.

Nor will manufacturing "economies of scale" leading to cheaper radios — the form of wishful thinking most prevalent among the industry operators — solve the problem alone. The chief factors limiting the growth of mobile telephony today, lack of system capacity and high charges for mobile service, have very little to do with the cost of the mobile unit.

"Competition" — the totem of the regulatory community — will not solve the problem alone. Nor will the suppression of competition. Both have been tried.

Broadly speaking, the solution will come in the form of "new technology," which will become the basis for the next generation of mobile telephone systems. Radio-communication techniques have progressed far since the standards for analog cellular radio were frozen. Some of the technical answers are at hand; others will emerge rapidly in the next few years. There is reason for considerable optimism on that score. Mobile telephony will have a second chance.

But the availability of vastly superior technical solutions will highlight the need for a vastly superior planning process, both among the industrial players who will develop, manufacture, and operate the new systems, and among the regulators who will develop the new rules. The failure of mobile telephony involves the interaction of industry decisions, regulatory policies, technological assumptions, technological realities. Many of the same pitfalls still gape at our feet.

There are new complexities. Will the next generation systems be developed separately from today's cellular systems, much as cellular systems were brought forth as a "clean slate" approach with no attempt to integrate them with previous mobile telephone systems? Or will new technologies be grafted somehow onto existing cellular systems?

The analog cellular process was driven largely (though not exclusively) by decisions made in the United States. American technical dicta were exported, along with some aspects of American regulatory philosophy. The next round will feature, I believe, three much more coequal centers of technological and regulatory influence: the United States position will be joined by a more or less united European position and by a very important Japanese position. In fact, the Europeans are already well ahead of the United States in thinking through the "next generation" questions, and it is quite possible that traditional American leadership will be ceded in this round. The real question, therefore, is: Will history repeat itself? Have we learned anything?

1.4 THE "DIGITAL" PANACEA

The next generation of cellular technology will almost certainly involve "digital communication" techniques. Over the past twenty-five years, digital systems have penetrated every segment of the network. Digital switches are revolutionizing telephone central-office and network control functions. Digital transmission systems — notably the so-called T-Carrier systems — are now installed on more than half of all interexchange telephone trunks. The use of digital microwave is increasing. In the local distribution segment (the local loop from the central office to the telephone subscriber's residence), hundreds of thousands of "digital" subscriber loops are being installed every year. "Fiber optics" — another tech buzzword for the 1980s and 1990s — utilizes digital technology. Digital techniques are invading other industries. Digital audio systems are rapidly displacing the standard analog media (LP records and analog tape cassettes). Digital television is on the horizon. Digital photography, digital x rays, are now being developed. Also, of course, the pervasiveness of the computer has transformed the workplace and brought digital electronics into millions of homes.

The digital evolution of the telephone network infrastructure is a fundamental, permanent trend that will transform the communication industry in ways that can only be partially foretold today. A vast, encompassing vision has begun to emerge, of an integrated digital network, piping bitstreams around with unheard-of flexibility and finesse. The boundaries between video and voice and data communication will continue to blur, as it becomes possible to handle all types of signals in a uniform digital format of 1s and 0s. Intelligence will permeate the network of the future.

Skeptics abound, of course. Whether digital techniques will usher in a qualitatively different kind of communication network, or whether they will "only" provide a much more efficient vehicle for current services, is hotly debated. But the growth of digital techniques is a pervasive trend.

Against this backdrop, cellular radio in its current form will almost certainly be the last major analog communication system ever deployed. Also, though it may be hard to accept for those who are committed to the development of today's systems, analog cellular radio finds itself in the late 1980s in a situation similar to that of the LP record industry a few years back, faced with the onset of digital compact laser disk media. (CD market share has rocketed from almost nothing in 1980 to overtake analog

LPs by 1988.) Communication engineers are beginning to talk about something called "digital cellular" in terms which suggest a similar upheaval of current industry patterns in mobile telephony.

Yet as one industry insider said in a recent interview, "Digital radio is the unknown, and it is a mysterious unknown [9]." The word "digital" carries an elusive magic that many in the industry find hard to grasp. What does "digital cellular" mean? Is it really a panacea for the industry's problems? How will it alleviate the economic difficulties of mobile telephone service? How will it address the capacity problems, the "spectrum shortage"? How will it solve the privacy problem? The data communication problem? A great deal of education will have to take place before the mobile communication community is ready to cope with the issues raised by the digital trend.

Beyond the specific questions, the advent of "digital cellular" will necessitate a fundamental change in mindset, in vocabulary, in analytical frameworks. Most of the digital systems that have been deployed so far are in the wireline segment of the network, and in the central office. The telephone world still has relatively little experience with digital radio. There are concerns about interference, about frequency reuse. The promise of tremendous economic advances has been held out. Cost trends in the allied digital field of data processing are startling. The costs of digital processors and memory devices are falling so rapidly that designers and users have grown accustomed to a sort of built-in future shock: today's technology is almost guaranteed to be technically obsolete by the time production systems reach the market, and functionally and economically obsolete (defined by the capabilities available to the user's competitors) within a very short time after that. Hardware costs, in particular, have almost ceased to have any importance at all in many types of applications. In place of steady incremental improvements, we have come to expect, and even demand, quantum leaps forward. It is an exciting time for the user, and a disturbing time for the manufacturer. The good old days of relatively long periods of technological stability (such as the forty-year reign of the LP record) during which manufacturing processes could be gradually refined and economics of scale fully exploited, have all but vanished for the foreseeable future. And even for the user, the risk of rapid technological obsolescence has complicated the planning process tremendously.

It is also clear that there is not one "digital," but many. "Digital cellular" is not a single step up from analog FM to some new monolithic standard. There will be many formats, and new formats every year. There will be many quantum leaps. This poses a serious challenge to the industry and to the regulatory community. It calls into question the feasibility of attempting to secure a single, permanent nationwide or worldwide standard for "digital cellular." Imagine trying to gain agreement on a single standard for microcomputers, to implement that standard by government fiat, and to freeze it for decades. Impossible? Inconceivable? Yet that is exactly what "digital cellular" will entail: the conversion of the mobile telephone into a kind of specialized, extremely powerful microcomputer. On the other hand, how can a communication network develop without standards? The decisions about "digital cellular" will be in many ways far more difficult than the choices that faced the industry in the previous generation, which carried out its design effort during the long technological stasis I refer to as the Age of FM.

REFERENCES

[1] F. L. Rhodes, *John J. Carty: An Appreciation,* New York: Privately Printed, 1932, p. 52.

[2] Herschel Shosteck, "Can Cellular Be Sold?," *Telephone Engineer & Management,* July 15, 1987.

[3] *Mobile Phone News,* November 14, 1984, p. 6.

[4] *Mobile Phone News,* July 31, 1986, p. 1.

[5] Martin Cooper, "Cellular Does Work — If the System is Designed Correctly," *Personal Communications,* June 1985, p. 41.

[6] "AT&T Gets out of Retail Cellular," *Industrial Communications,* September 26, 1986, pp. 7–8.

[7] Herschel Shosteck, *Cellular Subscribers & Brand Sales,* Vol. I, No. 3, published by Herschel Shosteck Associates, September 1986, cited in *Industrial Communications,* October 17, 1986, pp. 6–8.

[8] Remarks of John DeFeo, President of NewVector (US West's cellular subsidiary), at a seminar on cellular radio sponsored by Donaldson, Lufkin, Jenrette, New York, June 3, 1987.

[9] Cited by Steve Titch, "Cellular Plight will Last until the 1990's," *Communications Week,* February 2, 1987, p. 30.

Part II
THE AGE OF FM

Cellular telephony today is the final flower of frequency-modulated analog radio. When it was first developed, FM was regarded as a technical marvel. FM was so superior to existing radio that early demonstrations sometimes engendered open disbelief that any wireless system could perform so well. It superseded all other techniques in almost every application. Without FM, mobile telephony would have remained a primitive experiment. With FM, we have reached the beginnings of the wireless network of the future. The problems of cellular radio today, however, are largely the inherent limitations of a technology that has been refined and matured over five decades, and may not have much more to give.

Chapter 2
MOBILE RADIO BEFORE CELLULAR: 1921–1968

The forty-seven years between the first experiments in communicating by radio with moving vehicles and the opening of the rule-making by the FCC that would lead to cellular radio can be divided into two roughly equal periods, marked off by World War II. During the first period, mobile radio technology was still quite crude, dominated by the mechanical problem of constructing a radio system that could survive the bumps and bounces of a moving vehicle. The initial experiments were based on conventional amplitude modulation radio. There was no commercial service, although in the late 1930s mobile radio began to establish its first real application: police dispatch. The invention of frequency modulation put mobile radio on a sound technological footing, and the war brought about tremendous improvements in design. Commercial mobile telephone service commenced almost immediately after the war's end. The 1950s and 1960s saw a steady refinement of the radio and interconnection techniques that would become the basis for the cellular proposal in 1968.

2.1 THE PIONEER PHASE: 1921–1945

The first period of mobile radio is marked by three themes:

1. The dominance of military or paramilitary (police) uses;
2. The technical challenge of building a radio transmitter capable of operating within the size and power constraints of a moving automobile; and
3. The crucial technology breakthrough: radio transmission by means of frequency modulation (FM).

Ever since the American Civil War — which brought the first large-scale military use of electrical communication — the mismatch between conventional wireline techniques and wartime communication needs has been recognized. Wireline telegraphy and, later, telephony were suited for permanent installations where the slow deployment was not such a handicap and the high costs could be amortized over long periods. A military force in the field, however, found itself engaged quite often in rapid, mobile campaigns with temporary encampments. Wire was exquisitely vulnerable to hostile actions; a single cut could disable a line hundreds of miles long, and it was easy to tap the enemy's telegraph. Of course, in naval communication telegraph lines were out of the question; communication with ships at sea at the end of the nineteenth century was little better than it had been at the time of Christopher Columbus. This spurred the continuing interest of the military in various types of "wireless" communication. Such ancient techniques as semaphore signaling and heliography remained indispensable well after the telephone had been perfected and were widely used as late as World War I. But such systems could only transmit digital code; the great need at the turn of the century was for some means of transmitting the voice over great distances without wires.

In fact, the first wireless transmission of the human voice did not use radio, but light — actually reflected sunlight — coupled to a photoelectric selenium receiver. Alexander Graham Bell designed the "photophone" (he later called it the "radiophone") in 1880 and was able to transmit intelligible speech over distances of up to 700 feet — 25 years before this was achieved with radio. Bell considered the photophone his greatest invention, more revolutionary than the telephone. He grasped immediately its military significance. "In warfare the electric communications of an army could neither be cut or tapped. On the ocean communication may be carried on between vessels [1]." In the following years the photophone was refined by AT&T, using artificial light sources, to extend the transmission distance to several miles. Around the turn of the century, the German Naval Command commissioned Professor Ernst Ruhmer of the University of Berlin in a series of experiments to demonstrate the viability of speech transmission by lightwaves to warships. Ruhmer achieved transmissions of more than 15 kilometers (close to the limits imposed by the curvature of the earth). The Germans retained an interest in photophony; during World War II they made limited use of several advanced devices for speech transmission by lightwaves [2].

The photophone was soon superseded, however, by the development of radio, which, although it could not transmit speech effectively at first, could achieve much greater range for wireless telegraphy and could penetrate rain, fog, foliage, and even most buildings where lightbeams could not go. The primary initial application was for naval communication. Marconi, De Forest, and many early radio entrepreneurs built their businesses on navy contracts. At first, commercial shippers and passenger vessels were slow to take up the new technology. The sinking of the *Titanic* in 1912, however, highlighted the importance of wireless communication on the seaways, and marine radio telegraphy became widespread.

In 1905 some of the early wireless operators along the eastern seaboard, straining to pick the dots and dashes out of the background of static, were startled to hear Fessenden's first transmissions of voice and music. Practical speech communication by radio, however, took another decade to perfect. In 1915 and 1916, the United States War Department, anticipating involvement in the European conflict, began a program to develop ship-to-ship and ship-to-shore voice communication. Although the results were too late to have a great impact on the outcome of the war, ship-to-shore radiotelephone service was initiated experimentally in 1919 for ships along the eastern seaboard. By 1929, commercial radiotelephone service was begun to ships on the Atlantic. It was expensive, but the affluent clientele of the transatlantic ocean liners was ready to pay for it. The radio gear was extremely bulky and power-hungry, but this posed little problem aboard a large vessel [3].

Downsizing and ruggedizing the radio for a land vehicle was quite a challenge. The persevering visionary of this period was William P. Rutledge, the Commissioner of the Detroit Police Department. In the early 1920s the Detroit police under Rutledge's leadership carried out pioneering experiments with broadcast radio messages to receivers in police cars. It was one-way only; the patrolmen had to stop at a wireline telephone station to call back in, similar to today's paging systems. Daniel Noble recounts the frustrations of their work:

> For a period of six years, Commissioner Rutledge and his organization tried to develop a practical system to provide satisfactory voice communication to moving cars. Their point of failure was the receivers. They could not build receivers which would work reliably in the police cars. Both voice and radio telegraph were tried, but the basic problem of receiver instability and lack of sensitivity limited the coverage. With each new year they found new approaches, but all were failures. The accumulation of frustration was so strong that, in 1927, the station was shut down and the radio room was locked up [4].

Fortunately, the following year a young Purdue University student named Robert L. Batts was able to develop a receiver based on superheterodyne design that would withstand the buffeting of mobile usage. Rutledge had to manage the construction of the radio equipment by his own department; there was no radio manufacturer ready to undertake the task. The first operational mobile radio system (still one-way only) went on the air on April 7, 1928, and its effectiveness was immediate and dramatic. Noble writes:

> The Detroit Police Radio System drew world-wide publicity, and visitors arrived from all over the world to inspect the system. Other city police departments planned radio systems, and like Detroit, they were forced to build their own receivers. In September 1929 the Cleveland Police Department was the second system to go on the air with a few cars. . . . Equivalent systems were established in many cities [5].

Early radio broadcasting stations often cooperated with police to carry bulletins to police cars in the field, interspersed with ordinary entertainment and commercial programming, and "Calling all cars!" became a familiar phrase to radio enthusiasts everywhere.

In the early 1930s, mobile transmitters were developed, and the first two-way mobile system was placed in operation by the Bayonne, New Jersey, police department [6]. The problems of operating a very powerful transmitter in close proximity to a very sensitive receiver, however, limited the first systems to half-duplex, or "push-to-talk" transmission. The bulky radio equipment occupied most of the trunk of the typical vehicle.

During this period, mobile operators first began to plumb the special mysteries of mobile radio propagation. "Fading," "skipping," "shadows," and "picket-fence effects" were discovered. In general, all of the propagation disturbances that afflicted ordinary radio with fixed receivers were also experienced by mobile operators, but all the processes seemed to be tremendously speeded up. Experimenters recognized that in part this had to do with the movement of the receiver through the environment and the everchanging nature of the exact transmission path. Moving behind a building or a truck or under a bridge or tunnel could drastically alter the received signal. Unlike the ordinary radio listener who could manipulate his antenna to improve his signal, the mobile user was subject to wildly fluctuating channel conditions. Performance was not always completely satisfactory. Nevertheless, the utility of two-way mobile radio for police and fire departments was immediately recognized, and the demand for such systems grew quickly. By 1934, there were 194 municipal police radio systems and 58 state police radio stations serving more than 5,000 radio-equipped police cars [7]. This provoked the first spectrum crisis for the recently established Federal Communications Commission. There were only 11 channels allocated for police use; clearly additional allocations would be needed. In 1937, after extensive hearings the previous year, the FCC granted 29 new channels to law enforcement agencies.

In 1935, Edwin H. Armstrong stunned the radio industry with his invention of frequency modulation. Because so much of what follows will focus on the shortcomings of FM, it is worth recalling for a moment just how great a breakthrough it was. In his biography of Armstrong, Lawrence Lessing describes the first public demonstration of the new technology, which, with typical dramatic flair, Armstrong had arranged for the plenary session of the annual meeting of the Institute of Radio Engineers in New York City on November 5, 1935.

> Armstrong stood at the lectern, delivering his paper, giving no hint of the coming demonstration . . . until he received a signal that all was ready. "Now, suppose we have a little demonstration," he drawled. For a moment the receiver groped through the soughing regions of empty space, roaring in the loudspeaker like surf on a desolate beach, until the new station was tuned in with a dead, unearthly silence, as if the whole apparatus had been abruptly turned off. Suddenly out of the silence came [his assistant's] supernaturally clear voice:
>
> "This is amateur station W2AG at Yonkers, New York, operating on frequency modulation at two and a half meters."
>
> A hush fell over the audience. Waves of two and a half meters (110 megacycles) were waves so short that up until then they had been regarded as too weak to carry a message across a street. Moreover, W2AG's announced transmitter power was barely enough to light one good-sized electric bulb. Yet these shortwaves and weak power were not only carrying a message over the seventeen miles from Yonkers, but carrying it by a method of modulation which the textbooks still held to be of no value. And doing it with a lifelike clarity never heard on even the best clear-channel stations in the regular broadcasting band . . . Music was projected with a "liveness" rarely if ever heard before from a radio "music box." The absence of background noise and the lack of distortion in FM circuits made music stand out against the velvety silence with a presence that was new in auditory experiences. The secret lay in the achievement of a signal-to-noise ratio of 100-to-1 or better, as against 30-to-1 on the best AM stations . . . Armstrong was not satisfied with a 100-to-1 ratio, and he shortly succeeded in raising this to 1000-to-1 . . . By all the rules that had been drilled into radio engineers for nearly a quarter of a century this was . . . fantastic [8].

Instead of varying the strength (amplitude) of the radio signal, FM varied the frequency of the signal. Armstrong had found a way to conquer the ubiquitous "static" — random amplitude noise from natural and man-made sources — which had bedeviled radio since the beginning.

The dynamic range was tripled (compared to contemporary AM), giving an unparalleled crispness suitable for broadcasting fine music. Most important for mobile radio, FM needed much less power, opening the way for feasible vehicular transmitters and sensitive receivers [9]. FM also exhibited other desirable properties, notably the "capture effect": an FM receiver tends to "lock in" on the stronger of two competing signals and to reject the unwanted signal almost completely (unlike AM where two competing signals will often be superimposed upon one another) [10]. Finally, FM proved to be much more resistant to the peculiar propagation problems of mobile radio transmission, especially the problem of "fast fading" or flutter which virtually destroyed mobile AM signals.

In the late 1930s, the Connecticut state police implemented the first two-way FM mobile system. It was immediately recognized as a "better mousetrap," and by 1940 almost all police systems in the country had converted to FM [11].

The advent of FM was the first great watershed in mobile radio. It proved to be a remarkably long-lived technology. "There is very little incentive for changing the present FM pattern of systems design," wrote Daniel Noble (then the Executive Vice-President of Motorola) in 1962 [12]. The same words could have been written in 1972, or even in 1982. Although there have been major advances in network-level architecture, improvements in components and in the network interface, *no fundamental breakthroughs in the technology of the basic mobile radio link have reached the market since the late 1930s.*

The Impact of the War

World War II was an enormous stimulus to the field. In 1940 and 1941, Bell Labs and Western Electric were commissioned to undertake the development of mobile radio-communication systems for military vehicles, including tanks and other land vehicles, as well as military aircraft. Initial theoretical studies did not show a clear choice between AM and FM, but field trials of several prototype systems in early 1941 showed a decisive advantage for FM [13]. Ultimately several hundred thousand mobile radios were built for the war effort; great strides were made in packaging, reliability, and cost reduction. FM was in fact one of the areas of crucial technological superiority of the United States over its enemies in the war. Armstrong donated the use of his patents royalty-free, and as a result the mobile communication capabilities of the American forces were "beyond anything either the enemy or the other allied nations possessed [14]." The American army was the only fighting force to employ FM; the

German panzers still used AM, which was vulnerable to American jamming. The official history of the signal corps identifies the allied advantage in communication as a key factor in the outcome of the Battle of the Bulge.

> The jammer effectively filled the German AM receivers with a meaningless blare, while the American FM sets heard nothing but the voices of the operators [15].

From our current perspective, the main lasting result of the war stimulus was the creation of a commercial FM manufacturing capability. When the United States entered the war, almost every manufacturer involved in commercial (AM) radio set production was converted to supply military requirements (mostly FM). Size, reliability, cost, and performance were all substantially improved through successive product generations throughout the war. By the end of the war, every tank, ship, and airplane was equipped with a radio-communication system. The "walkie-talkie" had become a ubiquitous part of the military machine. The radio industry had been forcibly introduced to FM techniques and the feasibility of large-scale one-way and two-way systems had been demonstrated. The way had been cleared for the development of a true commercial mobile-communication market in the United States.

2.2 THE COMMERCIAL PHASE: 1946–1968

The next phase of the development of the mobile telephone industry saw the expansion of true mobile telephone service into the commercial arena [16]. Demand was strong and the beginnings of the chronic "spectrum shortage" were soon apparent. Technical improvements were oriented toward two chief goals, both of which were driven by the desire to improve spectrum utilization and increase carrying capacity.

1. The reduction of transmission bandwidths — channel splitting;
2. The introduction of automatic trunking into radio systems.

On the business side, the most striking development was the introduction of competition into the radiotelephone business by the FCC — the first and, for many years, the only crack in the monolithic facade of the AT&T communication monopoly.

During the 1940s, requests for spectrum for two-way mobile radio had increased dramatically. In 1945, Report No. 13 of the Radio Technical Planning Board of the FCC recommended mobile radio spectrum allocations for a wide range of private sector uses: police, fire departments, forestry services, electric, gas, and water utilities, transportation services, including railroads, buses, streetcars, trucks, and taxis. In 1949, the FCC officially recognized mobile radio as a new class of service. The number of mobile users exploded from a few thousand in 1940 to 86,000 by 1948, 695,000 by 1958, and almost 1.4 million by 1963 — the vast majority of these users were *not* interconnected to the telephone network [17].

True mobile telephone service — the interconnection of mobile users to the public telephone landline network to allow telephone calls from fixed stations to mobile users — was introduced in 1946, when the FCC granted a license to AT&T to operate in St. Louis. Within less than a year, mobile telephone service was being offered in more than 25 American cities. At last it was possible for someone to make a phone call from a moving automobile. These systems were based on FM transmission and utilized a wide-area architecture: a single powerful transmitter provided coverage out to 50 miles or more from the base, quite enough to encompass most metropolitan areas. The service charges were reasonable, about $15.00 a month and 15 cents a minute, and the systems were quickly oversubscribed. While the first systems involved an operator at the base station who manually patched in the radio call to the wireline network, the achievement of automatic, direct-dial service was quickly realized. By 1948, the first fully automatic mobile telephone system was in operation in Richmond, Indiana. Manual systems, however, remained the dominant mode of operation for many years.

Demand grew quickly and stayed ahead of available capacity in many of the large urban markets. The mobile telephone business was soon discovered to be quite different from dispatch-type mobile systems. In a dispatch mode of operation, each communication normally lasted only a few seconds. The amount of traffic or circuit usage generated by each mobile unit in a dispatch system is small. Thus, each dispatch channel can support a large number of mobile units, up to 100 or more, without serious blocking. In a mobile *telephone* service, however, the normal phone call may run for several minutes. The amount of circuit usage per customer may be ten times or more what it is for a dispatch circuit. This reduces drastically the number of customers who can share a single channel without serious blocking. In fact, in conventional wireline-telephony facilities planning a *concentration ratio* (the ratio of customers to available circuits) greater than 5 or 6 to 1 will usually result in degraded service. Yet mobile telephone subscribers were normally signed up far in excess of the effective traffic capacity of the early systems. Loading of 50, 100, or more subscribers per channel were common. Service was terrible. Blocking probabilities (the likelihood that a subscriber would not be able to obtain a circuit within a specified short period of time, such as ten seconds) rose to as high as 65% or more. The actual usefulness of the mobile telephone decreased as users found such blocking often completely prevented them from getting a circuit during the peak periods (rush hour). The situation ultimately became so extreme that in 1976 Bell mobile service for the entire metropolitan area of New York, more than 20 million people, constituted 12 channels, serving 543 paying customers, with a waiting list of about 3,700 [18]. Rates were high and customers were chronically dissatisfied, according to many surveys conducted over the years.

This placed mobile telephony in the center of the postwar spectrum struggle. A handful of channels would not be enough for true mobile telephone service to develop. Large blocks of spectrum would be needed to satisfy the demand in urban areas like New York — large enough that mobile telephony became competitive for spectrum with the other new and equally spectrum-hungry technology of the late 1940s, television.

In this context, FM suffered a serious disadvantage: it required much more bandwidth than AM. Each radio channel had to be wider, and the guardbands separating it from the next channel had to be wider, so that there could be fewer such channels carved out of a given portion of the radio spectrum.

The range of human hearing extends from very low tones around 20 cycles per second, or 20 Hertz in the terminology of the telephone world, to perhaps 15,000 cycles per second, or 15 Kilohertz (kHz). The bulk of the energy in the human voice is concentrated, however, between about 300 Hz and 3,300 Hz, and wireline telephone circuits are generally designed to pass only this band of frequencies, ignoring and in fact filtering out all higher and lower frequencies. The *bandwidth* of a voice circuit is therefore generally held to be about 3 kHz. Theoretically, AM might need a band only 3 kHz wide, although early systems were wider. FM, by comparison, requires theoretically an *infinite* bandwidth to achieve the maximum signal-to-noise ratio. In practice, Armstrong's first *FM broadcasts* required five times the spectrum of conventional AM stations [19].

The original *FM mobile telephone* channels required 120 kHz of spectrum to transmit a voice circuit with an effective bandwidth of only about 3 kHz. Such gross inefficiency indicated that there was considerable room for improvement, and in fact by 1950 the FCC decided to split the original channels into 60 kHz channels. FM receivers, however, had still not been refined sufficiently to handle the narrower bandwidths, so that at first only every other channel was allocated in a given city (much as adjacent TV channels today are not allocated in one broadcast area) [20]. When the FCC authorized new mobile telephone channels in the UHF band (around 450 MHz) in the mid-1950s, a bandwidth of 50 kHz was specified. By the early 1960s, however, FM-receiver design had been further improved, and channel bandwidth was halved again, to 30 kHz, and it was possible to use adjacent channels in the same city. The channelization for the UHF channels was cut to 25 kHz. The spectrum efficiency of analog FM systems was effectively quadrupled between the end of World War II and the mid-1960s. Without such improvements, the cellular radio proposals of the later 1960s could probably not have evolved; they would have required simply too much spectrum.

This is an important point. It is easy to be critical of the FCC's "inadequate" allocations for mobile radio during this period. Yet at 120 KHz per channel, or even 50–60 KHz per channel, the allocation of large numbers of radio channels was impractical. Moreover, it was clear that the trend of FM-system design would soon bring narrower bandwidths within the realm of feasibility. It would appear, in hindsight, that the FCC was justified in not undertaking a major spectrum allocation until the 1960s. Even allowing for "chicken-and-egg" effects (i.e., an adequate spectrum allocation would have stimulated technical improvements more rapidly), technology was probably more the pacing item than regulatory policy in determining when cellular radio could come into being.

The other important technological development during this period was the invention and application of automatic trunking radio systems. In the earliest radio systems, both transmitter and receiver were designed to operate on a specific fixed frequency. Each radio channel was dedicated to a specific user, or to a group of users who shared it like a party line. In a trunked radio system, by contrast, a group of channels are made available to the entire group of users. (See Figure 2.1.) If Channel A is in use, the next caller will be assigned to Channel B, and so on. Each user may find himself actually using any one of the group of channels for a given call. No channel is dedicated solely to his use. Instead all channels are available to all users.

The total traffic (measured in call-seconds) that can be borne by a trunked radio system can be *significantly higher* (under certain conditions) than the traffic that can be carried by an equal number of nontrunked, dedicated channels. For example, a single mobile channel can support only about two or three mobile telephone users (with a 10% chance of blocking and an average of 150 call-seconds per customer per hour). Twenty such channels, not trunked, could support only about 50 customers, evenly assigned to different frequencies. If, however, those 20 channels are trunked together such that any user can access any idle channel, then more than 420 customers can be supported with the same grade of service and the same calling behavior. This is referred to as *trunking efficiency.* (See Figure 2.2.)

Trunking is another way of increasing total system capacity and improving spectrum efficiency. It does increase the cost of the mobile unit

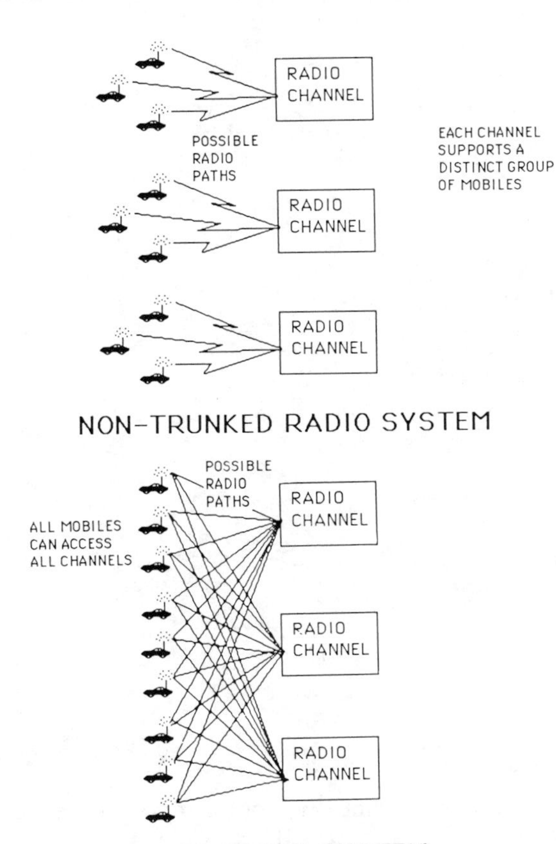

Figure 2.1 Trunked and Nontrunked Architectures.

considerably, however, by requiring that the unit be able to tune efficiently to a variety of different frequencies. In early sets, multiple crystal elements were employed. Each tunable frequency required "two quartz crystals and a position on the channel selector switch [21]." In the modern trunked radio systems which began to emerge toward the end of this period, the different frequencies were synthesized electronically. But efficient telephone (as opposed to dispatch) service virtually requires multichannel trunking.

The first mobile telephone systems were manually trunked systems: each caller had the ability to search through the available channels manually, determining by listening to see which ones were occupied and selecting an unused channel for his call. Later, more advanced systems

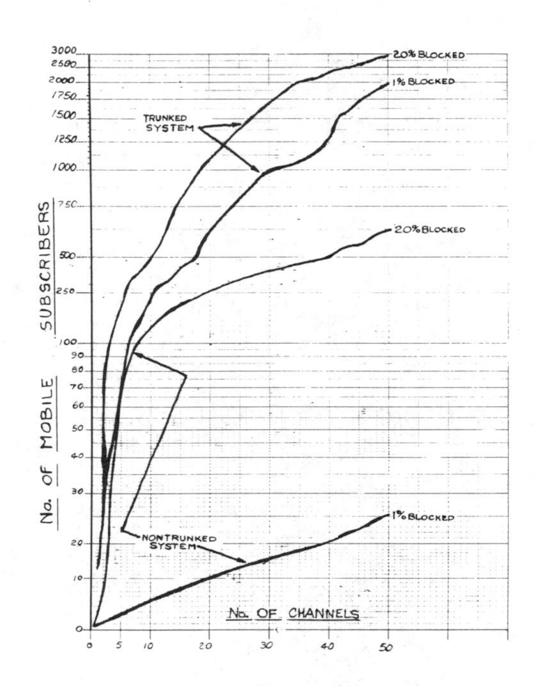

Figure 2.2 Illustration of Trunking Efficiencies

performed this search automatically. For example, in some systems fielded in the 1960s, a special tone was placed on all idle channels. The idle mobile units would automatically tune to this tone, thus placing them immediately in a position to make a call when the user so desired.

In the mid 1960s, a package of enhanced features, including automatic trunking, direct dialing, and full-duplex service (instead of "push-to-talk") was marketed under the rubric of Improved Mobile Telephone Service (IMTS). IMTS was designed to simulate the features and convenience (if not the grade of service) of conventional wireline telephony. Initial IMTS field trials were carried out in Harrisburg, Pennsylvania, from 1962 to 1964, and commercial IMTS service had been introduced into many urban centers by 1965. This repackaging of FM-based mobile telephony coincided with the final reduction in FM channel bandwidth to 25–30 KHz [22].

IMTS was the culmination of three decades of analog FM development. In fact, IMTS was the direct technical precursor, and in some ways the prototype, of cellular radio. It demonstrated the fundamental viability of the narrowband FM channel, automatic interconnection, trunking, and other key "modern" features, and seemed to justify the FCC's plans to scale up this technology to much larger systems which would, once and for all (it was hoped), eliminate the spectrum shortage and allow mobile telephony to become a high-quality service available to large numbers of people.

Industry Structure

In 1949, the Justice Department launched an antitrust proceeding against the Bell system. The suit dragged on and was finally settled by a Consent Decree signed in 1956 by the Justice Department and AT&T. Essentially, by the terms of this Consent Decree, Bell escaped with the telephone monopoly intact (for another 28 years, at least). There was a price to be paid, however, and one of Bell's concessions was its agreement to get out of the mobile-radio manufacturing business. This paved the way for Motorola to become the dominant North American supplier. It also strengthened the somewhat shaky position of one of the most interesting groups in the telecommunication business: the Radio Common Carriers, or RCCs.

Right from the beginning, against AT&T's pleadings, the FCC had decided to authorize independent operators to offer mobile telephone service. The argument that telecommunication is a "natural monopoly" did not seem to hold in the realm of radio, where there was no expensive fixed-wire plant to construct. Competition in the broadcast side of the radio world was vigorous, to say the least. Thus the FCC felt that competition could and should be allowed in mobile radio. The non-Bell carriers who applied for mobile telephone licenses were first called Miscellaneous Common Carriers, and were subsequently known as the RCCs.

The RCCs posed no threat whatsoever to Bell's basic business. There was not enough spectrum allocated to mobile telephony to allow them to grow very large. The typical RCC was a local small businessman who offered service to at most a few hundred mobile subscribers in an urban center, who might employ a half a dozen people, and who, if he were very successful, could glean several hundred thousand dollars a year from the business.

But the RCCs were significant, certainly in hindsight. They were the snake in the garden. They represented the first intrusion of competition into the telephone world since the Kingsbury Commitment of 1913 (which had settled a previous antitrust action against Bell). That the FCC had allowed them to exist was also an implicit acknowledgement that communication systems based primarily upon radio, without the need for extensive physical plant, were inherently amenable to competition and would operate by different rules. The importance of this reasoning would become much greater as microwave radio technology revolutionized long-distance telephony in the 1950s and 1960s, and the same opportunities for competition in the large and extremely lucrative long-haul segment began to become apparent to independent entrepreneurs such as those who built MCI. Also, when ideas for a new type of much-expanded mobile telephone service began to be put forward, the fact of preexisting competition in mobile telephony would have tremendous and unforeseen consequences.

REFERENCES

[1] Robert V. Bruce, *Bell: Alexander Graham Bell & the Conquest of Solitude,* Boston: Little, Brown, 1973, p. 337.

[2] Marvin K. Simon, Jim K. Omura, Robert A. Scholtz, and Barry K. Levitt, *Spread Spectrum Communications,* Vol. I, Rockville, Maryland: Computer Science Press, 1985, p. 48.

[3] Heidi Kargman, "Land Mobile Communications: The Historical Roots," in Raymond Bowers, Alfred M. Lee, and Cary Hershey, eds., *Communications for a Mobile Society: An Assessment of New Technology,* Beverly Hills, California: Sage Publications, 1978, p. 24.

[4] Daniel Noble, "The History of Land-Mobile Radio Communications," *Proceedings of the IRE, Vehicular Communications,* May 1962, p. 1406.

[5] *Ibid.,* p. 1407.

[6] Kargman, *op. cit.,* p. 26.

[7] *Ibid.,* p. 27.

[8] Lawrence Lessing, *Edwin Howard Armstrong: Man of High Fidelity,* Philadelphia: Lippincott, 1956, pp. 208–210.

[9] *Ibid.,* p. 210.

[10] William C. Jakes, ed., *Microwave Mobile Communications,* New York: Wiley, 1974, p. 383.

[11] Noble, *op. cit.,* p. 1408.

[12] *Ibid.,* p. 1413.

[13] M. D. Fagen, *A History of Engineering & Science in the Bell System: National Service in War and Peace (1925–1975),* Bell Telephone Laboratories, 1978, pp. 319 ff.

[14] Marvin K. Simon *et al., op. cit.,* p. 49.

[15] G. R. Thompson and D. R. Harris, *The Signal Corps: The Outcome (mid 1943 through 1945), United States Army in World War II,* Vol.

6, Part 5: *The Technical Services,* Vol. 3, cited in Marvin K. Simon *et al., op. cit.,* p. 49.
[16] Important sources for this section include Claude Buster, "The Beginnings of the Story: Some Historical Developments in Mobile Telephony," published by the Rural Electrification Administration, 1983; Heidi Kargman, *op. cit.;* Daniel E. Noble, "The History of Land-Mobile Radio Communications," *op. cit.;* David Talley, "A Prognosis of Mobile Telephone Communications," *IRE Transactions on Vehicular Communication,* PGVC-11, July 1958.
[17] Noble, *op. cit.,* p. 1410.
[18] William C. Y. Lee, *Mobile Communications Engineering,* New York: McGraw-Hill, 1982.
[19] Lawrence Lessing, *op. cit.,* p. 211.
[20] E. F. O'Neill, ed., *A History of Engineering and Science in the Bell System: Transmission Technology (1925–1975),* AT&T Bell Laboratories, 1985, p. 408.
[21] W. R. Young, "Advanced Mobile Phone Service: Introduction, Background, and Objectives," *Bell System Technical Journal,* Vol. 58, No. 1, January 1979, p. 7.
[22] E. F. O'Neill, *op. cit.,* p. 412.

Chapter 3
THE CELLULAR IDEA: 1947–1982

3.1 THE EVOLUTION OF THE CELLULAR IDEA

It is important to recognize that today's analog cellular radio is not so much a new technology as a new idea for organizing existing technology on a larger scale. In terms of the radio link, cellular systems do not differ significantly from the IMTS systems that preceded them. The backbone remains FM transmission. At the level of the individual radio-telephone circuit, the performance and economics of the two FM generations are similar.

The critical innovation was the "Cellular Idea." Cellular represented a very different approach to structuring a radio-telephone network. It was an idea that held out the fantastic promise of virtually unlimited system capacity, breaking through the barriers that had restricted the growth of mobile telephony, and it did so *without* any fundamental technological leap forward — simply through working smarter with the same resources. Indeed, "cellular architecture" was a system-level concept, essentially independent of radio technology. It appealed to mobile system engineers, because it kept them on relatively familiar hardware ground. It appealed to businessmen and entrepreneurs, because it seemed to open the path to a really large market; by the application of the cellular idea, mobile communication could become another first-class growth industry, like television or radio, or the telephone itself. The cellular idea also appealed to regulators, because it seemed to break out of the "zero-sum" spectrum game that had created terrific political difficulties for more than twenty years.

The cellular idea is elusively simple; one writer concluded that it seemed "to have materialized from nowhere [1]." It began to appear in Bell system proposals during the late 1940s. It had occurred to people that the problem of spectrum congestion might be alleviated by restructuring the coverage areas of mobile radio systems. The traditional approach to mobile radio viewed the problem in terms similar to radio or television broadcasting: it involved setting up a high-power transmitter on top of the highest point in the area and blasting out the signal to the horizon (as much as 40 or 50 miles away). The result was fair-to-adequate coverage over a large area. It also meant, however, that the few available channels were locked up over a large area by a small number of calls. For example, in New York City in the 1970s (a megalopolis of nearly 20 million people over a thousand square miles or more) the Bell mobile system could support just twelve simultaneous mobile conversations. The thirteenth caller was blocked.

The cellular idea approached the coverage problem quite differently. It abandoned the broadcasting model. Cellular called instead for *low-power* transmitters, lots of them, each specifically designed to serve only a small area, perhaps only a couple of miles across. Instead of covering an area like New York City with a single transmitter, the city would be blanketed with small coverage areas, called *"cells."* (See Figure 3.1.)

By reducing the coverage areas and creating a large number of small cells, it became possible (in theory) to *re-use* the same frequencies in different cells. To understand how this changes the total picture, imagine that all the available frequencies could be reused in every cell. If this could be done, then instead of 12 simultaneous telephone circuits for the entire city there would be 12 circuits for every cell. If there were 100 cells (each about 10 miles across), there would be 1,200 circuits for the city, instead of only 12.

Alas, it was not quite so neat. Early calculations indicated that, because of interference between mobiles operating on the same channel in adjacent cells, the same frequencies could not be used in every cell. It would be necessary to skip several cells before reusing the same frequencies. (See Figure 3.2.) But the basic idea of reuse appeared to be valid. The system engineer could, in effect, create more than one usable mobile telephone circuit from the same channel, reused in different parts of the city.

Moreover — and here lay the real power of the cellular idea — it appeared that the effects of interference were not related to absolute distance between cells, but to the *ratio of the distance between cells to the radius of the cells.* The cell radius was determined by the transmitter power; in other words, it was under the system engineers' control. It was within their power to decide how many circuits would be created through reuse. If, for example, a grid of 10-mile-radius cells allowed reuse of the frequencies in cell A at a distance of 30 miles, then a grid of 5-mile-radius cells would allow reuse at 15 miles, and 1-mile-radius cells would allow

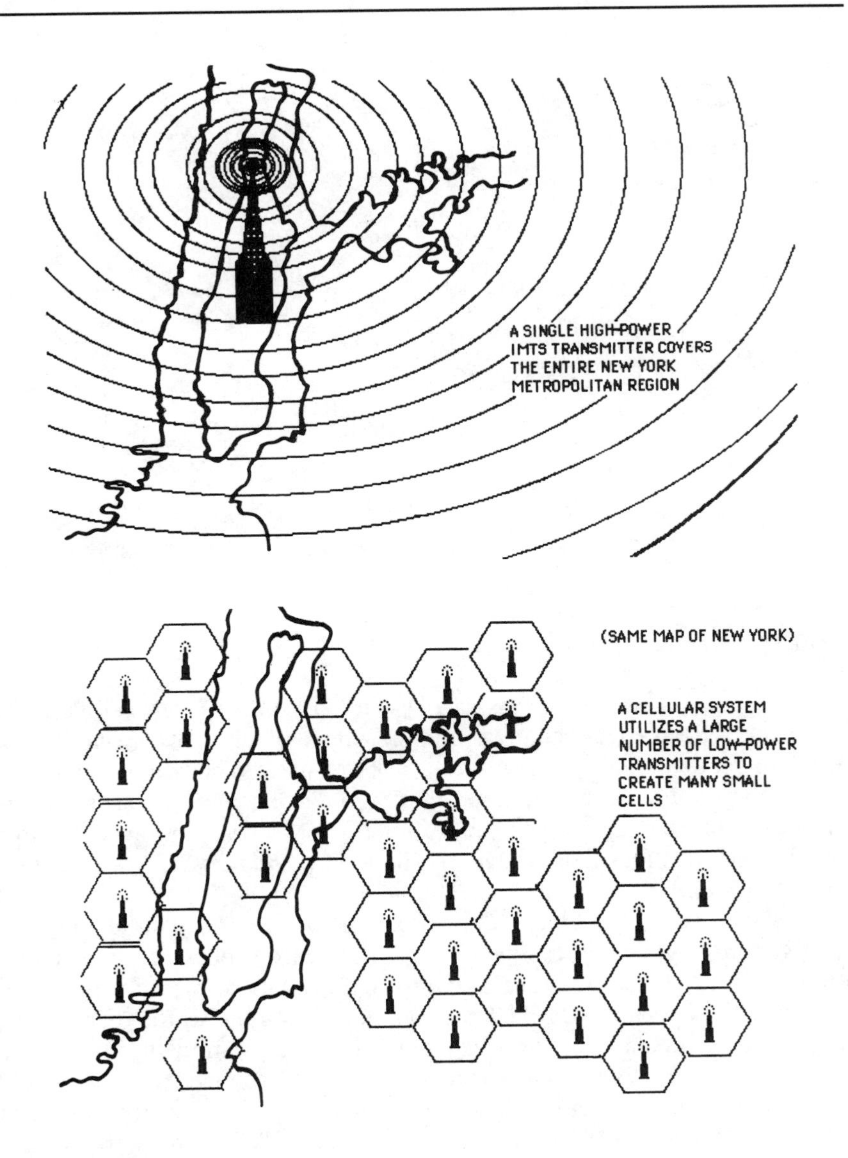

Figure 3.1 Cellular Architecture.

reuse at 3 miles. Because of the fabulous cellular geometry (based on the πr^2 rule for the cell coverage area), however, each reduction in cell radius by 50% led to a quadrupling of the number of circuits per megahertz per square mile. A system based on 1-mile-radius cells would generate one

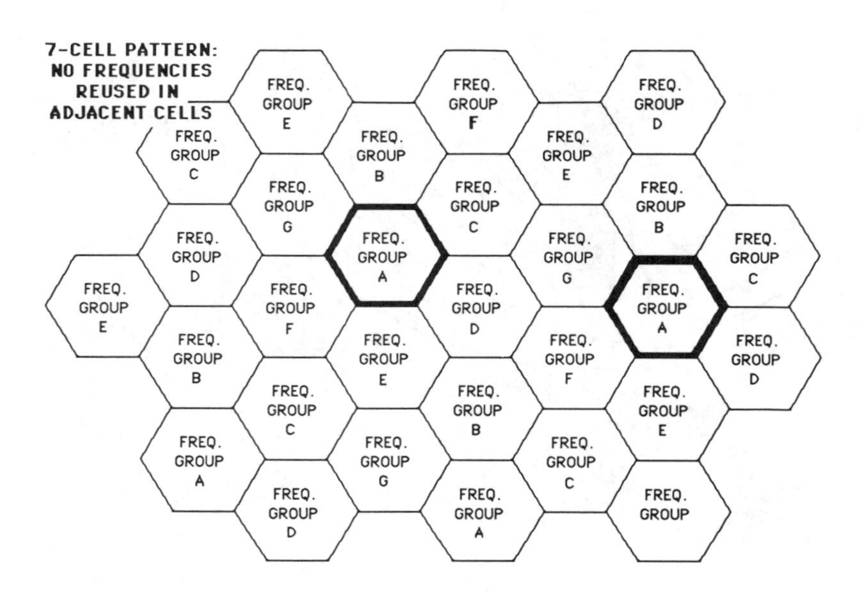

Figure 3.2 Frequency Reuse.

hundred times as many circuits as a system based on 10-mile-radius cells. Engineers began to speculate about the effect of using cells only a few hundred feet in radius, a few city blocks in area. They found that (on paper) they could create systems with thousands of cells and huge numbers of circuits per megahertz per square mile, capable of carrying the traffic of hundreds of thousands or even millions of paying customers in a metropolitan area.

Of course, it would have been enormously expensive to build thousand-cell systems right from the beginning. It appeared, however, that large-radius cells could *evolve* gracefully into small-radius cells over a period of time through a technique called *cell-splitting.* When the traffic reached the point in a particular cell such that the existing allocation of channels in that cell could no longer support a good grade of service, that cell would be subdivided into a number of smaller cells — with even lower transmitter powers — fitting within the area of the former cell. The reuse pattern of radio frequencies could be repeated on the new, smaller scale and the total capacity multiplied for that area by a factor equal to the number of new cells. When, in time, the smaller cells were saturated, still smaller cells could be created. Even without going to ultra-small cells only a few blocks across (later called microcells), the cellular architects conservatively forecast at least three rounds of cell-splitting. (See Figure 3.3.)

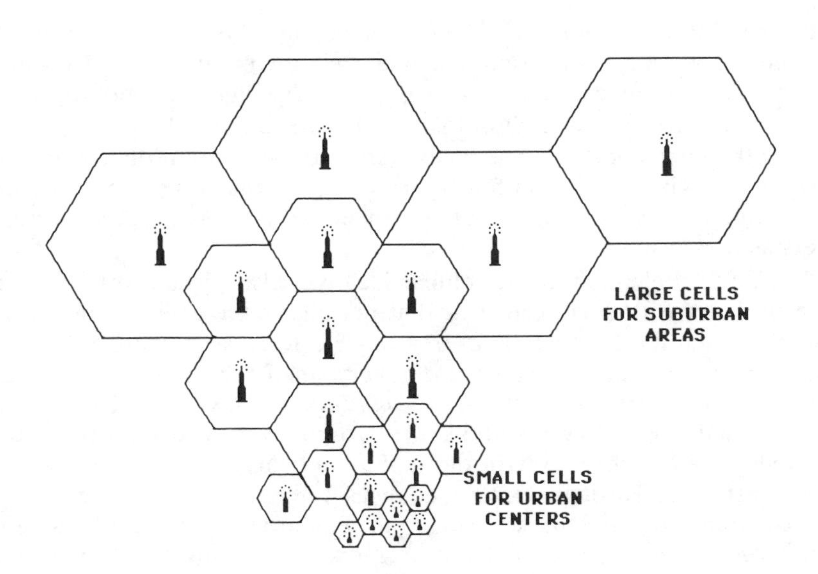

Figure 3.3 Cell-Splitting

Cell-splitting seemed to offer many advantages. It allowed the financial investment to be spread out as the system grew. New cells would only be added as the number of revenue-generating customers increased, providing the cash to support the continued investment. Moreover, it was thought that new cells could be created without scrapping the existing investment in the large-radius cell-site equipment; those transmitters would simply be powered down to fit within the new scale. Finally, cell-splitting could be applied in a geographically selective manner; the expense of smaller cells would only be necessary in the high density traffic centers. In outlying areas, larger-radius cells would suffice to carry the lower traffic densities in more suburban regions. Cell-splitting was flexible in the dimensions of time *and* space. In the minds of cellular architects, it was the ideal surgical technique for boosting capacity precisely where and when it was needed.

Frequency reuse, in a gross sense, had been applied before. For example, a TV station in Boston can use the same frequency as one in San Diego without mutual interference because of the geographical separation. But *cellular reuse* involves more than just taking advantage of the natural attenuation of a signal over great distances; it uses the much-reduced power levels of the cell-site transmitters to achieve reuse of frequencies at very short distances, in many cases only a few miles away. By reducing cell sizes, more and more circuits could be created. It was envisioned as a kind of spectrum machine: turn the crank, shrink the cells, and produce as many circuits as needed. In the light of modern disappointments with cellular radio, it is worth remembering the power of this idea in the minds of those who conceived it. It *was* revolutionary! For the first time, mobile engineers felt the conquest of the congestion problem was within their grasp. They could engineer the systems to serve as many users as they wanted.

The final element of the cellular idea was also (deceptively) simple. The problem with small cells was that not all mobile calls could now be completed within the boundaries of a single cell. A car moving at freeway speeds might zip through half a dozen very small cells in a single conversation. To deal with this, the idea of *hand-off* was invented. The cellular system would be endowed with its own system-level switching and control capability — a higher layer in the mobile network, operating above the individual cells. Through continuous measurements of signal strength received from the individual cell-sites, the cellular system would be able to sense when a mobile with a call in progress was passing from one cell to another, and to switch the call from the first cell to the second cell "on the fly," without dropping or disrupting the call in progress. This required new techniques for determining which of several possible new cells the mobile unit had strayed into, as well as methods for tearing down and reestablishing the call in a very rapid manner, which posed some modest challenges to the cellular architects. The concept of hand-off, however, seemed entirely feasible. Small cells would work. (See Figure 3.4.)

These were, and are, the essential principles of cellular architecture:

1. Low power transmitters and small coverage zones or cells;
2. Frequency reuse;
3. Cell-splitting to increase capacity;
4. Hand-off and central control.

Certainly it was an ambitious architecture, but in principle it seemed no more challenging than other contemporary telecommunication developments. There was an obstacle, however. To realize a workable cellular system, additional spectrum would certainly be needed. The handful of mobile telephone channels available in conventional IMTS allocations was far too small to permit the realization of the cellular dream. The cellular idea might be able to "manufacture spectrum," but it needed a critical mass to start with. Alas, obtaining the initial allocation would not be easy.

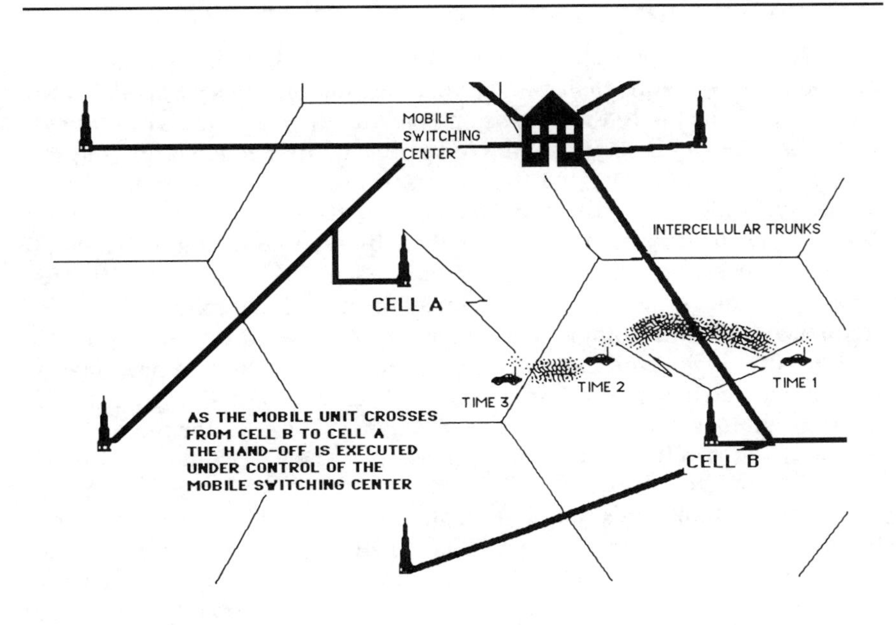

Figure 3.4 Hand-Off.

3.2 THE STRUGGLE FOR SPECTRUM: 1947–1970

The first proposal for a large-capacity mobile telephone system was put forward by Bell engineers in 1947, only a year after the introduction of mobile telephone service. In connection with the FCC's Docket 8658, a rule-making proceeding for the allocation of television spectrum, Bell suggested the creation of 150 two-way channels, utilizing 100-KHz channel spacing. The proposal was not acted upon [2].

Two years later, in 1949, Bell prepared a more elaborate plan in connection with a new Docket (No. 8976) [3]. The occasion was a momentous one: the FCC was in the process of creating a new and much expanded set of television allocations, in what would become known as the UHF band, encompassing more than 400 MHz of spectrum located between 470 MHz and 890 MHz. Bell argued that mobile services were also entitled to additional spectrum and should not be squeezed out by the broadcasters.

Television would stand astride the road to the realization of the cellular idea for more than twenty years. In the late 1940s and 1950s the nation was tuning in to the Television Age, and the visionaries of that new and profitable technology were not to be denied. TV needed large amounts of spectrum — a picture is worth a thousand words, and images need much more bandwidth in transmission than voice. Each TV channel utilizes 6 MHz of spectrum, enough for more than 100 mobile telephone channels (using today's channel spacing). In fact, each TV channel represented about three times the total spectrum of all the precellular mobile telephone frequencies. On the other hand, each TV channel provided service to millions of people, while a mobile telephone circuit benefited only an elite few. Measured by the utilitarian yardstick of "the greatest good to the greatest number," TV was viewed as a superior service. For the next two decades, the spectrum allocation policies of the FCC would be strongly skewed to the broadcasters.

The Commission's decision on the 1949 docket was a total victory for the broadcast interests. They rejected Bell's ideas and refused to allocate any portion of this valuable spectrum to mobile telephony. Seventy channels of UHF TV were created. In truth, it was probably the correct decision technologically. The elaboration of a full-scale cellular system in that era of 100-KHz FM channels and barely tunable quartz radios would have been awkward, if not impossible. The cellular idea was not yet mature.

Over the next two decades, however, a steady pressure built up to force a reconsideration of this position. Several trends were responsible for swinging the balance back toward mobile radio. First, mobile technology was improving: narrower channels, automatic interconnection, trunking for spectrum efficiency, and improved synthesizer technology (for frequency agility across large numbers of channels inexpensively) began to make large-scale mobile telephony look more feasible. Second, demand for mobile services grew robustly. A few more channels were allocated; but even with channel-splitting to 60 KHz and then to 30 KHz, the available mobile telephone allocations were, as indicated previously, woefully inadequate. Third, the UHF TV industry grew very slowly for a variety of reasons, including the reluctance of the set manufacturers to increase the price of their sets by adding the UHF tuning elements. As a result, in most cities and rural areas the vast ocean of prime UHF spectrum lay largely unused.

Change did not come quickly. In 1957, the FCC initiated an inquiry into the allocation of nongovernmental frequencies between 25 MHz and 890 MHz (Docket 11997) [4]. In response, the Bell system prepared another proposal which called for 75 MHz of spectrum in the 800-MHz region to be set aside for mobile telephony. But the Commission was still more concerned with encouraging the development of UHF TV. To this end it had secured the enactment of the All Channel Receiver Act of 1962, which required that all television sets marketed in interstate commerce be equipped with receivers for both UHF and VHF signals. There was hope that this would prove to be the boost that the UHF operators needed, and the Commission was accordingly reluctant to cast any cloud, however small, over the future of that spectrum for television use. In 1964, in their second major decision in the TV *versus* Mobile Radio struggle, the FCC once again rejected the idea of assigning any of this spectrum to mobile telephony [5].

The broadcasters, however, although helped by the 1962 legislation, still came nowhere near using all that spectrum. Meanwhile the pressure for relief of the mobile-radio congestion crisis was unrelenting. The FCC was forced to address this pressure by commissioning a new study group, the Advisory Committee for Land Mobile Radio Services, to make a thorough evaluation of possible solutions to the congestion crisis in mobile telephony. The regulators were apparently hoping that some "technological solution" could be found that would leave the established order of spectrum allocations intact and avoid the political storm that was sure to ensue if any spectrum were pried loose from the broadcast interests [6].

But the Advisory Committee was forced to conclude that no solution would work without substantial new spectrum allocations for the land mobile services, and it was equally clear that this spectrum would have to come from the existing TV allocations. They were influenced by proposals being circulated by Bell based on the cellular idea — now nearly twenty years old — which promised a spectrum breakthrough, if only the initial critical mass of frequencies were allocated. Bowing to the findings of the Advisory Committee, the FCC was forced to think the unthinkable. In the words of one group of historians:

> Confronted on the one hand . . . with the seemingly nonviable UHF [TV] system and on the other with the insatiable demand for land mobile services, the commission announced in April 1967 that it would study the feasibility of reassigning certain UHF channels to land mobile services [7].

The gauntlet was officially and publicly thrown down with the publication of the Advisory Committee's report in March 1968. A number of options were analyzed, including: (1) reallocation of the lower UHF channels 14 through 20 to mobile telephony; (2) sharing of all UHF channels depending upon patterns of geographical TV usage; and (3) the reallocation of the upper UHF channels 70 through 83 (the 800 MHz band) to mobile services.

Meanwhile, the mobile radio interests had also begun to exert pressure through the legislative and executive branches to overcome the perceived broadcasting bias at the Commission. In 1968, a House of Representatives Select Committee began hearings on the "crisis in land mobile communications." Observing that broadcast interests had been allocated 87% of the available spectrum below 960 MHz, compared to only 4% for mobile communication as a whole and less than one-half of 1% for mobile telephony, the House Committee called for "ample, additional, usable frequency spectrum to be allocated without delay for this means of communication [8]." In the same year, a Presidential Task Force Report on Communications Policy was issued, which diagnosed the chronic congestion and poor service in the mobile communication industry as the result of "spectrum mismanagement by the FCC [9]."

Spurred by such developments, in 1968 the Commission opened two new rule-makings. Docket 18261 called for comment on the proposed sharing of all UHF channels with mobile services, as an immediate relief measure, with specific allocations to be determined on a geographical basis. Docket 18262, which became known as the Cellular Docket, proposed an outright reallocation of UHF channels 70 through 83.

The "cellular rule-making" became one of the most contentious proceedings in FCC history. More than 110 parties filed public comments. There were two days of public hearings at which more than 40 parties offered their views to the Commission. On one side stood the long-suffering mobile radio community, led by the largest mobile operator, AT&T. They based their case on the demonstrable congestion in all existing mobile services, especially mobile telephony, and on the exciting promise of the cellular idea. Arrayed against them were the broadcasters who argued the perpetual need of television for spectral *Lebensraum*. The political dimension of the dispute was keenly apparent to all parties. Yet the arguments were often couched in esoteric technological jargon which seemed beyond the reasonable comprehension of many participants. This would be a recurring pattern.

After lengthy and often contradictory technical presentations, the Commission concluded by a 3–2 vote that only additional spectrum could solve the congestion problem. Two of the Commissioners still objected strenuously in favor of television uses for this spectrum. The swing vote was cast by Commissioner Johnson, who, in what one historian has characterized as a "reluctant concurring opinion," harshly criticized the Commission itself for failing to develop any "consistent, rational policy of spectrum management [10]."

Under such inauspicious circumstances, the modern era of mobile telephony began. Initial progress was so slow that it might have appeared as though the broadcasters had won yet another round. Docket 18261 resulted in very limited "emergency" spectrum relief measures: mainly the authorization of very limited sharing of small portions of one or two UHF channels in the ten largest cities only. The result was another two dozen potential mobile channels.

The Commission, however, had turned the corner. It was clear that the rule-making opened under Docket 18262 would lead to the long-sought broadband allocation of the 800 MHz frequencies. Indeed, the First Report and Order of May 1970 pronounced the tentative allocation of 115 MHz to mobile services, including 75 MHz to common carrier (telephony) systems! (This was later pared back to 40 MHz for immediate allocation, with additional spectrum in reserve.) At last there would be room to develop the cellular idea.

The decision on Docket 18262 was the second great watershed (after the invention of FM) in the development of mobile telephony. Despite continuing uncertainties about how the new spectrum would be administered — it would take another dozen years to unravel these uncertainties sufficiently to commence commercial service — it was clear that a very sizable new industry was going to be created. In 1968 AT&T — by far the largest radio common carrier — served only about 34,000 customers nationwide, about 50% of the total. The revenue base was negligible compared to the more than 100 million telephones in the wireline network. Growth in the wireline network, however, was "mature"; dramatic increases were unlikely. On the other hand, projections for cellular mobile service reaching millions of customers nationwide began to look like a real business with exciting growth potential, even to the jaded mega-managers of the "world's biggest company."

3.3 INDUSTRY POLITICS: 1970–1982

After the breakthrough of Docket 18262, cellular technology should have reached the marketplace by the early 1970s, but it was not until the mid-1980s that the first commercial systems became operational in the United States. This delay of from ten to fifteen years has been blamed on many factors, but mobile industry executives commonly cite "regulation" as the source of their woes. The hapless *FCC licensing* policies of the past several years (see Section 4.4) have tended to color our view of the prelicensing period. Yet from the First Report and Order on Docket 18262 in 1970, which established the claim to the spectrum and cleared the broadcasters out of the way, until the decision on Docket 79-318 in March 1982, which established the licensing procedures, cellular radio passed through twelve tortuous years of further inquiries, petitions, comments, judgments, challenges, reconsiderations, and lawsuits before the technology was even in a position to be licensed. These delays were not primarily caused by regulatory actions *per se*. Regulation tended to follow, not lead, the process. In the early and mid-1970s, the FCC was more a mirror of industry politics than an activist agency working through its own agenda. Kenneth Hardman, an attorney with considerable experience in the mobile industry, has written a convincing reconstruction of cellular politics of the 1970s [11]. In Hardman's view, the delays prior to 1982 can be traced primarily to the political wrangling among the industry players. Much of this section is based upon Hardman's analysis.

In 1970 it was assumed by almost everyone that the new mobile telephone service would be operated as an extension of AT&T's wireline telephone monopoly. The initial decision of 18262 specifically allocated the new mobile telephone spectrum to the wireline telephone companies *only,* which at that time meant AT&T for about 85% of the American population and for almost all the major cities. This seemed quite natural. AT&T had developed the cellular idea in the first place, which it now elaborated in a series of technical proposals that became known as the AMPS architecture (for Advanced Mobile Phone Service) [12], which were strongly shaped by the assumption of monopoly service. The AMPS approach certainly had the look of a conventional "utility-like" telephone network: the AMPS systems would be heavily dependent upon a skeleton of switching and call-processing technology linking up the dozens or hundreds of cells into an integrated metropolitan network under centralized control. In 1970, only AT&T had the expertise to build and operate such a system.

The AMPS architecture was very much a product of its techno-historical moment. Consider the technical environment within which the cellular technical standards evolved. In 1971 when the specific AT&T proposals for cellular radio were formulated, there was no such thing as a digital telephone switch (it would be five years before the first digital central office was installed in the United States, and it was not installed by the Bell system). In 1971 there were no digital microwave systems, no digital satellites. These backbone radio technologies of the 1980s were completely unknown in the telephone world. In 1971 there was no such thing as a microprocessor.

In 1971 the world was well into the fourth decade of the FM epoch. There seemed no reason to suspect or foment a revolution on the radio side (which was not Bell's strong suit in any case). The problems of mobile radio did not seem to be related to the radio technology. High mobile costs — as much as $1,000 or more — could be attacked through scaling up production; costs would fall as volumes rose. A large market would stimulate competition, encouraging manufacturing efficiency and aggressive pricing, to the benefit of consumers. (AT&T welcomed competition in the production of the mobile units, because they were barred from participating in that end of the market.) FM was not a proprietary technology. This gave it a perceived advantage. Any competent manufacturer could produce FM cellular radios. The FCC accepted the idea that competition and volume manufacturing should be encouraged by choosing a radio-link standard that was, in effect, in the public domain. Thus there were many arguments in favor of standardizing on FM — but the most compelling was that no one could really offer an alternative.

The result was a cellular architecture that was based upon:

1. A complex and expensive (some would later say "top-heavy") network architecture which overlaid the wireline network and to some degree duplicated it; this added considerably to fixed costs and weighted the investment heavily upfront, paralleling the "utility" cost structure of conventional telephony;
2. A "Big Machine" approach to system control (in the words of one participant), whereby processing was centralized in large, expensive switches instead of being distributed to "smart" cell-sites and mobile units as much as possible;
3. A "Least Common Denominator" radio link technology that was fraught with problems, not fully recognized at the time, for the scaling up of the relatively small and primitive *radio* systems of the precellular era into true wireless *telephone* networks.

Had AT&T been allowed to proceed as it intended, this architecture might have been brought into being relatively quickly through brute economic force. Unfortunately, the easy assumption that AT&T would, and should, monopolize wireless telephony as it then monopolized wireline telephony was contradicted by an obstinate fact: there had been head-to-head competition in the mobile telephone business for twenty years. Ever since the 1940s, there had been another side to the peculiar mobile industry: the Radio Common Carriers. In 1968 there were more than 500 of them, and while none could remotely compare with AT&T, together they served almost as many mobile customers as AT&T. Nevertheless, in the eyes of AT&T and, apparently, those of the FCC as well, the RCCs were an anachronism. Their existence was made possible only by the primitiveness of mobile radio at that time. The cellular architecture would involve a much higher level of infrastructure complexity, technical and operational sophistication, and capital. It was "obvious" that the average

"Mom and Pop" RCC serving a few hundred subscribers with antiquated equipment was in no position to capitalize on the cellular opportunity.

The story of spectrum management in the United States, however, is a story of entrenched interests, of living fossils and splinter groups, which are capable of fighting tenaciously to retain their hold on any spectrum they have once managed to sink their taproots into. In the face of FCC skepticism, the RCCs maintained their right to participate in the cellular bonanza. After all, they reasoned, they were already competing in mobile telephony — why should this not continue? They received support from the Justice Department, perennially suspicious of AT&T's monopoly powers [13]. In 1971, the Commission was forced to recognize their existence by redefining the eligibility requirements for providers of cellular service. The allocations would be open to "any qualified common carrier." Technically, the RCCs were eligible to apply. The plan, however, still called for the allocation of only one cellular license per city or region. It was almost impossible to imagine that AT&T would lose in any major urban market. The Commission openly admitted as much: "If we [are] correct that the wireline carriers [are] the only ones *capable* of cellular service offerings, then giving them *exclusivity* in this service would be superfluous [14]." (Emphasis in the original.)

Perceiving that their inclusion was more a fiction than a reality, the RCCs took their battle to the courts. According to Hardman, the "major players" at first presented a common front. Motorola, the largest of the manufacturers, sided with AT&T against the RCCs. Then, says Hardman, AT&T made a tactical error. In 1975 they announced their plans for a cellular test system in Chicago.

> The uneasy truce between AT&T and Motorola proved to be short-lived . . . AT&T announced the winning contractors for the first phase of its developmental cellular system, proposed for Chicago, Illinois. The mobile units were to be supplied by E. F. Johnson Company . . . and OKI Electric Industry Company, Ltd., a Japanese manufacturer of electronic equipment. Obviously outraged by its exclusion, Motorola changed its position in the court appeal, attacked the cellular allocation, and urged the court to set it aside. Less visibly, Motorola increased its research and development efforts on its own systems design . . . Motorola also joined with the RCC interests in urging denial of AT&T's application for its proposed developmental cellular system . . . (see [15], p. 388).

Whether or not the attributed motives are entirely accurate, the fact is that Motorola began to play a much more important role in the cellular process in the mid-1970s. This had a significant impact upon the ultimate outcome. Motorola was the major supplier of mobile radio equipment of all kinds, a business they had "inherited" when AT&T had been forced to get out of the radio manufacturing business by the 1956 Consent Decree. The advent of cellular radio betokened an upheaval in Motorola's traditional markets, and through the 1970s they fought to gain parity with AT&T as a potential supplier of cellular *system* equipment, in addition to their expected position as the market leader for the mobile units. It was natural that they ally themselves with their traditional customers, the RCCs. In 1977, Motorola teamed with one of the larger RCCs to file an application for a second cellular developmental system in the Washington area. Motorola's support gave legitimacy and strength to the RCC's claim on the cellular spectrum.

Finally, without quite going so far as to dictate policy to the Commission, the Appeals Court for the District of Columbia Circuit admonished the FCC against promulgation of a policy that could damage the competitiveness of the mobile communication market.

> AT&T [wrote the Court] is the likely recipient of a virtual monopoly in the operation of cellular systems [which would] result in significant increases in its market power, where cellular systems are operative . . . AT&T appears likely to dominate substantially the field of radio telephone service which has heretofore been occupied in significant part by small radio common carriers operating in a highly competitive environment . . . it would weaken incentives for development of improved mobile radio systems . . . it would enhance the already enormous economic power of the Bell System [16].

The rules were changing.

Meanwhile, as the litigation delayed the projected start of cellular service, technology was opening up new Pandoran boxes. New digital switching systems were emerging and becoming cheaper and more "turnkey"; by the early 1980s, it was possible indeed for a "manufacturer" with very little technological depth to purchase the necessary switching components in "kit" form for assembly into cellular base stations. Much of this technology was coming from outside the Bell system. By the end of the 1970s, it was clear that AT&T no longer had the monopoly on switching technology and know-how that it had enjoyed in 1971. In short, it began to dawn on people that an RCC could actually set up and operate a cellular system.

At the same time, as Hardman further points out, the RCCs were beginning to develop into substantial businesses, mainly on the strength of the rapidly growing paging business. RCC paging subscribers went from about 50,000 nationwide in 1970 to more than 600,000 in 1978. RCCs were developing the management and operational talent to play in the cellular game, and play well. They were also developing access to, and expertise in, using new funding sources. The capital to build the systems was becoming available to them.

In short, one of the chief effects of the litigation was that it bought time — a crucial five to eight years — during which the RCC industry transformed itself from an embarrassing anachronism into a viable competitive alternative to the wireline operators. Competition began to look like it might just be workable, even for the complex cellular architecture. This would now provoke further delays while the industry and the Commission wrestled over how best to implement competition (see Section 3.4).

But something even more fundamental was going on. The technological stasis of conventional mobile telephony was beginning to break up. The long-term viability of the cellular standards was coming under a cloud, at first no larger than a man's hand, but growing. This would have ominous consequences in the 1980s when the systems actually became operational.

Unfortunately, while at another moment a delay between standardization and commercialization might not have meant very much in terms of the viability of final technical standards — both television and FM broadcast radio had experienced such delays — the cellular architects were freezing their design concepts and basic technology choices just at the moment of the great "take-off" in communication and information processing technology. The "digital revolution" was already coming into the core of telephone switching and transmission. As we shall evaluate more analytically in Chapter 4, the shortcomings of today's cellular systems are largely rooted in the obsolescence of the technical standards established fifteen years too early (or a system implemented fifteen years too late).

3.4 COMPETITION AND THE CHANGING REGULATORY AGENDA: 1970–1982

During the 1970s, while the debate over cellular radio ebbed and flowed, a deeper set of changes were taking place within the American telecommunication industry. In 1970, the institutional landscape had changed very little since the Kingsbury Agreement of 1913 (in which AT&T had settled the first government antitrust suit with Woodrow Wilson's Attorney General, establishing the monopoly framework for the telephone industry). Telephony was still a monolithic network, a public utility protected from competition and guaranteed a reasonable rate of return. For practical purposes, there was only one long-distance carrier: AT&T. In each city there was only one local telephone company, and about 85% of the American households were linked into the Bell network. There was only one supplier for AT&T — Western Electric — for most kinds of telecommunication equipment, from central office switches to residential telephone sets. Except for the one tiny corner of mobile telephony, with its "anachronistic" RCCs, there had been no breath of competition in any segment of the telephone world within the memory of anyone in the business.

Nor did it seem likely that there ever would be again. Under the Kingsbury rules, the industry had reached a plateau of general satisfaction. "Universal service" had been virtually achieved. Telephone rates were reasonable. Service in the United States was widely regarded as far superior to that in most other countries. The idea that there could ever be a return to the wild and woolly era of wide-open competition that had existed prior to 1913 seemed as remote and as bizarre as the thought that people might suddenly turn off their electric lights and go back to reading by kerosene lamps.

By 1980, everything had changed. The long reign of the monopoly was ending. Competition had been legalized in the long-distance segment of the business, and was apparently going to work. Competition had transformed the telecommunication equipment business. Not only had "legitimate" competitors like Northern Telecom entered Western Electric's private reserve to sell more advanced digital switches than Bell could field, but a horde of smaller, entrepreneurial upstarts had torn huge chunks out of Bell's 100% market share for data modems, PBXs, and microwave radio. The concept of "bypass" was sending shivers through the corridors of the Bell system: the idea that large individual communication users, such as Fortune 500 companies, might set up and operate their own communication systems, saving money by "bypassing" the AT&T network, seemed to attack the foundations of the monopoly.

Moreover, AT&T was once again embroiled with the Justice Department in a massive antitrust action. Having already lost numerous preliminary skirmishes, the incredible possibility of a Standard Oil type of breakup of the Bell System was beginning to be pondered. Although the climax of the drama — the dismemberment of the AT&T system into eight separate companies — would not come about until January 1, 1984, the drift toward all-out competition was unmistakable. What had seemed unthinkable only a few years before became not only thinkable, but seemingly inevitable.

The role and goals of the Federal Communications Commission were also changing. In 1970, the FCC and AT&T had worked hand-in-glove in many areas of telecommunication policy. While the Commission was often critical of AT&T's position on economic matters such as tariffs and rate structures, it still usually acknowledged the paramount authority of the Bell system on technical matters. Also the basic principles of the Kingsbury Agreement — that the telephone network was best managed as an integrated, nationwide public utility and that competition in most aspects of the telephone business was inherently "uneconomic" and therefore undesirable from the consumers' standpoint — were still generally accepted by state and federal regulators.

This attitude was evident in the early phases of the cellular proceeding. The Commission not only turned to Bell for technical proposals concerning the cellular architecture, but explicitly assumed that AT&T would develop and operate the system nationwide. There would be only one cellular operator in any given city, with untrammeled access to the entire cellular spectrum. Under such a regime, there would be no real need for a drawn-out cellular licensing process. The Commission could rely upon AT&T to ensure high-quality service and adherence to the appropriate technical standards (which had emerged from Bell system plans in the first place).

At first, therefore, the Commission was spiritually and practically opposed to the RCCs' insistence on competitive procedures. Not only did the FCC doubt the abilities of the small "Mom and Pop" mobile operators to build and operate complex cellular systems, the Commission staff dreaded the prospect of comparative hearings to decide among competing applicants. Such hearings always introduced delays; no matter how unequal the applicants or how obvious the choice among them, the demands of due process would have to be met. This could disrupt the orderly roll-out of cellular technology.

As the RCCs were able, however, to prevail in part in the courts, and especially as they began to develop and present a credible competitive alternative, the thinking at the Commission began to change. Hardman believes that initially this change was reactive. In a single-operator-per-market arrangement, which the Commission still favored, it was recognized that comparative hearings would become the rule instead of the exception as capable RCCs brought forth competitive proposals on the basis that they had "nothing to lose." Each cellular license would become a tangle of controversy and litigation. The FCC's license awards would be subject to further court challenges — with attendant further delays in the start of service. And the test of competitiveness applied to the FCC's procedures would dictate in any case that AT&T could *not* win in all markets. This brought the Commission back to square one on the question of competition.

True competition meant allowing for multiple cellular operators in any given market. Obviously the total available spectrum would have to be divided among them somehow. At this, AT&T balked. They pointed out to the FCC that the original cellular design was based on the assumption that a single operator would have access to a large block of spectrum, at least 40 MHz. The system overhead, the frequency allocations and reuse plans, and the control protocols were all designed and optimized for a certain scale. Cellular could not operate efficiently in a downsized mode for a competitive environment. Dividing the spectrum up among even two operators would result in higher user costs. In fact, AT&T calculated that the costs of operating in a two-system mode with 20 MHz for each system would be 30% higher than if the entire allocation were licensed to a single operator [17]. They also predicted that there would be severe near-term capacity shortages if the spectrum were split up among two or more operators [18].

To resolve the problem of whether, and how, to introduce competition into cellular licensing, in 1980 the FCC began yet another cellular rule-making, Docket 79–318, to evaluate alternative licensing approaches. Three options were considered:

1. Maintaining the single-operator monopoly concept favored by AT&T and incorporated into the previous rule-making (18262);
2. Allowing "open entry" by an unlimited number of competitors, the "let the market sort it out" approach;
3. Creating two systems per market, each with half the spectrum.

A new round of public comments now ensued, drenched in technical esoterica. AT&T bolstered its arguments for a unified 40-MHz allocation. The Justice Department argued for modified "open entry," with allocations of perhaps as small as 5 MHz per operator [19]. Motorola had previously brought forth its own studies to show higher costs for smaller systems [20]. Paradoxically, Motorola had supported an initial spectrum allocation limited to only 12.5 MHz, with more to be granted as needed. Some observers attributed this to Motorola's interest in "protecting its substantial share of the dispatch equipment market" by making sure that cellular would not become too inexpensive and threaten to woo away traditional dispatch users [21].

In the end, the FCC resorted to the Solomonic solution. They substantially accepted the Bell system arguments that smaller systems were inherently less efficient and more costly. Nevertheless, they decided to split the total allocation into two operating licenses for each market, each to receive 20 MHz. The principle of efficiency — which is the traditional monopoly virtue — was to be sacrificed for the principle of competition.

The Commission wrote:

> Even the introduction of a marginal amount of . . . competition into the cellular market will foster important public benefits of diversity of technology, service, and price . . . The commission is unpersuaded that these benefits would be outweighed by the benefits associated with the increased efficiency of a 40-MHz system over that of two 20-MHz systems . . . [22].

It is easy to dismiss AT&T's warnings as self-serving. As discussed in Chapter 4, however, many of these predictions proved to be accurate. System costs were higher than expected, and capacity was inadequate. The warnings were sound, based on the architectural and technological unsuitedness of the cellular standards for a competitive service.

The FCC also addressed the question of how the licenses should be awarded and to whom. It was decided that in each market one license would be set aside for the local wireline telephone company. The other would be reserved for companies other than telephone companies, the "nonwireline carriers." This "fence" was not necessarily envisioned as a permanent fixture of the license; it was instituted to offset the perceived handicap of the nonwireline companies in a head-to-head comparative hearing with the wirelines.

Incredible as it now seems, the FCC actually saw the two-operator arrangement as a way of streamlining the licensing process by reducing or eliminating comparative hearings. It was assumed that by "pre-sorting" the applicants into two groups, the number of competing applications would be reduced. Where, despite this, there were competing applications, the Commission would decide among them through a full-scale comparative evaluation of the proposals. Applicants were instructed to provide ample and detailed engineering, marketing, and financial information. It was apparent that a proper application would cost quite a bit to prepare. This would further discourage large numbers of applications and simplify the licensing process.

Finally, on March 3, 1982, the Commission issued its final rules for the cellular licensing process and indicated that it would accept applications for the top thirty markets on June 7, 1982. The long wait for cellular radio seemed nearly over.

Since the licensing process subsequently degenerated into almost total chaos (see Section 4.4), it is relevant to ask whether the FCC was wrong in embracing, but only partially embracing, the idea of competition. Would things have worked out better had the cellular industry opted for either of the other alternatives, monopoly franchises or more open entry?

The question is probably impossible to answer. Yet given all the criticism which has been heaped upon the FCC for its halfhearted choice, it is worthwhile pausing to consider why this conclusion was reached.

Initially, at the beginning of the process, the Commission appears to have been relatively conservative, passive, reactive. After two decades of holding the line against any additional spectrum allocation for mobile radio, the FCC was compelled to act in the late 1960s, as we have seen, not so much by any new procompetitive regulatory "theory" emerging from within the Commission itself as by the pent-up forces of the mobile radio industry which was successful in persuading a growing number of people within the government that mobile radio had indeed been given short shrift at the hands of the broadcast interests. The Commission undertook its momentous move in Docket 18262 with a palpable reluctance, and without any real "theory" or plan. "This scheme [Dockets 18261 and 18262] is simply a response to a political situation," wrote Commissioner Johnson whose vote "swung" the issue. "Our response is not a well thought-out plan; it is not the result of planning; rather it is an immediate response to an immediate need [23]." Indeed, the initial character of the cellular allocation was unmarked by any pronounced bias toward competition. As we have seen, the Commission actually took a step away from a procompetition position by assuming a monopoly service under the auspices of AT&T, rather than a continuation of the precellular competitive markets.

The procompetitive ideas came from outside at first. One such influence was a 1973 report of the Office of Telecommunications Policy (OTP), an agency of the Executive Branch which interacted closely with the Commission staff in the early 1970s. The OTP formulated in its own submission on 18262 the competition issue: they asked whether "increased availability of mobile communications services is best achieved by a regulatory technology or by the creation of a diverse competitive environment." The implicit answer was the latter [24]. Another external influence was the Anti-Trust Division of the Justice Department. Anti-Trust was perhaps the most hostile and well-informed opponent of the conventional monopoly thinking which so dominated everyone in the communication industry, including the FCC. In 1970 and again in 1973, they urged the Commission to allow for competitive forces in the mobile telephone market. They supported the RCCs at a time when the RCCs probably lacked credibility with most of the FCC staff. They also urged the Commission to prohibit the wireline operators from "owning, manufacturing, supplying, or maintaining user mobile equipment [25]."

There was a keen awareness of the potential for cross-subsidization of cellular from wireline profits and of the *de facto* monopoly resulting from AT&T's dominant market position. The cellular proceeding was important in pointing the way toward a wider competition, particularly since radio-based systems did not fit the "natural monopoly" arguments that tended to buttress the *status quo* in conventional wireline telephony. As some analysts have observed: "The Department of Justice viewed the docket as providing the commission with a unique opportunity to expand the role of competition in the communications industry [26]." Yet it must be remembered that these comments were intermixed with hundreds of other pleadings. The handful of government policy-makers who were beginning to believe in the totem of "competition" were more than offset by the conventional industry players, especially AT&T, who were not interested in promoting competitive thinking at all. The Commission in the late 1960s and early 1970s was far from leading the charge for competition and deregulation in the mobile telephone field, although they were responsive to persuasive reasoning along these lines.

By the early 1980s, in contrast, the FCC had armed itself with a much more definite set of ideas about the proper structure for, and the role of, competition in telecommunication as a whole, and in mobile telephony in particular. Several years of largely favorable experience with deregulation in other industries (such as trucking and the airlines) had convinced many people that deregulation and competition could work in telephony, too. Because of the antitrust suit brought by the Justice Department, however, even at this stage the policy direction was not entirely in the hands of the Commission. In the field of mobile radio, therefore, a set of cautious policies for "controlled competition" was implemented. The Commission still approached the problem with the idea that competition had to be "created." The thought that it might simply be "allowed" was still too radical; there was still in the early 1980s a strong, lingering distrust of unbridled market forces.

The Commission was also changing in other ways during this period. It was evolving from an old-style politicized body which doled out rights to a scarce resource (the spectrum) on the basis of the relative strengths of the interest groups which might at one time or another exercise political influence and becoming a more modern, "scientific" agency which attempted to work out and implement a theory of spectrum management that would ensure the most efficient utilization of the scarce resource. There was a growing interest in market mechanisms; the Commission even explored the idea of "auctioning" the spectrum to the highest bidder, in much the same way that offshore oil leases were sold. The courts, however, later determined that the power to sell spectrum rights lay outside the Commission's charter.

To have followed a more procompetitive path would have required a much stronger hand than the FCC could play in the 1970s. Had they exercised that strong hand, the delay might have been reduced. But in the absence of an internal "theory" for the implementation of competition in cellular service, the Commission was paralyzed by conflicts among the major interest groups. In the end, the decision was to straddle the options.

The real tragedy of the cellular delays, however, was not the postponement of service by a few years. It was the fact that the technical and regulatory environments both underwent such tremendous changes during those years: by the time the problem of the industry structure was "solved" in the early 1980s, a technological gap of far greater proportions had emerged. In effect, cellular technology had become obsolete even as it was reaching the marketplace. It is a commonplace in our era that by the time any new technology reaches the market, a better, faster, cheaper version is already well established in the laboratories. This is taken into account in normal product planning cycles. The obsolescence of analog FM-based cellular radio was far more fundamental. By sheer bad timing, the ten-year delay in cellular deployment straddled what will come to be seen as one of the great "revolutions" in communication technology of this century.

As the cellular proceedings developed against a background of sweeping procompetitive changes in the telephone world, the stance of the FCC began to shift from reactive to proactive. It had begun to seem more and more that competition would work for cellular radio, and the new economics of the 1980s certainly favored the reintroduction of competitive processes wherever possible. By early 1982, the AT&T monolith was already programmed for dismemberment. Seven Bell operating companies would soon inherit the operating licenses for the wireline allocations (along with other local telephone companies). The vision of a single nationwide operating entity for cellular radio, so fundamental to cellular planning in 1968, had itself become an anachronism. The Commission had finally moved to a fully procompetitive posture in the early 1980s as cellular systems were just about to come to the market after a ten-year delay.

REFERENCES

[1] W. R. Young, "Advanced Mobile Phone Service: Introduction, Background, & Objectives," *Bell System Technical Journal,* Vol. 58, No. 1, January 1979, p. 7.

[2] *Ibid.,* p. 6.

[3] Mark Nadel, Robert E. Glanville, and Philip L. Bereano, "Land Mobile Communications and the Regulatory Process," in Raymond

Bowers, Alfred M. Lee, and Cary Hershey, eds., *Communications for a Mobile Society: An Assessment of New Technology,* Beverly Hills, California: Sage Publications, 1978, p. 69.
[4] Young, *op. cit.,* p. 6.
[5] Nadel *et al., op. cit.,* p. 68.
[6] *Ibid.,* pp. 68–69.
[7] *Ibid.,* p. 68.
[8] *Ibid.,* p. 76.
[9] *Ibid.,* p. 78.
[10] *Ibid.,* p. 70.
[11] Kenneth Hardman, "A Primer on Cellular Mobile Telephone Systems," *Federal Bar News & Journal,* Vol. 29, No. 11, November 1982, pp. 385–391.
[12] *Bell System Technical Journal,* Vol. 58, No. 1, January 1979.
[13] Nadel *et al., op. cit.,* p. 77.
[14] Cited in Hardman, *op. cit.,* p. 388.
[15] *Ibid.,* p. 388.
[16] Cited in Hardman, p. 389.
[17] Harold Ware, "The Competitive Potential of Cellular Mobile Telecommunications," *IEEE Communications Magazine,* November 1983, p. 18.
[18] *Ibid.,* p. 19.
[19] *Ibid.,* p. 18.
[20] Erwin A. Blackstone and Harold Ware, "The Emerging Mobile Communications Industry: Structure & Regulation," in Raymond Bowers, Alfred M. Lee, and Cary Hershey, eds., *Communications for a Mobile Society: An Assessment of New Technology,* Beverly Hills, California: Sage Publications, 1978, p. 345, Note 45.
[21] *Ibid.,* p. 371.
[22] Ware, *op. cit.,* p. 19.
[23] Nadel *et al., op. cit.,* p. 79.
[24] *Ibid.,* p. 77.
[25] *Ibid.,* p. 77.
[26] *Ibid.,* p. 77.

Chapter 4
CELLULAR REALITIES

The Chicago cellular system began operations on October 13, 1983. AT&T, the prime mover behind the cellular idea, became a cellular operator for exactly seventy-nine days. (On January 1, 1984, the Bell system was broken up into eight separate companies. AT&T withdrew from the operating business, and the Chicago system was taken over by a subsidiary of Ameritech.) By the end of 1984, cellular service had come on line in twenty-five American cities [1]. After fifteen years of planning, research and development, political wrangling, and regulatory delay, cellular radio finally reached the market.

It was soon clear, however, that something had gone very much awry. The industry was beset by problems that seemed increasingly intractable. Costs were high, service was flawed, customers reluctant, profits thin or nonexistent, and the industry structure was in a shambles (at least in terms of the expectations of those who had started out so optimistically in 1983–84). Some were beginning to question cellular's fundamental technological viability.

The problems went deeper than the normal difficulties of developing a new market. Compare, for example, the historical timetable for the commercial introduction of broadcast television in the United States with the cellular radio experience. (See Figure 4.1.) Initial market penetration was slow to develop. At the end of Year 1 of television (1946) there were scarcely 8,000 American homes that boasted a set [2]. In Year 3 the number of TV-equipped households was up to 172,000 — still well under 1%. Then came the explosion: the compound growth rate was 130% per year for the next six years. By Year 9, TV had reached 55% of American homes, more than 26 million households. By Year 18 (1963), penetration had reached 90%.

Mapping the cellular pattern onto the same relative timeline discloses a growth pattern that is even faster — at first. Cellular leapt out of the

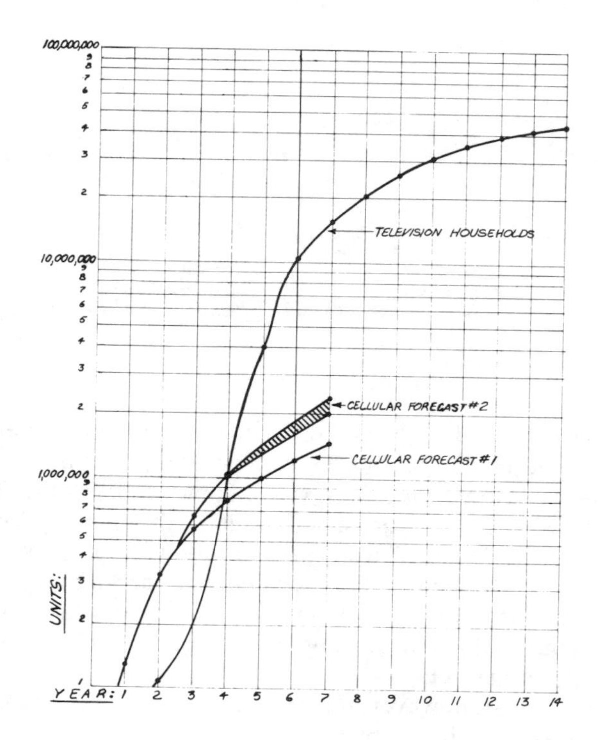

Figure 4.1 Relative Time Lines for Consumer Penetration of Television and Cellular Radio.

Sources: U.S. Census; Frost and Sullivan; Herschel Shosteck.

blocks in Years 1 and 2 (1984 and 1985) on the basis of pent-up demand in the major cities. The one millionth subscriber was projected to be signed on during Year 4 or Year 5. Cellular has kept pace, more or less, allowing for the different measures of market penetration. So what is the problem?

First of all, many market analysts believe that by any measure the peak of cellular sales (in units and in dollars) may already be past, having occurred in Year 3 or Year 4. (See Figure 4.2.) The rate of growth in the number of subscribers is slowing, and forecasts for the smaller markets are especially gloomy. (See Figure 4.3.) The trends portray a deceleration typical of a saturating market in the mature stages of the product life cycle. Can it be true, however, that the market for mobile communication is saturating at around 2%? Is the cellular market maturing at the age of six or seven?

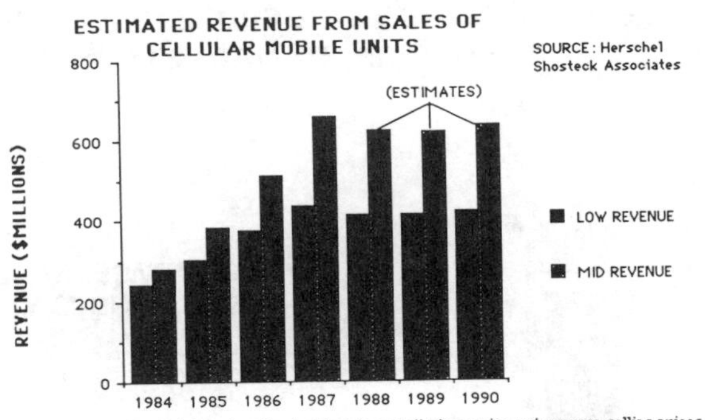

This estimate is based on Shosteck's data on units in service and average selling prices. The "Low Revenue" is based on Shosteck's "Average Lowest Price" obtained from dealer surveys. "Mid Revenue" is based on "Average Midpoint Price" from those surveys.

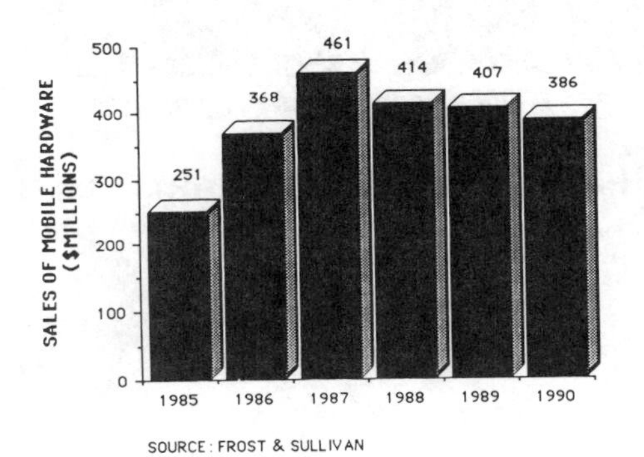

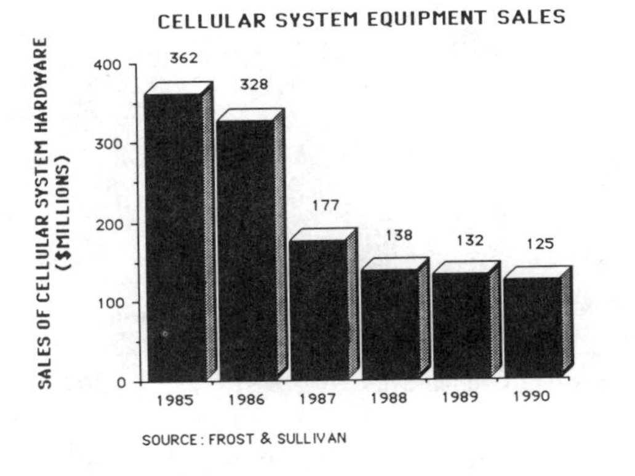

Figure 4.2 Estimated Sales of Cellular Equipment.

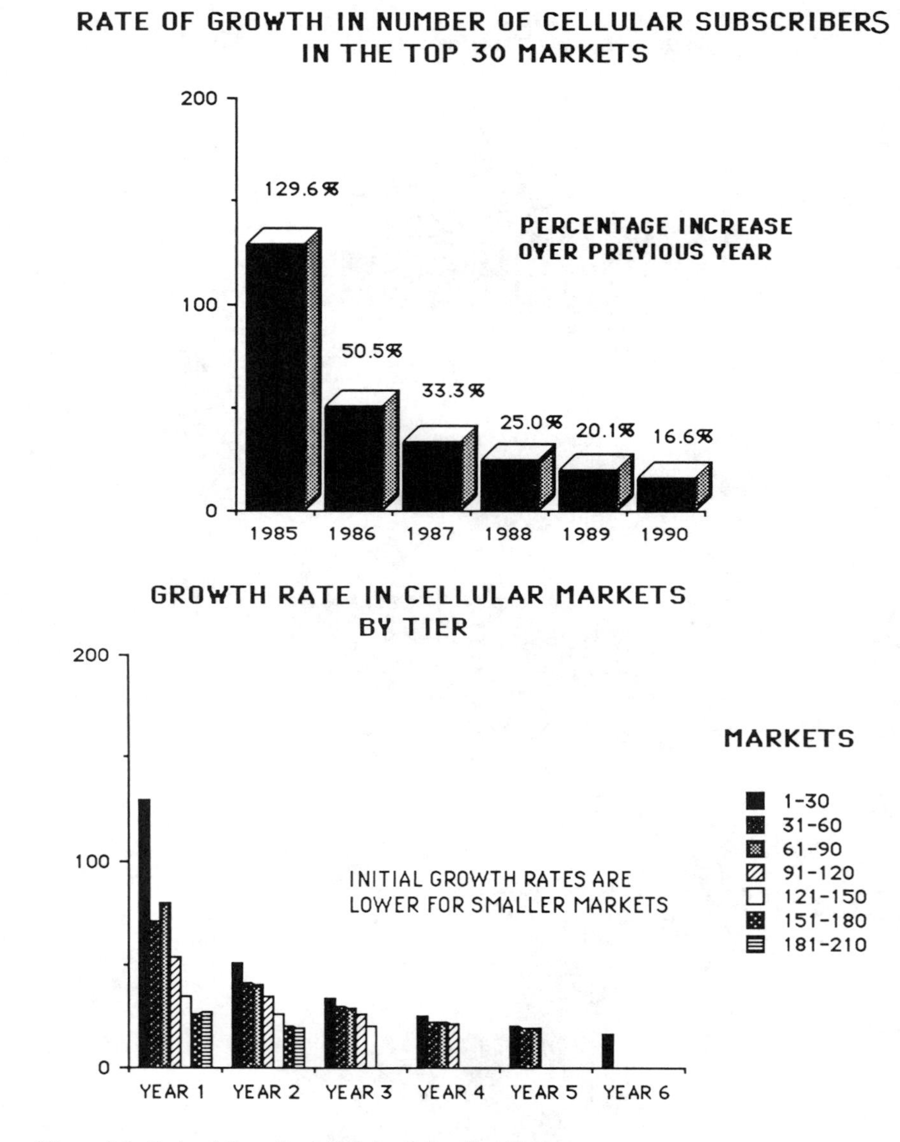

Figure 4.3 Rate of Growth of Cellular Subscriber Base.

It is generally acknowledged that it will be simply impossible for cellular radio to keep pace with the television growth path beyond about Year 5, assuming today's technology. At the point at which television was entering its true growth phase, becoming established as a mainstream consumer industry, cellular radio is reaching its maximum. In the macroperspective, it does not matter a great deal whether the ceiling is 1% or 3% of the population; either case represents a stalled takeoff for a would-be mass consumer product.

Second, the basic standards for broadcast TV remain intact in Year 40 of television. TV was a standard that worked; it created a stable technical environment within which manufacturers could steadily reduce their costs and expand their markets.

By comparison, Year 4 of cellular radio (1987) saw the mobile telephone industry beginning serious discussion of not merely the modification, but the abandonment of the established technical standards. The FCC has recently proposed scrapping existing technology restrictions and compatibility requirements and opening the cellular spectrum to "new technology [3]." The next technological generation is already being designed in Europe. The current forecasts as to when that generation will actually be deployed range from about Year 7 to Year 12 (1990 to 1995). Some observers have forecast *two* new technology generations before Year 17 (2000) [4].

This is not a case of planned obsolescence. If a new generation of technology is to come in so soon, the massive investment in today's analog cellular technology (including development costs and system hardware) will probably not be fully recovered. Instead of reaping its profits, the cellular industry will embark on an even more expensive round of technology development; some estimates run to more than $1 billion to be spent worldwide just on R&D. The changeout of system hardware will be a staggering expense for hard-pressed operators. Cellular subscribers will be saddled with obsolete and costly mobile units. Clearly no one sought to bring about this state of affairs.

What went wrong? A system old before its time, its growth capped at very low levels, a technology standard that must be changed even before it has been fully deployed — all this speaks of megamiscalculation. After thousands of man-years of planning and development invested by hundreds of private companies and government planners, how could the result have been so far off the mark? How could an entire industry converge on the same wrong answer?

The purpose of this chapter is to review and analyze with a cold eye the realities, the problems, of cellular radio in the mid-1980s in the United States. In some respects the debacle is uniquely American; in most, it is not. It can be broken down into several distinct syndromes.

First and foremost, the economics of cellular operations never came into line with either the designers' expectations or the requirements of the market. Second, the cellular system suffered from unyielding performance problems — often the structure simply did not work as planned. A third set of difficulties relates to the spectrum efficiency question: the spectrum management scheme envisioned by AT&T in the original cellular idea, including the multiplication of the spectrum through cell-splitting, did not evolve as anticipated. Fourth, there was a virtual breakdown of the regulatory and licensing process, which dimmed the prospects of realizing the FCC's vision of a healthy, competitive cellular industry. Let us consider each of these syndromes in turn.

4.1 THE COST SYNDROME

The economic difficulties that have afflicted cellular radio can be differentiated into three categories:

1. Mobile-unit cost problems;
2. System cost problems;
3. Operating cost problems.

The total cost of a cellular telephone is composed of these three elements. In a deregulated environment, the potential customer must of course purchase or rent his own mobile radio unit. This is a direct outlay from his own pocket which constitutes the initial barrier to his recruitment as a subscriber to a cellular service.

The cost of the system is defined by the operator's investment in cell-site base stations and other system overhead equipment (including the mobile telephone switching office, which controls the metropolitan cellular network). If we assume a large cellular system operating at or near capacity, it is meaningful to talk about an average system cost per subscriber, which is calculated simply by dividing the total investment in system equipment by the number of subscribers that can be served at maximum capacity. The monthly charges are directly correlated with system cost per subscriber.

The final element of the costs of cellular service is derived from the ongoing costs of operating the cellular system. These include the conventional costs for operating any type of communication service business, such as administration, sales, maintenance, and so forth. Pricing of the service must be sufficient to cover these ongoing expenses.

4.1.1 Mobile Equipment Economics

The cost of the mobile unit was identified early in the planning process as the key target for cost reduction. In the early 1970s, the 800-MHz spectrum band was still *terra incognita* for most radio engineers. There was concern over the high cost of 800-MHz components. Forecasts for the initial cost of the mobile radio unit ranged between $2500 and $4000, which was felt to be the most serious impediment to the development of a broad market.

To bring costs down, large production runs would be necessary, or so it was felt by early system planners. The chicken-and-egg problem of price and volume emerged: high volumes could only be sold if the price of the mobile unit were low, and the price would only be low if manufacturers could gear up to produce high volumes. The cellular industry became convinced that if manufacturers planned for high volumes of sales and production, the resultant pricing would create the market they needed. A forecast of rapidly falling prices became one of the mainstays of cellular promoters.

The faith in this process was often quite naive, based upon rather simplistic views of production, derived more from economic theory than from real industry experience. The following typified the early optimism (Bowers, Lee, and Hershey, *Communications for a Mobile Society,* 1978, p. 355.):

> It appears that the initial market is apt to be substantial enough to allow equipment prices to fall. There are several reasons for this conclusion: such a market makes it feasible to introduce specialized, automated production techniques; experience in producing many similar units reduces costs along "learning curves"; input costs may decline when purchases are in bulk; and the knowledge that there is a large market may promote greater expenditures of research and development (R&D) funds to increase efficiency.

As the market developed, mobile equipment prices did begin to fall. In 1984–85, mobile prices hovered around $2000 to $2500. By 1987, prices had dropped by about 50% to around $1000 to $1200. Some retailers were selling bargain-priced equipment for as low as $800. On the surface, it was a dramatic decline.

Much of this reduction, however, came not from "learning curves" and other such semimythical factors, but from old-fashioned competition and price-cutting under conditions of severe oversupply. Manufacturers had geared up to produce far more units than the market could absorb. A glut developed almost immediately. At the end of cellular Year 1, there were two mobile phones in inventory for every one that had been sold [5]. Many equipment suppliers reconsidered their commitment to the market; the shakeout took place much earlier, at unit sales levels much lower than would normally have been expected. Although profit margins on mobile equipment are difficult to determine, it also seems clear that most manufacturers were shaving their margins to the bone to stay in the game and avoid drowning in inventory. One analyst from Arthur D. Little, Inc., summarized the turmoil of the first year as follows:

> In the year since cellular service first became commercially available, the retail price of subscriber equipment has fallen from the $2500 range to as low as $1500 . . . This dramatic decline has resulted from dealer competition — dealers have been forced to accept low or nonexistent margins . . . [6].

Another analyst estimated in 1985 that "although U.S. production capacity is currently upwards of 400,000 units annually, demand is forecast at approximately 150,000 units for 1985, a 38% capacity utilization rate [7]."

The entrance of Japanese manufacturers also upset the market plans of many of the more traditional suppliers. The Japanese have proved that they can capture and dominate many markets by their emphasis on long-term planning, on production expertise, on pricing for market share (labeled "dumping" by some), and on an impeccable reputation for quality. The result was particularly unsettling for Motorola, which had been accustomed to "owning" the North American mobile radio market, with an approximately 65% market share prior to the cellular introduction [8]. Due to recurrent quality problems (Motorola units were reported to show defect rates 10 to 20 times higher than units from Japanese competitors) and what many regarded as shortsighted pricing and distribution practices, Motorola found its market share in the cellular mobile equipment market in North America slipping [9, 10]. The company responded by bringing dumping charges against its Japanese competitors [11] and patent infringement allegations against others, including Novatel [12]. Although the dumping allegations in particular have been sharply criticized by some economists [13], the Commerce Department ruled in favor of Motorola. It may have been a Pyrrhic victory. Far Eastern suppliers appear to have conquered the mobile equipment market anyway. In 1986, according to dealer surveys, the top four suppliers, controlling 46% of the market, were non-US companies: Panasonic (12% market share), Audiotel — a Toshiba distributor (11%), Novatel (11%), and OKI (11%). Motorola was in a dismal fifth place (9%), followed closely by Mitsubishi, NEC, and General Electric (7%) [14]. By 1987, Motorola had slipped to ninth place (according to Shosteck's dealer surveys) with about 4.8% of the market [15]. Today, even some of the nominally "North American" suppliers (like the Canadian Novatel) are little more than transhippers from factories in Hong Kong, Taiwan, or Korea.

This is a very different industry structure from that envisioned by the founders of cellular back in the 1970s. Many past regulatory actions were predicated upon the desire to foster a strong domestic mobile equipment industry. Today the US manufacturers probably account for less than 20% of the cellular mobile market. The control of subscriber technology has emigrated offshore.

FM and the "Commoditization" of the Mobile Industry

In hindsight, the insistence on a single technology standard which virtually mandated the choice of a "least common denominator" technology such as analog FM, was the chief cause of the destruction of the domestic equipment industry. In most industries, proprietary technology is an important part of the competitive picture. The desire to obtain a proprietary patent, which affords the company a period of time during which it may recoup its investment and enjoy a competitive advantage, is an important motivator. The period of real advantage is usually considerably less than the life of the patent — truly fundamental patents capable of controlling entire industries for the full seventeen years of a patent's life are extremely rare. But even a shorter-lived edge is nevertheless the main incentive to drive the R&D investment that is the engine of technological progress. Even aside from the absolute, although temporary, monopoly afforded by a patent, companies still find an incentive to develop new technology in the lead time they may gain over their competitors, even where the competitors are more or less free to duplicate the new techniques. Engineering know-how at the innovating company, combined with inertia among the company's competitors, may mean that a new development can provide several years of competitive advantage even without any patent protection whatsoever.

In some communication markets, however, notably broadcast radio and television, a policy of fixed transmission standards has been adopted. All manufacturers must build to the same transmission specifications. In almost every case, this has decimated the domestic equipment suppliers, because it means that they cannot use new technology — the traditional American edge — to gain a competitive advantage. There may or may not be compelling reasons to retain such standards for broadcast services. The assumption by mobile radio planners that the same approach would have to be applied to cellular radio is more open to question. In any case, by specifying analog FM, a technique so mature that it virtually assured that no company could develop a meaningful competitive edge based upon proprietary technology, the FCC and the EIA and others unwittingly created a commodity-like market for cellular mobile equipment. All mobile units were fundamentally similar, regardless of the source. Differences in "features" were superficial and short-lived. Consumers began to buy mainly on price. It is the iron law of commodity-like markets that sales will tend to go to the producer with the lowest prices, or, to put it another way, that the prices of all producers will tend to become equal, and will tend to be equal to costs. The profit margins will be squeezed to the bare minimum. The most successful supplier is the supplier with the lowest prices, who will be the supplier with the lowest costs (hence the movement to lower-cost offshore manufacturing) and the lowest margins (hence the long-term perspective and the so-called "dumping" tactics of the Japanese).

The most serious effect of "commoditizing" the cellular equipment market, however, is the disincentive it creates for further fundamental research and development. The technological stasis tends to entrench itself.

This impact of "commoditization" on the cellular mobile equipment market may be fully appreciated by contrasting the cellular market with the microcomputer market, which is not as highly commoditized. The domestic computer industry is very robust, and highly competitive. Because the competition is based as much on capabilities as on price, technology investment is an advantage, not a handicap. There is a high level of expenditure on R&D. Foreign competition exists at a healthy level, but is not such that domestic producers are driven offshore or destroyed altogether.

Of course, the reader undoubtedly recognizes that the commoditization problem is inextricable from the question of transmission standards. Commoditization would seem to be an unavoidable consequence of fixing a single standard. Could the industry have developed — can it develop in the future — without a single standard? We shall return to this question in Part VI.

4.1.2 Cellular System Economics

In spite of the devastation of the domestic mobile-equipment *suppliers,* falling prices for mobile hardware could be viewed as a positive development for the cellular *operators.* Since cellular users purchase their own radiotelephones, lower equipment prices ought to make the recruitment of new customers that much easier for the operators. This, however, has not happened. Growth of the subscriber base is decelerating. (See Figure 4.3.) It has become increasingly difficult, and more costly, to sign up new cellular customers. In fact some of the retail price declines have actually been the result of "bounties" paid by cellular operators to marketing agents and distributors, cash bonuses of up to $800 for each new customer. Dealers have sometimes passed this through to the end purchaser in the form of a further deep discount, creating situations where the mobile equipment has actually been priced below cost [16]. From the operators' standpoint, they are subsidizing the mobile phone and hoping to make money on the service.

In reality, the cost of the mobile unit is no longer the chief economic obstacle to the development of cellular radio. Cellular planners were very worried about consumer resistance to a $2000 car phone. Yet even at $2000, the mobile unit was no more expensive (in constant dollars) than radio or TV sets were in the first years of those respective industries. Today, as hardware costs have dropped to half that or less, the *real* economic problem of cellular radio stands out more clearly: the problem of system costs.

Cellular radio is by nature a system with very heavy overhead. High fixed costs, a high proportion of which must be invested up front, create two problems for the cellular operator. First, the operator must be able to attract a relatively large critical mass of customers to share the heavy fixed costs and so reduce service prices to reasonable levels. Unlike other types of communication services, cellular radio is difficult to build up in small steps; it must be established from the very beginning on a scale far larger than anything ever seen before in mobile telephony. The subscriber population required to reach the break-even point for a single large metropolitan cellular system (e.g., Los Angeles) is probably greater than the entire population of mobile telephone users nationwide during the IMTS era!

Second, the operator must be able to survive the relatively long start-up period before reaching this critical mass of subscribers. Large amounts of capital are needed not only to build the system but also to sustain the operating losses until revenues catch up. It is difficult to gauge how long this period may be. It is probably at least a couple of years in most markets; some analysts are now predicting that it may be sometime in the 1990s before many of today's systems turn a profit. It becomes imperative to minimize these costs by speeding up the achievement of the critical mass. This "sprint for break-even" has engendered extraordinary and expensive marketing strategies (such as the use of bounties).

There is also concern over the steady state economics of today's cellular systems. Even if prices are optimistically set based on the assumption of a fully loaded system, the price of cellular service may be too high ever to attract that critical mass. For example, let us assume that a given cellular operator has determined his maximum system capacity (without additional investment in cell-splitting) to be 10,000 customers. Further assume he has determined, based on his fixed costs, his operating costs, and his bare minimum profit requirement, that at maximum capacity he needs an average of $150 revenue per customer per month. Are there enough people in his market willing to pay $150 a month for cellular service (on top of the cost of the equipment, installation, and insurance)? Are there still enough after we give half of them to the competing operator? If not, then the system cannot achieve economic viability.

4.1.2.1 System Overhead Costs (Hardware)

The high fixed costs of cellular systems are the result of three factors:

1. The cellular architecture itself, especially the proliferation of cell-sites;
2. The traffic-engineering objectives of cellular service;
3. The inefficient radio-link technology (our recurring theme).

Cost Structure of Cellular Architecture

As a system for providing mobile telephone coverage for a metropolitan area, a cellular system is much more expensive per square mile than the older IMTS systems. It also requires a *much* larger up-front commitment, with a riskier growth path which usually dictates a more expensive financing strategy.

(All cost numbers in this chapter are essentially hypothetical unless otherwise indicated. Equipment costs are constantly changing, generally downward. The objective here is to illustrate the cost relationships that determine the economics of cellular systems.)

Consider first the structure and growth pattern of a conventional, single-cell IMTS radio-telephone system. The IMTS operator can enter the market with a relatively small amount of capital. Let us assume that he begins with five radio channels. He covers the metropolitan area with one transmitter; he needs only one base station, one piece of real estate, one antenna tower. Let us assume that each channel costs him $40,000 for radio hardware — his initial equipment investment is $200,000.

His growth strategy is simple. He signs up customers until he reaches a level that he has determined as 100% capacity, according to the traffic and blocking standards he has set for the system. This determines the cost basis of his service pricing. If, for example, he determines that 100% capacity on five channels is equal to 100 customers, then he must allocate

$2000 in system costs to each customer. The $2000 system cost per customer becomes his basis for setting his prices (ignoring other costs for the moment). Let us assume that the operator wants to recover his base station investment in five years. To do so he must receive $33 a month from each of his 100 customers. He sets his base rate prices so that the average revenue per customer is $40 a month (excluding usage or toll charges). At 100% capacity, his monthly income statement looks like this:

Revenue	$4000 ($40 per month from 100 customers)
System Hardware Amortization	$3333 ($200,000 divided by 60 months)
Profit (Before All Other Expenses)	$ 667

When this hypothetical IMTS operator begins to exceed capacity, resulting in a degradation in service, he adds another channel, costing another $40,000. At the moment he adds the channel, his system cost per customer jumps to $2400. His monthly income statement is now:

Revenue	$4000 (Pricing is unchanged)
System Hardware Amortization	$4000 ($240,000 divided by 60 months)
Profit (Before All Other Expenses)	$ 0

As he signs on new customers to return to 100% capacity, his revenues increase and his costs per customer come back down to the target level of $2000 upon which his pricing is based. His growth path is thus relatively manageable; his fixed costs per customer are never more than 20% higher than his target. Moreover, each increment is small, which increases the chances that growth can be financed from operations and reduces the likelihood of the operator ever having to suffer a prolonged period of unprofitable operations. (See Figure 4.4.)

This model is characteristic of many "simple" communication networks, such as paging systems. Start-up costs are small. Growth is easily managed, incremental, and low-risk. Investments are closely matched with immediate revenue prospects. Most operators are actually able to gauge the pent-up demand from a waiting list of potential customers and are assured before they make expenditures for each new increment that the new customers will be there. This pattern allowed for healthy competition in mobile telephony throughout the precellular years. RCCs flourished

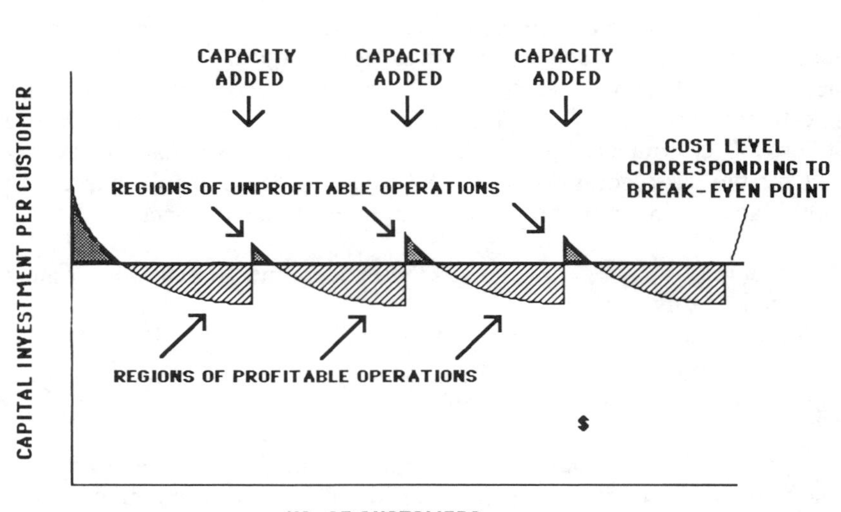

Figure 4.4 Economics of IMTS-Type Mobile Operations.

without any monopoly or duopoly franchise, without the benefits of lower equipment prices from megavolume production lines, indeed, without any of the special features that regulators attempted to design into the cellular market to "ensure" adequate competition.

The cellular architecture is very different. To cover his metropolitan area, the cellular operator is forced to make a much larger up-front investment. Instead of a single transmitter, he must construct dozens of cell-sites. Every cell-site requires real estate, tower, and considerable hardware. Basically, the cost of the IMTS operator's single site is multiplied by the number of cells in the initial system configuration. Most cells must be operational before service can commence. The cellular operator must also build or lease transmission facilities to connect all the cells into the mobile switching center.

The cellular ante is much larger. Currently the estimated system hardware and facilities costs appear to be running between $500,000 and $1 million per cell for urban systems. A system for a major metropolitan area can cost from $10 to $20 million or more before the first revenue-producing call is placed. While this investment buys much more capacity than an IMTS system and (based on these factors only with all else being equal) the system cost per subscriber would be about the same as for IMTS, the matching of capacity to demand is much "lumpier" for a cellular system (particularly the first "lump"). (See Figure 4.5.) The cellular operator is far more exposed financially than his IMTS counterpart.

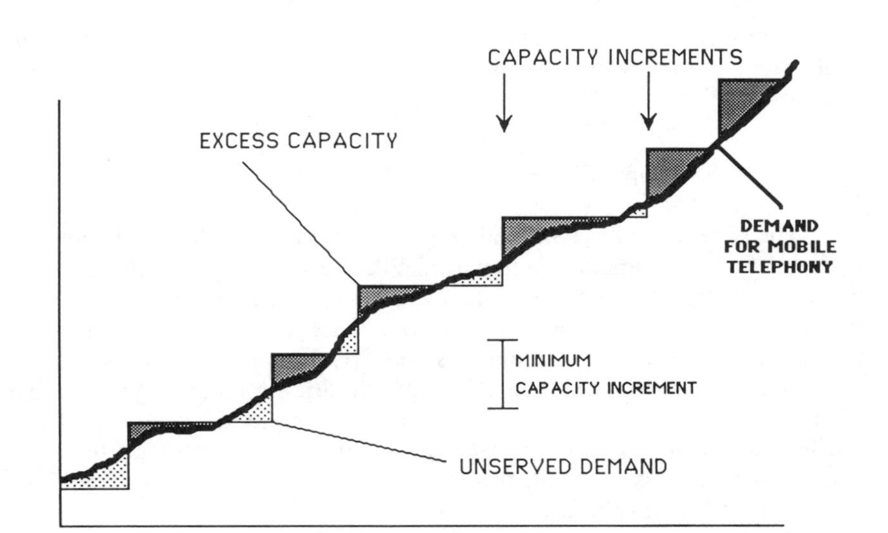

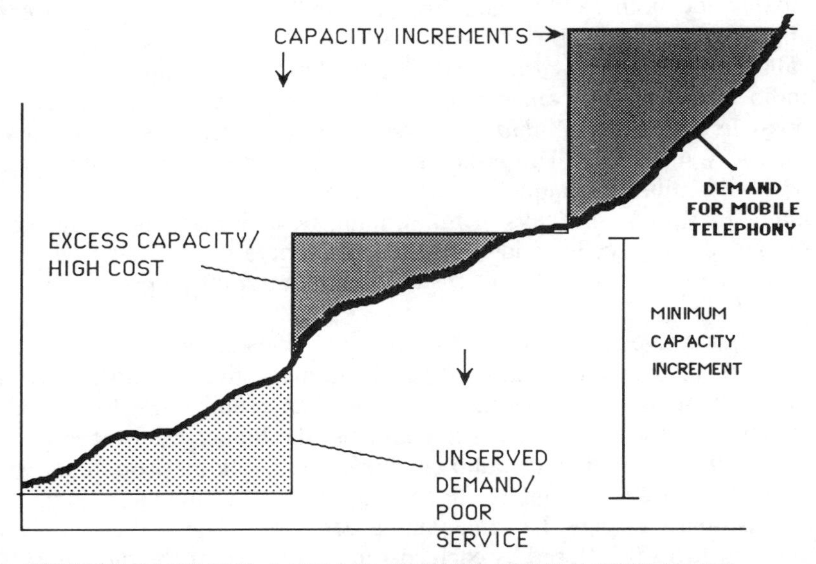

Figure 4.5 Growth Paths for IMTS *versus* Cellular Mobile Operations.

In this sense, cellular radio truly is "big business." In its cost structure, it resembles less a paging system or an IMTS operation than, say, a nuclear power plant. It is highly leveraged, both operationally and financially. Its economics belie its origins under the aegis of the AT&T monopoly.

Traffic Objectives of Cellular Service

The objective of the original architects of cellular radio was to achieve a *grade of service* comparable to wireline telephony, both in terms of voice quality and circuit availability. The availability of circuits in earlier IMTS systems had been a particularly sore spot. One measure of availability is the *blocking probability*. Cellular systems are mandated to achieve blocking of 2% or less: no more than 2% of all calls attempted during the busiest hour (e.g., rush hour) shall fail to obtain access to the network. Even during the period of heaviest use, 98 out of 100 calls must go through. This contrasts with blocking probabilities for IMTS systems of 50% or more during the busy hour.

Designing for a given blocking probability depends upon the amount of calling traffic that is "offered" on the average by the cellular subscribers (to include both calls to and from the cellular customer, which can be characterized by both the number of calls and the average length of the calls from an average customer during the busy hour), and upon how many radio channels are available for a given population of subscribers. It is possible to design a *nonblocking* system, in which there is a radio channel for every subscriber. This is expensive overkill, however, since most users will only utilize a channel for a few minutes out of the busy hour, and many may make no calls at all. Calling behavior can be analyzed and predicted statistically, and this allows the operator to get by with fewer radio channels, since not all customers will attempt to place calls at the same time.

For a given number of subscribers offering a known amount of traffic (seconds of usage, or "Erlangs") and sharing a given number of trunked radio channels, the blocking probability can readily be calculated. The higher the ratio of customers to channels, the higher the blocking. If 100 customers are sharing 10 radio channels, the blocking probability will be lower than if 200 customers are sharing those 10 channels (assuming that all customers display the same calling behavior). The higher the average amount of traffic offered by each customer, the higher the blocking. If 100 customers are sharing 10 channels and each customer tends to spend an average of 10 minutes each busy hour on the phone, the blocking probability will be higher than if each customer spends an average of 2 minutes on the phone.

Traffic engineering is a well-established discipline which is essential for many areas of communication design (including switching systems, long-distance trunks — any shared network facility). As a general rule, assuming that the size of the calling population and its behavior are fixed, a higher grade-of-service objective (lower blocking probability) means a *reduction* in the ratio of customers to trunked circuits (in this case, radio channels). To put it another way, more base-station equipment is needed for each customer.

Thus, the establishment of 2% blocking for cellular radio automatically increases the system overhead, the system cost per customer, compared to precellular systems. It is the inescapable price of better service.

The impact on system economics can be surprisingly large. The difference between a 2% blocking standard and a 50% blocking standard works out to about two and one-half times more equipment required by the cellular system per subscriber, even at the single-cell level. Consider an example derived from Dr. W. C. Y. Lee [17]. A cell with 45 voice-channel frequencies (ignoring control-channel overhead) can support about 1200 users with a 2% blocking probability, assuming that each user generates an average of 105 call-seconds of traffic in the busy hour. If we allow an IMTS-style 50% blocking, the same 45 channels could support more than 3000 users with the same calling characteristics. The total hardware cost of the 45 channels at the cell-site is the same in either case; but the cost per customer is much higher for the cellular case. In the 2% blocking situation the cost is shared by 1200 users, while in the 50% situation it is shared by 3000. The system cost per user is about 2.5 times higher for the cellular customer *based on this factor alone.* (See Figure 4.6.)

Actually, the situation is likely to be worse than this. One of the characteristics of true telephone-grade service is a longer *holding time,* i.e., calls are more frequent and longer. A very short call in the context of wireline telephony is 105 seconds. A more typical busy-hour figure would be something like 360 seconds (6 minutes, or 0.1 Erlangs) of traffic per customer at the busy hour. If cellular radio really does begin to approximate landline service, customer usage will tend to increase (setting aside the effect of pricing for the moment). This further reduces the *concentration ratio,* the ratio of subscribers to circuits, and increases the cost per subscriber.

To illustrate, let us return to Dr. Lee's example. At 105 seconds of traffic per customer, 45 channels can support over 1200 users, a concentration ratio of about 27 users per channel. For simplicity, let us assume that each RF channel has a cost of $27,000 (which is approximately within

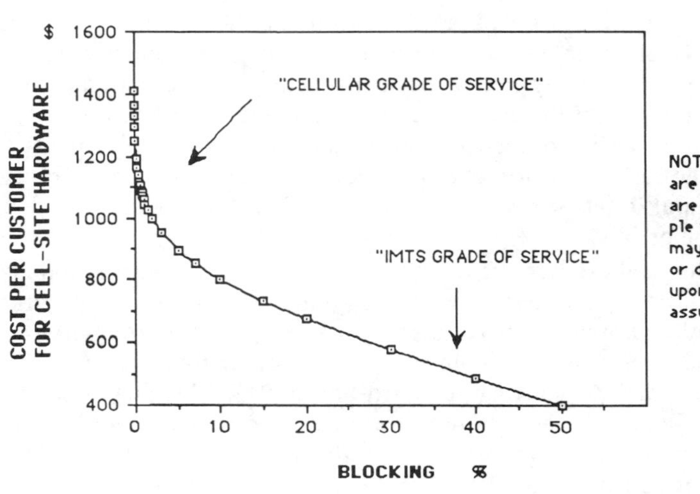

Figure 4.6 Cell-Site Hardware Costs as a Function of Blocking Objectives.

the range of today's equipment prices). The system cost per subscriber would be $1000. (For comparison, the cost per subscriber for a cell operating at an IMTS-style 50% blocking would be about $400.)

Now compare this with a system where each customer offers 6 minutes of traffic at the busy hour. The 45 channels can then support only about 356 users, or 8 per channel. The system cost per subscriber is about $3375. (See Figure 4.7.)

In fact, the situation may be worse yet. Cellular radio is by definition targeted at communication-intensive businessmen and businesswomen, people who so need to use the telephone that they are willing to pay well over $1000 to install a mobile telephone in their car. Is it reasonable to believe that such an individual will want to use that expensive device for only about 50 seconds during a typical 30-minute rush hour commute (which corresponds to Dr. Lee's assumption)? Yet such heavy users will soon be frustrated, because the system must charge them heavily to recoup the cost of supporting their calls. For example, an individual who uses his cellular phone for 5 minutes during that 30-minute commute, would generate 10 minutes of call traffic per hour. Dr. Lee's 45 channel system could support only 214 such users, and the shared system cost per subscriber would be almost $5700 (still assuming the nominal $27,000 cost per RF channel).

These figures are only illustrative, but they indicate the bind that cellular operators find themselves in. By aiming for high service standards

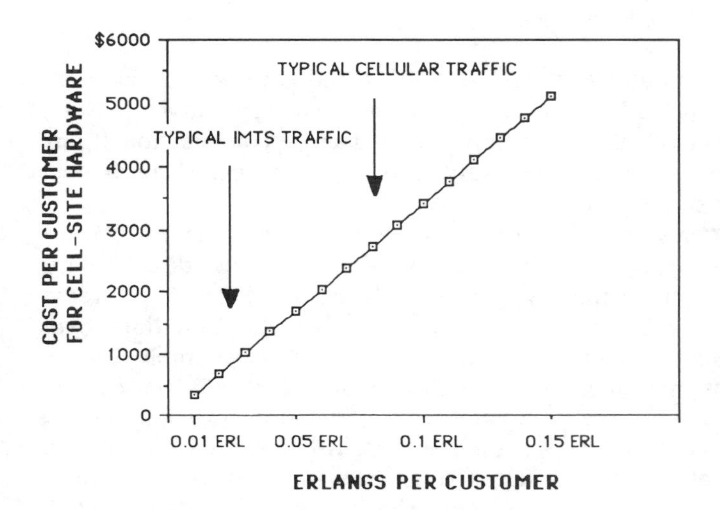

Figure 4.7 Cell-Site Hardware Costs as a Function of Erlangs per Customer.

and "mainstream" telephone users with calling patterns similar to wireline customers, cellular radio incurs enormously increased system costs per customer within the cell-site itself. There is something illogical, almost self-contradictory, about this situation. Cellular operators are trying to persuade prospective customers that they "need" a car phone because they have so many important calls to make, and have specifically targeted people who use the telephone a lot more than normal wireline customers. Moreover, the peaks of mobile telephone usage would probably tend to be more sharply concentrated during the morning and evening rush hours than landline calling, which is spread more evenly throughout the business day. Yet if these customers really use the system heavily, they drive effective capacity down and system cost per customer up, perhaps to intolerable levels. Some cellular operators are indeed almost schizophrenic in their marketing efforts. As recruitment of new subscribers becomes more difficult (see Section 4.1.2.2 below), there is greater emphasis on increasing the revenue per subscriber — getting him to use the phone more. On the other hand, to counteract the shock of high usage charges which may discourage many new customers, some companies advise customers to use the phone efficiently, "make the call and get off [18]."

Inefficient Radio-Link Technology

A third contributor to high system overhead is the radio-link technology itself. Analog FM is inherently a *single channel per carrier* (SCPC) system. Each radio channel can support only a single conversation at any one time. That one customer effectively bears the entire cost of the channel for the duration of the call.

A basic premise of this book is the need for a new, more efficient radio-link technology. This issue will be examined in greater detail in Part V. Without specifying technology, however, it should be clear that a radio link capable of carrying, say, four telephone calls at the same time, over the same base-station channel equipment, would be considerably less expensive in terms of system cost per customer. For example, the current projected standard for the next generation of European digital cellular systems is eight-circuits-per-carrier, that is, eight telephone calls handled simultaneously by each base-station transmitter-receiver. At any given moment, there would be eight customers sharing the cost of that channel. Even if the eight-circuit-per-carrier channel equipment were 50% more expensive than today's SCPC FM channel equipment, *the system cost per subscriber would still be more than five times lower.*

This is obviously a very promising avenue to reducing system cost per subscriber. Today's SCPC radio architecture can be extremely expensive, as we have seen.

What Are the System Costs?

How high are system costs per subscriber in today's cellular systems? Even holding technology constant, the answer is not easy to obtain. It depends upon:

1. The cost of the cell-site equipment;
2. The number of cells;
3. The actual traffic characteristics of cellular users;
4. The *true* blocking and grade-of-service standards (which may be quite a bit worse than the FCC 2% standard).

The cost of cell-site equipment has been falling. In the first systems, the cost per RF channel per cell was perhaps $50,000 or more. It is projected that prices of $20,000 to $30,000 will soon be reached. No doubt further improvements may be expected. Nevertheless, many of today's systems have already made the bulk of their investment at higher prices, and these costs must form the basis for their pricing. Also, of course, the costs of towers, real estate, and so forth, are not likely to fall.

According to some industry sources, the actual system cost ranges between $1000 and $2000 per subscriber (assuming 100% capacity) [19]. These figures may be optimistic, however, since they are based upon the assumption that the traffic offered by each customer will be low, compared to wireline telephone-usage patterns. In other words, "100% capacity" may turn out to be less than was assumed when the system was engineered. Cellular engineers often assume average busy-hour traffic of 0.01 to 0.03 Erlangs per customer (about 36 to 105 seconds per hour). I believe this is far too low *for design purposes.* If the mobile users were able to use their telephones as much as they wanted to (like landline telephone customers with flat-rate, unlimited calling-area service), they would generate much higher traffic, probably at least equal to landline averages.

Here we encounter one of the fundamental contradictions of cellular system economics. Cellular radio is purportedly designed to provide telephone service *comparable to wireline.* Yet if wireline calling statistics are used, the resulting system cost per customer will be so high that usage charges become prohibitive. In a number of analyses involving the application of cellular in situations where wireline statistics are assumed, the system cost per customer has been in the range of $4000 to $5000. Of course it is unlikely that such traffic will actually be generated, because cellular users could not afford to pay for it. If a cellular customer, however, cannot afford to use his car phone for more than a few minutes of commuting per week, how much utility does it really provide him?

This raises another another "steady state" viability issue. The business customer performs some type of cost-benefit assessment on cellular service, as on any other business expense. To make the phone in the car worthwhile, he must be able to use it fairly freely. But liberal usage will generate high monthly charges. Cutting back on calling can control expenses, but raises the question of what the value of the car phone really is, if it cannot be used very much. A few people are on the road all day, but most potential cellular customers are commuters who have the option of waiting a short period (or even stopping) and placing their call from a wireline telephone. Car-calling is a convenience for most users, not a necessity.

4.1.2.2 Start-Up and Operating Costs

Start-Up Costs

A cellular operator generally bases his pricing on an assumed target capacity utilization. That is, the fixed, up-front system costs are spread across a certain number of assumed subscribers for pricing purposes. For example, Shosteck has estimated that the average cell in the top 90 cellular systems will require at least 550 subscribers, generating $100 a month each, to cover current costs (including depreciation), and to achieve the operating break-even point [20]. Simplifying somewhat, this implies a revenue requirement, i.e., fixed costs, of $660,000 per cell per year.

The system, however, does not start out with 550 subscribers. At some early stage, there may be only 440 subscribers per cell signed up and generating revenue. The system is then at only 80% of break-even capacity. If the total fixed costs are still $660,000 per cell, there will be a shortfall of $132,000 (on a yearly basis), or about $25 per month per current customer. For every month that the system operates at less than break-even capacity, the cellular operator is losing money through unpaid-for fixed costs. This becomes an additional capital requirement.

High Marketing Costs

This makes it imperative for the cellular operator to sprint flat out for the break-even line, to load the system as fast as he can. A large city cellular system certainly requires many thousands of customers to reach break-even capacity. Unlike the IMTS operator who usually operated in a seller's market, the cellular operator has to do whatever is necessary to bring the subscriber base up to cover his fixed costs. The IMTS operator is like the builder of small office buildings who never needs to find more than a few paying customers to cover the costs of his modest up-front investments. By comparison, cellular radio is a towering skyscraper. It takes a lot of money to put the huge building in condition even to take the first tenant, and, once built, it must be tenanted as quickly as possible to begin covering the enormous finance and operating charges.

This urgency leads to an accelerated marketing effort. It is typically found that pent-up demand unsatisfied by precellular systems is insufficient to reach break-even capacity. Moreover, the high cellular service charges have made it difficult to expand the market rapidly. By most accounts, cellular is a "hard sell." Operators have found it necessary to develop aggressive marketing programs. High-pressure sales tactics have been adopted. Alluring pricing policies have been developed — "a cellular phone for only $29.00 a month" — and as we have seen some operators have been willing to offer bounties of up to $800 to retailers and distributors to sign on new customers. In some cases, the bounty reportedly has exceeded a year's gross profit per customer [21].

The high costs of marketing cellular service have come as something of a shock to many who had seen the rapid market successes of CB radio, cordless phones, and paging services. Whereas these earlier, less sophisticated forms of wireless communication had practically sold themselves (so much so that they rapidly glutted the airwaves), cellular operators have found that it can cost as much as $2000 *in selling costs* to recruit a new cellular subscriber [22]! In effect, this is an additional capital investment of $2000 per subscriber, over and above the system hardware and start-up costs. Of course these costs, too, must be factored into cellular service pricing, driving the rates still higher, making it even harder for salesmen to recruit new customers.

This has the makings of a vicious circle: high cellular rates (based on high hardware costs) make it hard to recruit the required critical mass of customers, which requires more intensive and expensive marketing efforts, in turn increasing the cost base, driving the prices up, driving marketing costs up, and so on, until the system economics choke off the market. Can this be happening?

The Costs of Overselling

It gets worse. Today many operators are facing cost problems that are the unfortunate but inevitable outcome of such recruitment practices. Many customers, subjected to a "hard sell" to bring them into the cellular network, may not be fully cognizant of the real prices for airtime when they sign on for a "low monthly charge." (Very few customers are accustomed to keeping track of the number of minutes per month that they use the phone. Forty cents a minute may sound inexpensive.) The industry is beginning to show the classic signs of such overselling: high rates of default and bad debts among customers, high rates of unauthorized and unbillable calls (in part due to the poor security features of the current cellular architecture), as well as incipient consumer boycotts among those who have already signed on (see Section 4.1.2.3). These costs must also be factored into the basis for service pricing.

Backhauling

Unlike IMTS single-cell designs, cellular systems require extensive networks of connecting telephone trunks to link the cell-sites back to the mobile switching center. Such "backhauling" circuits must be either constructed by the cellular operator or leased from the local telephone company. They represent an infrastructural requirement made necessary, in part, by the cumbersome and overly centralized control architecture of today's cellular systems. Backhauling charges constitute a large component of operating costs for cellular telephony.

It is difficult to produce an accurate and general accounting of cellular operating costs. Many operators are reluctant to share such information, understandably for an industry that is dependent upon rapidly reaching critical mass. Shosteck has estimated that for cellular operators in the top 90 urban markets in the United States, annual operating costs average $450,000 to $500,000 per cell (on top of about $1 million per cell in hardware investment). Since Shosteck assumes an average of 635 subscribers per cell (based on data from the ten largest systems), this works out to be from $700 to $800 per year per subscriber. It is not clear whether this includes a portion of the up-front marketing costs [23].

4.1.2.3 Cellular Pricing

The cost structure of cellular radio determines the pricing to the end user, which defines the size of the potential market and the scope of the business opportunity. Like anything else, demand for mobile telephony is elastic: the lower the price, the larger the market. Cellular service prices are high and likely to remain so; more than any other single factor, this has hampered the growth of the industry.

For the individual customer, the cost of owning a cellular telephone includes both the monthly service charges and the cost of the mobile telephone equipment (plus installation, insurance, *et cetera).* As we have indicated, the equipment cost does not appear to be as much of a barrier as once thought. Falling retail prices have helped bring down the amortized cost of the mobile hardware to around $35 a month [24]. In itself, this is not onerous for the presumably upscale, urban, professional and business segments that constitute the main markets for cellular telephony. Moreover, there is some reason to believe that consumers may view the purchase of the mobile set as an impulse decision that, once made, is regarded as a sunk cost and excluded from ongoing cost-benefit calculations. Unless financed through a lease or rental, the monthly amortization of the mobile equipment becomes an abstraction that may not loom very large in the minds of many users. Cellular radio is probably similar in this respect to other "options" acquired as part of a package in a new car purchase. As part of a $20,000 automobile, a $1000 cellular phone may seem no weightier than, say, a sunroof.

The real problem remains the high monthly cost of cellular service, based on high system costs for the cellular operator. This is the "front line" of the customer's cost-benefit analysis, revisited with every monthly bill.

How high are typical usage charges? Let us consider a cellular user operating at the "economic threshold": an individual who uses his car phone prudently and for justifiable business purposes, such that if he cut back his airtime any further the service would cease to have sufficient economic utility. We assume, in other words, that to be justified economically, i.e., other than as a prestigious car ornament, the cellular phone must be used for at least some minimum number of minutes per week. Let us assume that the threshold is 10 minutes of calling time per working day, 20 working days per month, concentrated during business hours, especially during the morning and evening commutes when the user is out of contact with conventional wireline telephones. This is not a lot of calling. It could represent, for example, the traffic generated by one 4-minute business call (about the average for business-related telephone calls) during the morning drive, one similar call during the evening drive, and a shorter 120-second call to office or home for scheduling purposes. Obviously, 10 minutes per working day is not enough to do much real business over the phone. A stockbroker, for example, who may spend several *hours* per day on his wireline telephone, would find such a limit constraining. Roaming (calling from "foreign" systems) and off-peak calling (nights and weekends) are not considered in this analysis.

The cellular customer at an economic threshold of 200 minutes per month would pay at least $100 per month in most American cellular systems (excluding any long-distance charges). Different operators have published different rate structures. Some have a relatively large fixed monthly charge and a moderate usage (per minute) rate. Others minimize the fixed monthly charge — "only $9.95 per month!" — but charge more for airtime. The threshold user would pay pretty much the same in most systems, regardless of the rate structure.

This hypothetical usage level apparently conforms fairly closely to actual usage rates among cellular customers. According to Shosteck [25], the "average" actual service charges in 1987 are around $120 a month. With equipment expenses factored in, the total monthly cost is averaging around $155.

This is far above the level at which anyone believes the development of a true mass market is possible. Shosteck has compiled an elasticity curve for *business* demand, adapted in Figure 4.8. At levels of around $150 a month for equipment and service, only a small percentage of the general business market will purchase cellular service. Anecdotal evidence and industry discussions indicate that Shosteck's views are at least generally

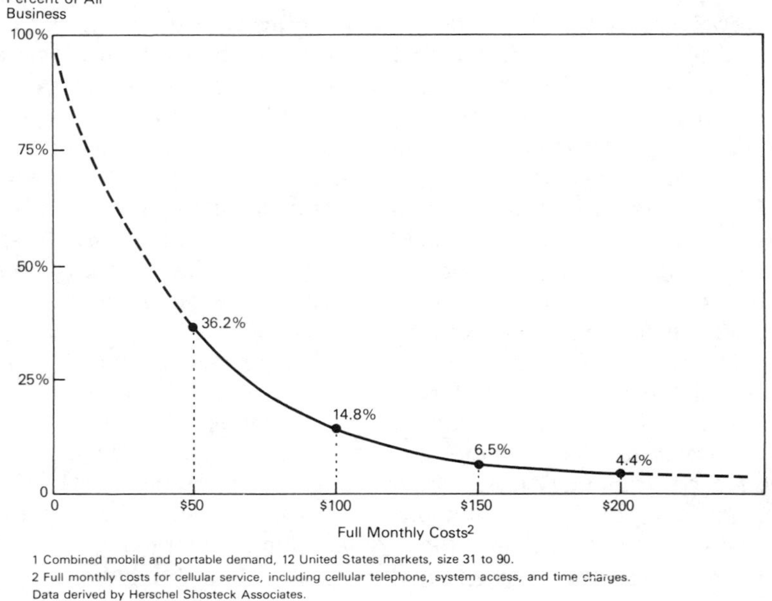

Figure 4.8 Elasticity of Business Demand for Cellular Telephone Service.
Source: Herschel Shosteck.

accurate. In fact, the current feeling among some cellular operators holds that cellular pricing is currently so high that demand is actually quite inelastic: modest decreases in monthly rates will not significantly increase the potential market [26]. Therefore, it is argued, operators should concentrate on *raising* revenue per subscriber by encouraging usage through enhanced services and equipment features such as hands-free mobile control heads for automobiles, electronic dashboard "Rolodex" devices, and so forth. This suggests that, if anything, Shosteck's curve may not be flat enough in the high-price region.

With falling equipment prices and competition in cellular rates the picture will moderate somewhat. It may not be enough, however. Arthur D. Little, Inc., has forecast that "typical cellular subscribers will continue to face bills of at least $100 per month for equipment and service through the early 1990s."

> Our research, which included interviews with over 10,000 potential cellular users, leads us to conclude that at these charges a true "consumer market" is unlikely to develop within a decade [27].

Shosteck is even more blunt: "In reality, the consumer market is non-existent [28]."

The microdynamics of cellular usage patterns are also troublesome. Customers who are wooed with low base rates tend initially to use the car phone in a somewhat lavish fashion, i.e., in a manner similar to the way in which they are accustomed to use wireline telephones. Then the first monthly bill comes in at $200 to $400, certainly a disconcerting experience. Their reaction is perhaps equally disconcerting to the cellular operators: a large number of them tend to discontinue service. Some of them actually disconnect from the service altogether. Others keep paying the "low" monthly fixed fees, but become more or less dormant, ceasing to use the cellular phone. According to the head of one major cellular operator, there is an annual drop-out rate of approximately 25% of the total subscriber base [29]! In other words, if the operator begins the year with 10,000 subscribers, 2500 of them will have discontinued service by year end. Another 10%, or 1000 users, will stay on the subscription list, but will cease using their phones.

This attrition may account for another mystery in the sales numbers for cellular mobile units. Herschel Shosteck compared production and sales of mobile units with reported growth in the active subscriber base of cellular users and found that:

> Half again more cellular telephones [are] being produced than are accounted for in net subscriber gains . . . Manufacturers may be producing 35–55% more units than show up in the subscriber base; retailers may be selling 25–30% more . . . These units are churning into a "black hole." That is, people who discontinue service or are dropped by carriers just keep the phones idle [30].

The urgency of the cellular marketing effort described in the previous section can now be more fully appreciated. The cellular operator is trying to fill a bucket with a large hole in it. According to these figures, if he begins the year with 10,000 subscribers and succeeds through strenuous marketing efforts in signing up 5,000 new customers, he will end the year with around 11,500 — a 50% increase in recruitment produces only a 15% increase in the subscriber base.

More disturbing still, each new cohort of subscribers appears to be more difficult to sell, more sensitive to pricing, and generates less billable airtime than the previous groups. The leader of one of the Bell operating companies recently noted that cellular usage per customer was below expectations:

> Our experience has been very similar to that of the national marketplace. The market in terms of the number of units is developing just about how we had projected it would, just a little under. But the usage of the line is lesser by quite a degree, 20% to 30% lesser depending upon the particular market, than we had forecasted originally. So the revenue stream is down. The units are about right, usage down, revenue stream down . . . [31].

As cellular salesmen exhaust the best prospects — those for whom the usage costs are least important, those who are best equipped and most inclined to pay for the service — new customers can only come from the more price-resistant segments of the market. As one analyst put it, "there are only so many Cadillacs and Lincolns out there." At some point, the cellular salesmen must reach down to the Buicks, Oldsmobiles, and Chevrolets. Price resistance increases, the likelihood of "bill shock" increases, and average airtime decreases. Indeed, according to the head of one major cellular operator, many operators are experiencing a *shrinkage* in average revenue per subscriber of from 1% to 2% *per month* [32].

Elasticity curves can be squirmy things, especially for new markets. There are reasons for modest optimism in a few large, relatively high-income markets such as Los Angeles. Yet everyone who has studied the problem agrees on several general points: (1) current prices are too high to attract large segments of the business market; whether the potential at current levels is 3% or 8% of all businesses, it is still disappointingly low; (2) current prices prohibit the development of a general consumer market; and (3) there *is* a breakthrough price level, below which the potential market for cellular services is believed to expand dramatically. Most observers believe that the breakthrough price level is around $50 to $60 per month (for at least 200 minutes per month).

There is also general agreement that current cellular systems based on analog FM technology probably cannot reach the breakthrough price level. Shosteck has estimated that to get to the $50 to $60 per month level, the cost of the mobile equipment would have to fall below $300 installed [33]. Considering that labor-intensive installation must be included, this forecast implies a very cheap mobile unit indeed. Cellular service charges would have to drop to less than $50 a month (for at least 200 peak-hour minutes of airtime), which seems out of the question given current technology and service standards (the 2% blocking objective).

The most nettlesome question of all is perhaps the one we raised earlier as a hypothetical question: Given current high price-floors, will cellular marketing efforts be capable of attracting the critical mass of customers to achieve operating break-even points? Are there enough Cadillacs and Lincolns out there? In a few markets — large, communication-intensive metropolises like New York, Chicago, and Los Angeles, and perhaps a few very upscale smaller communities (Colorado Springs? Las Vegas?) — the answer appears to be a tentative Yes. But the infrastructure costs of cellular radio do not decrease linearly as the size of the market decreases. Certain fixed costs may even *increase* on a proportional basis in smaller markets. In cities like Richmond, Virginia, or Birmingham, Alabama, there is room for real doubt as to whether the market potential exists at currently achievable price levels to allow operators to reach the break-even point.

In recent discussions, the projected payback period for cellular systems in such smaller markets has been the subject of reevaluation. The emerging, sober view is that the payback for cellular systems in smaller markets may be "long to very long," as one cellular operator recently put it. Paybacks of ten years or more are being forecast for some of these markets. Competition among over-leveraged competitors will further depress business prospects and stretch out paybacks. The viability of the two-operator competitive framework laid out by the FCC appears to be in jeopardy in such markets. Some operators are beginning to discuss the possibility of sharing central system equipment as a way of reducing costs and consolidating markets. Such long periods begin to approach the useful life of the equipment base and are considerably longer than the rate of generational change in mobile-radio technology at this time.

What can be done to alleviate this situation? The high system costs, resulting in high service charges, are the very crux of the cellular impasse. Within the current framework, it appears that little can be done — because it is that framework itself which produces the problem.

We could scrap the cellular architecture with its high fixed costs and high critical mass of subscribers and return to single-cell IMTS-like designs. This would shut off the opportunity to realize high-capacity systems and would doom mobile telephony to a perpetual marginal status in the communication world. As such, it seems undesirable.

We could scrap the traffic and blocking standards and allow cellular to degrade to an inferior IMTS-style service, with high-blocking, high costs, and self-limiting performance and economics. Once again, this closes off the possibility of developing a large-scale, mass market for mobile communication.

The third option, the only real alternative, is to attack the root of the problem: the obsolete and inherently high-cost technology that forms the basis of today's cellular standards.

4.2 THE PERFORMANCE SYNDROME

In the first years of cellular operations, performance often fell short of the standards set by the architects of the cellular idea. There were too many dropped calls, hand-offs that were misconnected to the wrong parties, poor voice quality in many areas because of inadequate coverage. Some of this was part of the natural process of proving in a new system, developing and honing new operating and engineering procedures. Cellular systems operate on a much larger scale than anything previously attempted in mobile telephony, with many more transmitter sites, more channels at each site, and vastly more sophisticated call processing. As the operators gained experience, some of the initial difficulties began to diminish. Voice quality was improved with better coverage and more cell-sites. Some call-processing problems were addressed with better software.

There remain a number of areas, however, in which the performance of today's cellular systems continues to fall short and which cannot be easily remedied by improved operating procedures or better planning. Like the dilemma of high fixed system costs, these problems are embedded in the basic way the system is put together and probably cannot be solved within the framework of today's technology standards.

Some of these performance problems are related to the difficulties of operating a telephone system in a radio environment. In hindsight, some of the design principles of today's cellular architecture seem to have been based on assumptions about the nature of the transmission process which were derived from the wireline world and today appear somewhat naive. It is the nature of wireline telephone engineering that the transmission environment — the wire or cable itself — can be specified and controlled by the engineer to a very high degree. Cost aside, a wireline circuit can be engineered to provide a relatively specific communication capability, with known and measurable characteristics, between two points. The engineer can specify almost any type of circuit he is willing to pay for. If the signal-to-noise ratio is poor, a larger gauge cable can be used or repeater spacing improved. Wireline engineers are accustomed to working with a relatively definite knowledge of the transmission path (at least for local or metropolitan transmission).

By contrast, mobile radio transmission is by nature much less predictable and certainly less controllable. These uncertainties are not really reducible at any cost; the mobile system must work around them. (See Chapter 8.) The radio engineer must more or less take the world as he finds it. Sometimes the too finely specified tolerances of multicell system designs cannot accommodate the sloppiness of radio reality. As one engineer once said to me: "Building a cellular system is like trying to put together a finely crafted jigsaw puzzle with a hundred tightly fitting pieces, except that all the pieces were soaked in a bucket of water for a week where they became warped and swollen so that they do not really fit cleanly anymore."

Other problems grow out of oversights in the original design. The cellular architecture is today more than twenty years old, and there is a degree of obsolescence in the performance specifications of the system. For example, no one really gave much thought back in the late 1960s to the need for the transmission of high-speed data to and from mobile users. As a result, the cellular architecture was developed in a way that makes it very hard to accommodate mobile telephony to the rapidly developing data capabilities of the wireline network with which it interfaces. A much more serious oversight was the lack of adequate measures to assure even a modest degree of conversational privacy for cellular conversations.

There are four areas in which today's cellular systems show serious performance shortcomings which do not appear to be remediable without a major technology shift.

1. Flaws in coverage;
2. Flaws in call processing;
3. The lack of communications privacy and security;
4. The awkwardness of data transmission.

4.2.1 Coverage Problems

Mobile-system engineering encompasses the art and science of defining for a particular system the best placement of transmitters to provide acceptable signal quality over a wide area, or, alternatively, to evaluate the coverage provided from a given transmitter location (since mobile operators often cannot erect their antennas at precisely the optimum site). The analysis takes into account factors such as the transmitter power output, the directional characteristics of the antenna, the height of the tower, the surrounding terrain, and so forth. It evaluates signal strength contours, potential cochannel and adjacent channel interference, projected availability of service, and other aspects of mobile system performance in the real world, to produce a system plan.

System engineering is a relatively simple and inexpensive process for an IMTS-type operator. IMTS systems generally use only a single high-power transmitter, and the goal is to cover as large an area as possible by putting out as much signal as possible. The siting of the tower is usually straightforward: the systems engineer looks for the tallest building or hilltop in the area and operates at the maximum allowable transmitter power.

The cellular engineer faces a much greater challenge. There are many cell-sites, and many more antennas, each with specific local constraints and nonoptimal characteristics. The cellular engineer's objective is not simply to blast out the maximum power in all directions. Instead, cellular architecture calls for the creation of controlled small-coverage areas (the cells), such that the transmission from one cell-site will attenuate sufficiently at a given distance for the frequencies to be reused. In other words, the problem of cochannel interference comes to the fore. Indeed, for the IMTS operator cochannel interference is really not a major limitation, since the FCC rules will restrict the reuse of frequencies to areas more than 70 miles distant, well beyond the curvature of the earth. His coverage area has a "natural boundary" at mobile frequencies. A cellular operator, on the other hand, may be reusing the same radio channels at a distance of only a few miles; he does not have a natural boundary to define his coverage area. The boundaries of each cell are largely defined by the system engineer himself, through the placement, power, and directionality of the cell-site transmitters.

In retrospect, the early architects of today's cellular technology held some surprisingly fastidious views of what they called the "cellular geometry," the way in which multiple low-power transmitters could be used to provide coverage over a large area. While acknowledging that actual cell boundaries were likely to be rather irregular, they were nevertheless captivated by the overarching rationalism of the cellular idea, which led them to postulate that the irregularities could be — indeed *should* be — ignored for practical purposes, and cellular designs laid out in regular hexagonal grids. The following comment is drawn from the *Bell System Technical Journal,* issue of January 1979, which presented the first comprehensive published overview of cellular technology:

> Amorphous cell shapes might be acceptable in a system where the initial system configuration . . . could be frozen indefinitely. In practice, however, the absence of an orderly geometrical structure in a cellular pattern would make adaptation to traffic growth more cumbersome than necessary. Inefficient use of spectrum and uneconomical deployment of equipment would be likely outcomes. A great deal of improvisation and custom engineering of radio, transmission, switching, and control facilities would be required repeatedly in the course of system growth.
>
> Early in the evolution of the cellular concept, system designers recognized that visualizing all cells as having the same shape helps to systematize the design and layout of cellular systems . . . If, as with present-day mobile service, omnidirectional transmitting antennas were used, then each cell's coverage area — bounded by the contour of constant signal level — would be roughly circular. Although propagation

considerations recommend the circle as a cell shape, the circle is impractical for design purposes because an array of circular cells produces ambiguous areas which are contained either in no cell or in multiple cells. On the other hand, any regular polygon approximates the shape of a circle and three types, the equilateral triangle, the square, and the regular hexagon, can cover a plane with no gaps or overlaps. A cellular system could be designed with square or equilateral triangular cells, but, for economic reasons, Bell Laboratories system designers adopted the regular hexagonal shape several years ago. [34]

Such idealizations even survived the first round of systems design. In the AMPS test bed in Chicago, a 12-cell system was developed to serve about 2000 customers in the late 1970s. Apparently on such a scale, at least for the flat Chicago landscape, there were no serious discrepancies. The AMPS concept was described by one of its designers as "approximately circular cells with no gaps and no overlaps [35]." (See Figure 4.9.)

Dead Spots and Hot Spots

Unfortunately, overlaying the clean circles and hexagons drawn by the early system architects onto the irregular realities of many urban terrains rapidly disclosed the weaknesses of the geometric method for detailed system engineering. Operators soon found that dead spots appeared in unexpected locations. Details of the local microgeography, such as tall buildings, groves of trees, and street orientations, produced unanticipated attenuations in signals. No model was complex enough, no database comprehensive enough, to allow such microfactors to be engineered on paper. Cellular engineers found it to be a widespread problem. In the words of one engineer:

> Virtually every city has one or two spots that are considered "call-dropping" areas . . . If you drive into such an area, you start getting a lot of noise and then the call drops . . . Some of these areas are as large as two miles or more. You'd be amazed at some of the spots that design engineers didn't anticipate would be as severely impaired as they are. At the time many systems were designed, operational experience from around the country wasn't available. Some engineers were a little more optimistic than they should have been [36].

It soon became clear that actual system engineering could not be completed in advance or optimized in any precise way. A great deal of ongoing "improvisation and custom engineering of radio, transmission, switching and control facilities" in fact became necessary to provide workable coverage. Even after several years of field experience, some of these

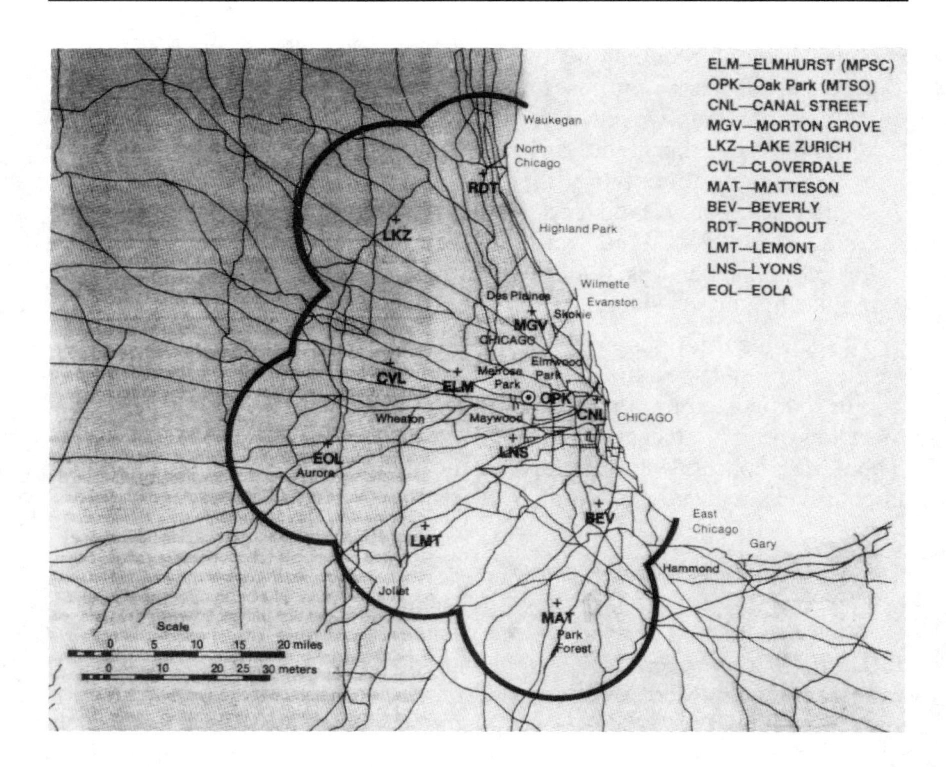

Figure 4.9 Coverage Boundaries for the AMPS Cellular Test in Chicago. (Note Circularized Cell Boundaries.)

Source: Nathan Ehrlich, *IEEE Communications Magazine,* 1979.

propagation anomalies resisted the corrective efforts of cellular engineers. For example, in Philadelphia there was for a long time a serious dead spot astride the Schuykill Expressway, a main commuter artery. "Bell engineers believe the problem is caused by heavy foliage," the press reported. "Autumn will take care of that [37]." The nonwireline operator in Washington, D.C., reported similar difficulties: "Coverage in areas that are normally trouble-free can temporarily become less than perfect when trees bud in the spring [38]."

The Schuykill Expressway anecdote indicates another breakdown in the mapping of cellular hexagons onto the real world: the idealized patterns often failed to take into account the placement of the traffic arteries and automobile flow patterns. Choke-points sometimes developed at unexpected locations where traffic jams tend to occur, again for reasons which may simply not show up in the radio engineer's database (such as where poor exit design tends to create backups during rush hours). A lot of cars stuck in traffic means a lot of drivers whose schedules are being delayed; this will tend to generate a mobile telephone "hot spot" that can overwhelm poorly placed cell-sites. Cellular system engineers soon learned that actual, anomalous phenomena of this kind can be far more important than average calling statistics in determining when and where a system will actually saturate.

What Is a "Cell Boundary"?

Retrofitting to fill in coverage gaps and supplement traffic spikes of course adds to the cost of cellular overhead. But the more fundamental performance problem with the cellular concept in practice became apparent as the true nature of the "cell" began to be appreciated. As is evident from the quotation above, the early systems engineers tended to view the cell as a stable, fixed entity with a coverage area "bounded by a contour of constant signal level." This assumption, which underlies every representation of cellular systems as a grid of circles or hexagons each with sharp fixed boundaries, was another legacy inherited unquestioningly from the precellular era.

Actually, the "boundary" of a cell is a statistical idealization. It is a spatial average of microscopic variations in signal strength; the actual signal level may vary by as much as a factor of 10,000 over a distance of just a few inches. This is a fundamental characteristic of the mobile radio environment, which will be discussed in greater detail in Chapter 8. If it were possible actually to plot the "contour of constant signal level," the cell boundary would look something like what is shown in Figure 4.10 (derived from a Motorola study of the Baltimore-Washington system) [39].

In fact, even this representation does not do justice to the problem of cell boundaries. The edges of the cells are not fixed in time, but are constantly quivering and shifting in an almost Brownian motion. The cause of this remarkable behavior is the phenomenon known as *multipath,* which is discussed in greater detail in Chapter 8. Because of the "fuzziness" of cell boundaries, it becomes increasingly difficult to predict mobile link quality and, indeed, the mobile unit's behavior at the edges of cells. In a normal single-cell IMTS system, the edges are also indistinct, but the mobile unit is expected to spend most of the time within the core coverage area. The system engineering for an IMTS operation simply does not concern itself too much with fringe conditions. An adequate database would have to encompass an enormous detail of microfactors, and such information is simply not available.

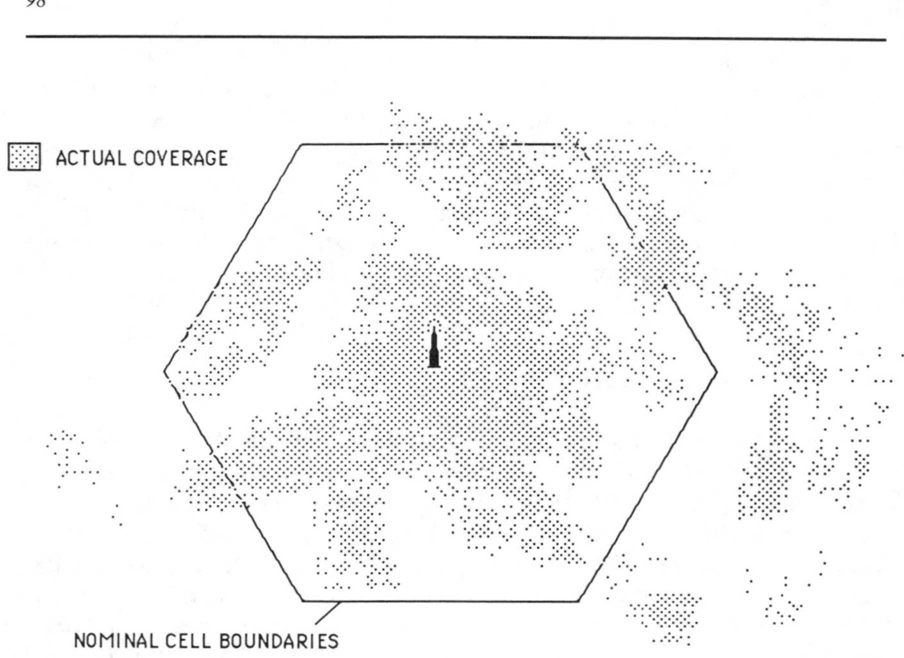

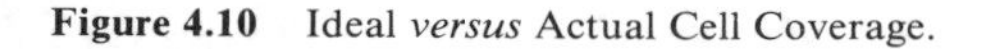
Figure 4.10 Ideal *versus* Actual Cell Coverage.

The effects of real-world irregularities can seriously reduce the efficiency of the cellular system. An AT&T study in 1986 reported that:

> Irregular traffic, terrain, and growth situations limit the spectrum efficiency of orthodox [cellular designs] to about half the ideal — for example, orthodox [reuse] plans result in about 20 to 25 channels per cell, as opposed to the 45 nominally available (under U.S. allocations). This is the consistent and repeated result of a series of detailed growth-planning case studies at AT&T [40].

The result is that cellular system economics are impaired even further. The same paper also offers evidence that the frequency of such problems increases as the cell size shrinks (through cell-splitting) [41]. In effect, the boundaries grow less distinct as the number of cells multiplies. This appears to be a fundamental physical feature of the multicellular radio environment (see Chapter 9).

4.2.2 Call-Processing Problems

The "fuzziness" of cell boundaries creates a particular problem for the cellular system in making the hand-off decision. When an IMTS mobile unit reaches the fringe of the coverage area, the signal simply begins to deteriorate; no action is taken. In a cellular system, when a mobile with a call in progress reaches the "edge" of its cell, the system has to hand-off the call to another cell. (See Figure 4.11.) Yet unlike the top-down view implied by the usual diagrams of this process, the cellular system actually does not know where the mobile unit is, which direction it is headed, how fast it is moving, or if it is moving at all, nor does it know which of the many adjacent cells to hand-off to. The cellular controller actually knows only that the signal level from the mobile has deteriorated below a certain threshold. If this condition persists for more than a short period of time, the system will decide to try a hand-off to another cell.

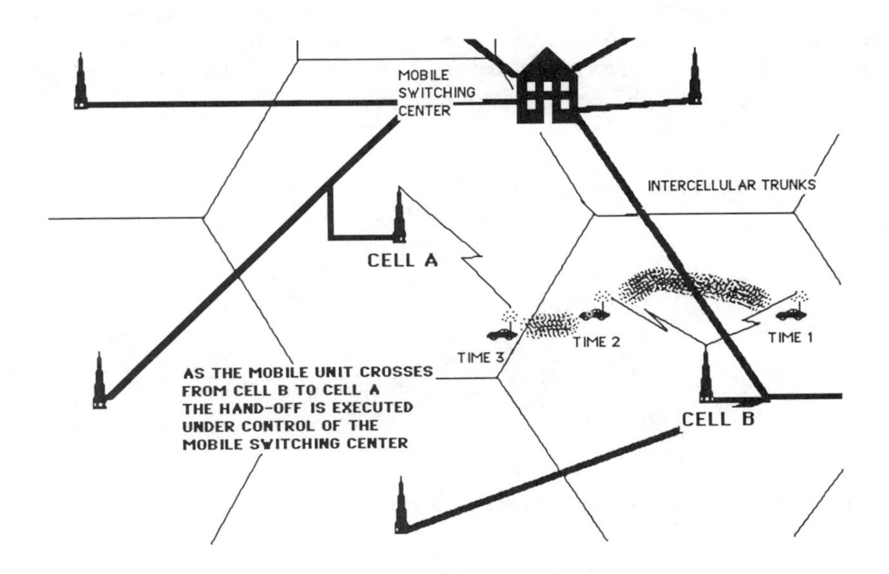

Figure 4.11 Hand-Off.

To determine which of several cells should receive the hand-off (without knowledge of the mobile's actual location) the cellular system initiates a complex hand-off processing routine. Each of the adjacent cell-sites makes a measurement of the received signal strength from the mobile in question and transmits this data to the mobile switching center. The mobile switching center then selects the cell reporting the strongest signal and allocates an idle channel in the new cell to receive the hand-off. An instruction is transmitted to the cell-site currently handling the call, indicating the new frequency to which the call will be shifted. This information is relayed to the mobile, which then executes the frequency shift to the new assigned channel and homes in on the new cell-site. (See Figure 4.12.)

In theory the process works smoothly — as long as the cell boundaries are sharply defined. What happens, however, if the mobile is located in an ambiguous zone between two potential hand-off cells. (See Figure 4.13.) This zone is full of small spots where the signal strength from the "right" hand-off cell will be overpowered by the signal from the "wrong" hand-off cell. (Why this is so will be discussed in Chapters 8 and 9; it is a fundamental characteristic of the mobile environment.) That is, the mobile may be located at a particular point where, especially if it is stationary at a stop light or in a traffic tie-up, the signal strength happens to be stronger from the farther of two potential hand-off cells. Because of these quirks of the boundary situation, the hand-off may be completed to the "wrong" cell. If this happens, the improper hand-off is likely to become apparent as the mobile moves out of that particular spot. A new hand-off will be needed to make the correct cell assignment. In the meantime, however, the mobile on the incorrectly assigned channel may also create severe interference within the cell it has moved into.

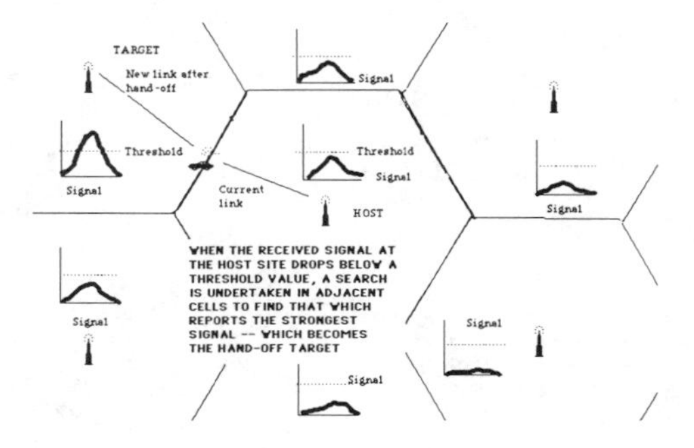

Figure 4.12 Hand-Off Procedure in Analog Cellular. Transmission.

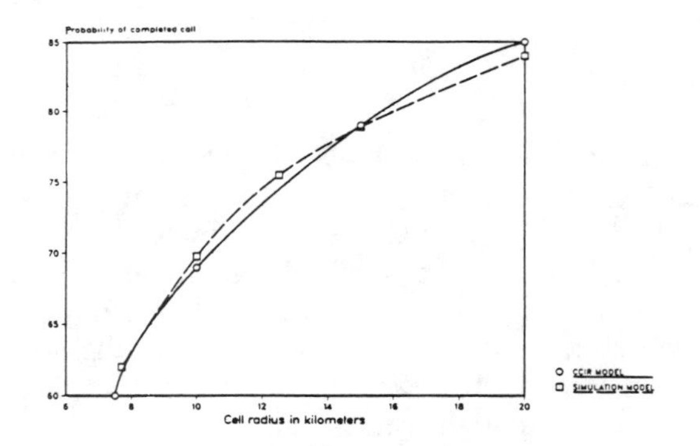

Figure 4.13 Probability of completed message before boundary crossing as a function of cell radius (100 km/hr; 180 s average call length).

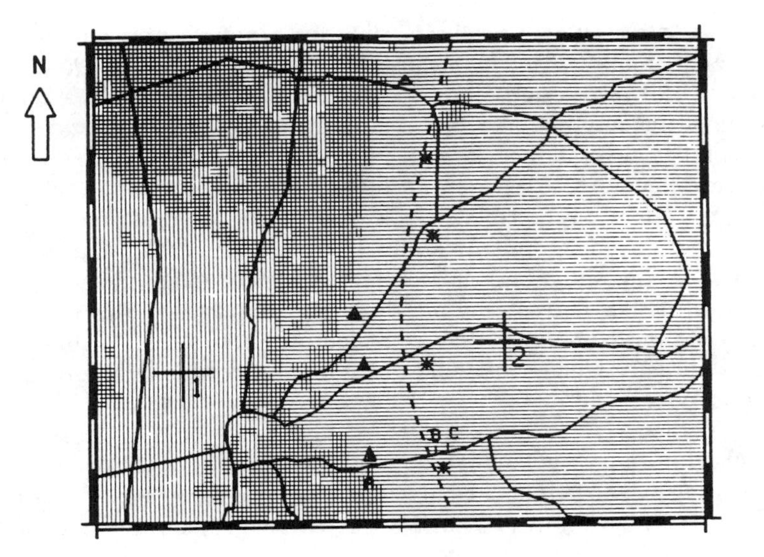

Figure 4.14 A Realistic Portrayal of a Mobile Hand-Off Situation.

Source: Greiner.

According to some cellular operators, there *is* a problem with incorrect hand-offs. For example, the cellular system in Los Angeles has experienced "considerable problems with crosstalk," according to an officer of a major cellular reseller there. "You get to a high elevation, your own signal weakens and you start hearing another conversation. Right after that, you lose the call [42]." The former director of cellular development for Motorola has written that:

> . . . anywhere from 10% to 25% of the time the system made decision errors when a subscriber moved from one cell to another. When the switch made an error it assigned the subscriber's call to a distant cell site rather than the closest one. The system would then start serving a different subscriber on the same frequency in the same cell as the first subscriber.
>
> The result was cellular disaster. When this situation occurs, subscribers hear other conversations on their channels. They get cut off or, if they are lucky [?], the channel gets noisy. This problem happens often enough to be infuriating [43].

Dropped calls and crosstalk may not be the worst of it. The actual margin between two cells is a very ragged and constantly changing interface. Figure 4.14 is drawn from a study of the actual boundary between two cells in a German mobile system.

It is at least theoretically possible for a mobile user moving along the fuzzy margin between two transmitters to "ping-pong" back and forth between the two through repeated hand-offs. The probability of this oscillation is increased by a stop-and-go traffic situation. The hand-off procedure takes several seconds from start to finish (although the actual frequency change, the last step in the process, is accomplished relatively quickly). Thus if a subscriber's car is stopped for a few seconds in a spot that represents a deep fade for Cell 1 and a correspondingly strong signal from Cell 2, this situation will be detected and the hand-off made to Cell 2. The subscriber moves on a little way and stops again, this time in the middle of a fade for Cell 2, and is handed off back to Cell 1. Under these circumstances, a single call could generate a very high hand-off load for the total system.

Whether such extreme effects as this "ping-pong" process actually take place in today's cellular systems is the subject of some debate. There is mounting evidence that current cellular systems are indeed experiencing considerably higher rates of hand-off events than theories of cellular geometry (based on "hard" cell boundaries) had predicted. Since each hand-off imposes a heavy processing load on the central switch, the excess hand-off rate has caused concern in some quarters that cellular capacity may be reduced. In other words, the actual capacity of the cellular switch may be lower, in terms of number of subscribers, than originally estimated. This perplexing phenomenon, named the Losee-Shosteck Effect by the trade press, has been attributed to poor installation of the mobile antenna [45]. I believe it is a more general result of the procedures for hand-off based on the misleading view of cell boundaries as sharp contours, rather than statistical idealizations. A number of proposals have been offered for solving the problem, based on more precise *location* of the mobile by the base, through the use of actual *ranging* information as well as signal strength to make the hand-off decision. This would allow the system to control false hand-offs by overruling signal strength data when necessary. Unfortunately, this requires a precise timing capability (to measure propagation delays from which distance information can be derived) which today's FM transmission cannot accomplish [46].

It is difficult to quantify the impact of the surplus processing load generated by the Losee-Shosteck effect (in part because the operators have so far refrained from publishing their data on the matter). Nevertheless, the problem is apparently significant enough to cause concern among cellular engineers. As Shosteck summarized in 1987:

> At this point, our data are insufficient to estimate either the extent of the Losee-Shosteck Effect or whether or not it may reduce switch capacity significantly. However, none of the cellular engineers with

whom we have discussed this phenomenon would discount the possibility that the effect of lowered switch capacity could be extensive and profound [47].

The Effect of Cell-Splitting on Hand-Off Rates

What is clear is that processing requirements will grow — leading to increased overhead costs — as initial large cells are split into smaller-radius cells to accommodate growth in the subscriber base. The average length of cellular calls, and the distance covered while the call is in progress, will tend to remain the same; decreasing cell radii mean that cell boundaries will be crossed more often, producing more hand-off events per call, and a higher processing load per subscriber. (See Figure 4.15.) This effect, which is a straightforward deduction from the cellular geometry, is likely to be far greater than the Losee-Shosteck phenomenon (if Losee-Shosteck actually exists). Some calculations indicate that a reduction in cell radius by a factor of four will produce a more than tenfold increase in the hand-off rate per subscriber [48]!

The implication is frightening: the processing load will tend to increase geometrically with the growth in the subscriber base. An increase in subscribers by a factor of ten may lead to a processing load one hundred times greater! (See Figure 4.16.) Moreover, this effect appears to be inherent in *any* cellular system, regardless of the radio technology employed. A new generation of more efficient radios may help the Losee-Shosteck effect but will probably not alleviate the geometrical growth in processing loads. Small cells will still mean more boundary crossings, and more hand-offs. Instead we must look to more efficient control procedures, such that hand-off can be accomplished without as much centralized control. Unfortunately, the over-centralized design of today's cellular systems precludes an easy implementation of less burdensome call-processing procedures. We shall see below that cell-splitting will prove to be unworkable and expensive (Figure 4.17).

4.2.3 Privacy and Security

Privacy has long been an important concern for the designers of radio systems. During World War II, the need for private radio communication was the direct stimulus for the development of two of the most important techniques in modern communication: pulse code modulation for digitized speech (see Chapter 6) and spread spectrum radio transmission (see Chapter 12). The scientists on both sides of the Atlantic who were brought

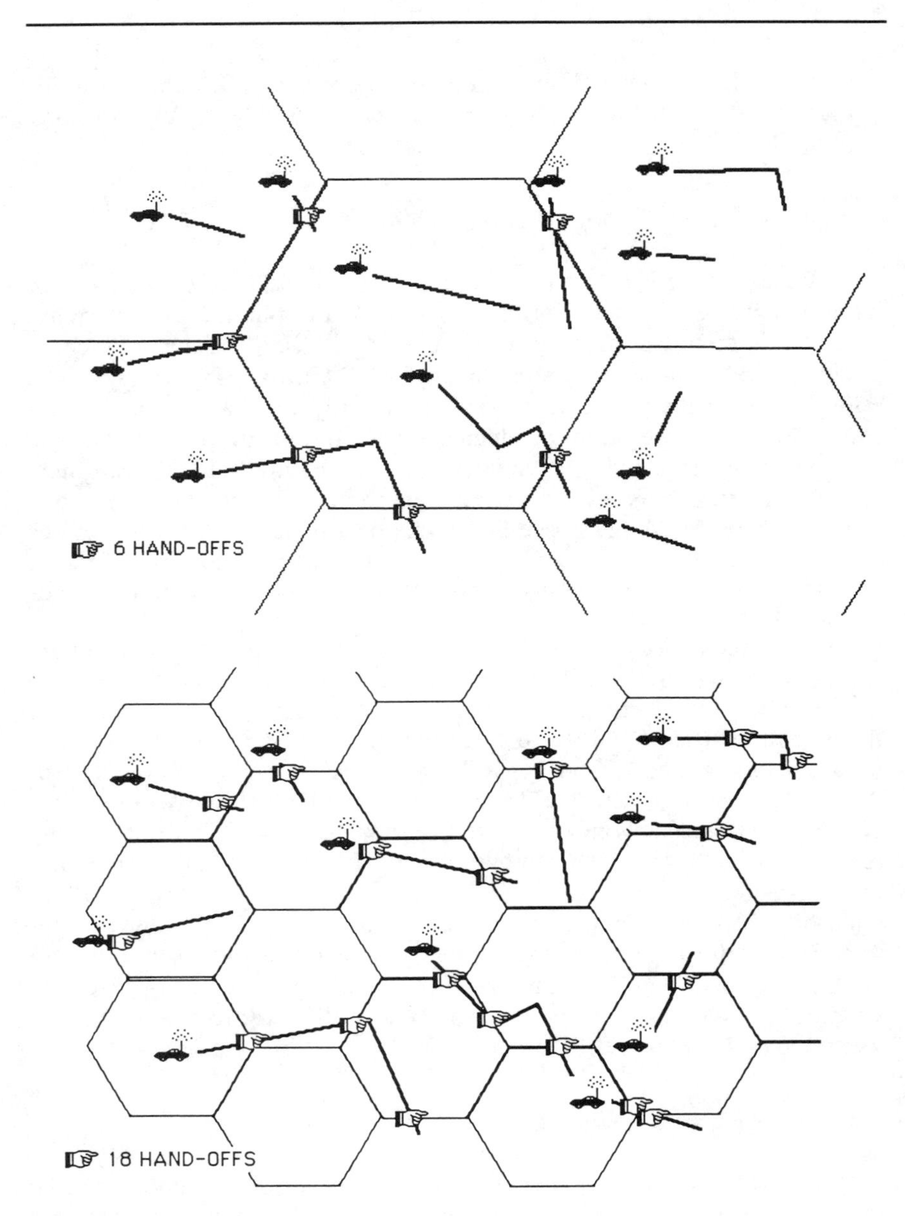

Figure 4.15 Relationship between Cell-Size and Rate of Hand-Off.

together by the war effort to lay the foundations of modern communication theory were motivated in large part by the problem of designing systems for coded transmission of speech and data.

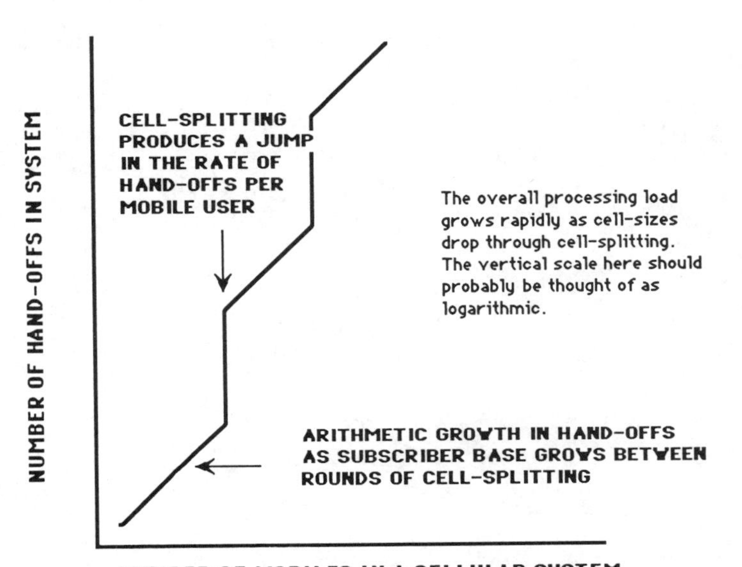

Figure 4.16 Idealization of the Relationship between Growth in Subscriber Base and Hand-Off Processing Load.

It may seem somewhat surprising, therefore, that privacy was never considered very important in commercial mobile telephony. In fact, unlike the wireline network where from the time of Alexander Graham Bell the principle of privacy was usually accepted (although not always honored), the ethos of radio tended in the opposite direction. The airwaves were viewed as a sort of public property, in the sense that anyone had the right to tune a receiver to pick up any signal that might be out there. Restrictions were placed on who could *transmit*, and how; but the "right" to listen in became enshrined in radio practice, and, to a considerable degree, in radio law. Indeed, the freedom of any citizen to enjoy unimpeded reception of radio signals has taken on a libertarian cast. The Soviet practice of jamming Western broadcasts is regarded by many as proof of the need to maintain the policy of open airwaves. This view is so entrenched in the United States that even the police and government authorities are unable to find legislative protection for their own communications. (In the 1986 Electronic Communications Privacy Act, discussed below, government uses of radio are explicitly exempted from any privacy protection.)

An industry has even emerged to outfit a curious public to take advantage of this freedom. A wide range of commercial scanners are marketed to consumers at very affordable prices, advertising openly the

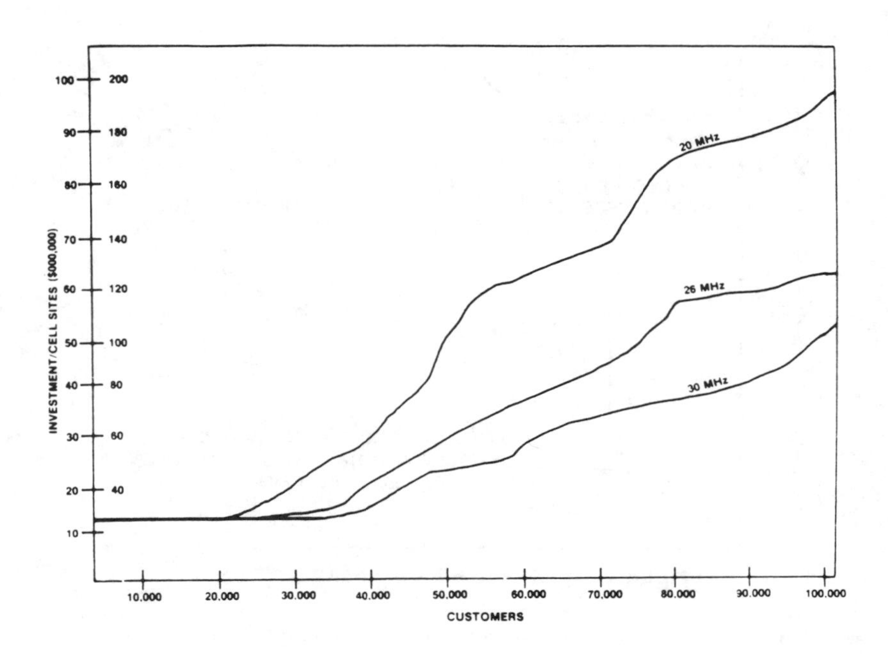

Figure 4.17 New York Cellular System's Estimated Costs of System Growth.

Source: Wholl and DiBella

entertainment value of monitoring police and mobile telephone transmissions — "soap operas without pictures," "more real life intrigue" than TV dramas. Books have been published to identify frequencies used by the police and other emergency services. Some amateur radio operators have even specialized in intercepting the transmissions to and from Air Force One; it is considered a fair sport to tape radio-telephone calls even involving the President of the United States and the members of his cabinet. It may surprise many people both inside and outside the United States to learn that there is no legal prohibition against much of this eavesdropping. The ethos of radio has always accepted such a lack of privacy as simply "part of the game."

Thus, it is really not so surprising that back in the 1960s and early 1970s the designers of cellular radio did not consider it necessary to enhance the privacy of radio-telephone communication. In fact, some even convinced themselves that cellular *would be* private, at least by comparison with other radio systems, since each call was assigned a unique radio channel (unlike the "party-line" circuits in CB radio and some older mobile phone systems) and since the call frequency might change during hand-off, presumably making it more difficult to track.

As cellular radio systems began to be deployed, however, concerns for the lack of privacy started to surface. It was soon apparent that there had been a major oversight in the original cellular specifications. FM transmissions are very easy to intercept. For example, even UHF TV sets and VCRs may be used to pick up today's FM cellular conversations. If wireline telephone standards are the model for cellular service, then mobile systems ought to be able to assure at least the same level of privacy that the wireline network offers today. The user should enjoy the reasonable assurance that he is not being routinely subjected to eavesdropping by his fellow citizens. The analog cellular user has no such assurance.

Cellular operators have tended to minimize the privacy deficit. Actually, however, some operators routinely monitor cellular communication themselves to determine the activities of competing operators in the same market. In December 1986, Metroplex Telephone Company, the non-wireline cellular operator in Dallas, filed a complaint against its wireline competitor, Southwestern Bell Mobile Systems, alleging "deliberate commercial spying" on Bell's part through eavesdropping on Metroplex's cellular communications. Southwestern Bell actually admitted that it monitored Metroplex's transmissions "to obtain an estimate of its market share [49]."

Some local regulatory bodies have begun to take a look at whether consumer safeguards are needed. Some have called for warning labels to be affixed to cellular equipment. This would not be adequate, however, to warn wireline customers on the receiving end of cellular calls. The California State Utility Commission recently proposed a number of measures to warn all parties to cellular communication that their conversations are not private. One idea called for a warning tone or recorded message at the beginning of each cellular call to indicate that the conversation was unsecure. The cellular operators argued strenuously against this, raising dire predictions that the additional processing overhead would cause the cellular switches to overload, costs to increase, and service to break down. (One must wonder whether the system is not operating on a rather too slender margin if the overhead of providing a short tone or a one-sentence announcement at the beginning of each call could actually cause such severe problems.) Clearly such an intrusive reminder of the privacy deficit would not assist the marketing effort.

The concern over privacy, however, is also coming from the marketers themselves. Privacy is particularly important to customers involved in sensitive business transactions as well as to many government users; the lack of conversational privacy has possibly impacted sales to these segments. The operators have begun to realize that some positive solution must be found. There are two lines of attack: legal and technological.

Legal Perspectives on Cellular Privacy: A Clash of Principles

The checkered legislative and judicial history of the principle of privacy of mobile telephone communication is marked by the growing intermingling of telephony standards (where privacy is sacrosanct) with radio standards (where freedom of the airwaves often takes precedence). The modern history of the privacy issue begins in 1968 with the passage by Congress of a law defining the citizen's right to privacy in telephone communication. The so-called "wiretap law" sought to "prevent or deter improper invasions of privacy," but limited the definition of "telephony" to "wire communications [50]." Subsequently, drawing in part on the implicit failure of the wiretap law to extend the privacy guarantee to radio communication, several courts held that there was "no reasonable expectation of privacy for calls transmitted over a mobile car telephone when the conversations could be easily overheard with an FM radio receiver [51]." This conforms to the conventional radio ethos.

What happens, however, when a car-telephone subscriber is talking to a wireline customer and someone listens in with a scanner? Does the wireline caller relinquish his normal protections? Some courts have attempted to rule that a radio-telephone conversation might be covered by the wiretap law if it connected to a landline telephone, but would not be so protected if it connected to another radiotelephone. Then, however, came the explosion of cordless telephones. Clearly cordless phones usually interconnect to wireline calls. The guiding principle became even muddier. Generally cordless phones have not been afforded privacy protection, again because interception is so easy that "no reasonable expectation of privacy" could be maintained. Indeed, even the most recent legislation to ensure privacy for radio specifically excluded cordless phones.

Cellular radio brought the issue to a head. In a notable case in Louisiana, a cellular telephone call involving one Marion Edwards was intercepted on a scanner by a private citizen. Edwards happened to be the brother of the governor of the state and also happened to be the target of a federal investigation on racketeering allegations. In his cellular phone call, Edwards was overheard making certain statements which the eavesdropper apparently felt were suspicious. The eavesdropper tape-recorded the call and later turned the tapes over to federal prosecutors. When he found out about it, Edwards himself then sued the eavesdropper in a federal district court alleging a violation of his privacy and of the 1968 wiretap law. The district court judge, however, threw out Edwards's suit, ruling once again that a radio conversation inherently has no "reasonable expectation of privacy [52]." In fact, wrote the judge, Edwards "had broadcast his conversation for all to hear." This was tantamount to a voluntary waiver of the privacy of the communication: "Disclosure of a public conversation cannot, under the law, amount to an invasion of privacy [53]." This ruling was later upheld by the federal appeals court [54]. In fact, the district judge went so far as to commend the eavesdropper for his actions: "Citizens in this country are encouraged to cooperate with and assist law enforcement officers in the detection and prevention of crime," the judge wrote [55]. One can surmise that his views would have been quite different (his ruling would certainly have been different) had the private citizen in question performed a vigilante wiretap upon Mr. Edwards's ordinary wire telephone.

The Edwards case points up the legal schizophrenia that has been engendered by the head-on clash of two "fundamental rights" — the right to privacy and the right to unimpeded radio reception — which in past times were segregated in different technological domains, but which cellular mobile telephony has brought together. I suspect that a wiretap of one citizen by another would be regarded by a vast majority of judicial and legislative authorities as well as the general public as an egregious violation of the wiretap subject's rights. Yet the interception of the same call, placed over a cellular radio link, would be regarded in a totally different light. There is increasingly little real difference between the two situations.

What is the basis of this hazy notion of "reasonable expectation of privacy"? It seems to involve the question of *enforceability*. Where privacy can be enforced easily, it tends to be sanctioned by law. Where it cannot be enforced, the law tends to take the view that the expectation of privacy is unreasonable. The reasonableness of the expectation is, I believe, implicitly referenced against the technological possibilities for interception and countermeasures against interception. In the wireline world, a wiretap involves (at least traditionally) a physical violation of someone else's property to place the tap. (Taps by the telephone companies themselves would fall outside this proscription and have always been the subject of controversy dating back to the earliest days of the telephone industry.) Interception of a radio signal is a passive process, involving no such trespass. It is, moreover, almost completely undetectable. The ease of interception and the difficulty of detecting and enforcing any anti-interception statute become the true test of "reasonableness." As Southwestern Bell argued in the case of competitive eavesdropping previously referred to, "transmissions that may be intercepted by the use of readily available scanning equipment are not protectible [56]."

To attempt to clarify this situation, a new privacy bill was introduced in the United States Congress in 1986, lobbied vigorously by the cellular operators (including, ironically, Southwestern Bell Mobility), and ultimately passed as the Electronic Communications Privacy Act. It extends privacy guarantees to cellular conversations, on a par with anti-wiretap protections in the 1968 law. Once again, however, the issue of enforceability was central. The Department of Justice (DOJ) initially opposed criminalizing "the recreational interception of cellular transmission" because of the extreme difficulty of detecting the act of interception [57]. After intensive behind-the-scenes negotiations between the DOJ and the bill's supporters, a tepid endorsement was obtained. "It was an incredible feat to get the DOJ to sign off on it," said one procellular lobbyist [58]. Yet it was clear that the DOJ continued to regard the law as unenforceable; one Congressman indicated that he had been informed that the DOJ had "absolutely no intention of enforcing that part of the bill [59]." In fact, it was widely conceded even by the bill's supporters that enforcement would be practically impossible. Why write such a law? "There are often bills passed that the Justice Department does not enforce," said one congressional staffer. "They use them as a message." "It will discourage an attitude," said a cellular lobbyist. "Laws are put on the books not only as deterrents but as reflections of societal expectations . . . [60]"

Nevertheless, other observers regarded the law with skepticism. They felt it was riddled with inconsistencies, ambiguities, and much careless wording. Even supporters of privacy guarantees lamented the botch that it became. "Congress," wrote one cellular advocate,

> in an attempt to update the wiretap act, has taken the ambiguities and inconsistencies of that 1968 legislation, about which the courts have repeatedly complained, and actually *expanded the scope* of its imperfections . . . [It is] a classic example of special-interest legislation whose privacy "cure" is worse than the disease [61].

In any case, it is well understood that such legislative measures are cosmetic at best, misleading at worst. The real answer to the privacy deficit will come from technology.

Technological Solutions?

The technological solution to the privacy problem has been known since the 1940s: digitize the communication and encrypt it. Yet once again the implementation of such a solution is frustrated by the choice of analog FM for the radio link. It is not possible to ensure transmission privacy in an analog radio system without significantly impacting cost and voice quality [62]. Analog scrambling techniques are not adequate. Of course it is possible to build over the existing analog system with enough digital technology to perform a digital encryption. Bell Atlantic's cellular subsidiary has experimented with a fully digital encryption process; the analog speech input is digitized, encrypted in digital form, reconverted to analog FM for transmission, with the same process performed in reverse at the base station. It is claimed that "the result is a good approximation of the original voice [63]." The cost: about $2500 for a "shoe-box sized" encryption device which must be interfaced to the user's mobile telephone transceiver, plus a $65 per month fee to cover the cost of the base station encryption equipment. For a market that is already in trouble because of high service costs, such a solution is not particularly attractive.

The real solution is to move to digital transmission techniques which are compatible with digital encryption methods already in use today. Privacy thus becomes one more factor leading us away from the analog standard, toward a new generation of radio technology for cellular systems.

4.2.4 Data Transmission

The current generation of cellular radio technology cannot really handle the developing requirements for data tranmission within modern communication networks. The difficulty is twofold.

The more immediate, and more solvable, problem is created by the method of cellular hand-off. At the moment of hand-off, the cellular unit, acting under base-station control, performs what is called a "blank and burst": the voice signal is shut off, blanked, and a burst of signaling is transmitted to accomplish the hand-off [64]. This burst is relatively brief, a fraction of a second; it is advertised as "unnoticeable" for voice communication. Nevertheless, any data message which is being transmitted in the cellular channel at the moment of hand-off is lost. The interruption is also long enough to disrupt most synchronous transmission schemes.

The available cellular data modems simply ignore the hand-off hiccup. The user is advised to avoid transmission when crossing cell boundaries. Since hand-off, however, is not under the control of the mobile unit or of the cellular user, predicting its occurrence may be difficult for a superimposed data-transmission protocol. In principle, this should be correctable without a major technology overhaul. It is likely, however, to add cost to the cellular circuit.

The second difficulty is fundamental: the analog FM channel is adapted for voice, not for data. The FM transmission standard cannot accommodate data rates higher than about 10 kb/s within the current channel structure. The high error rates often found in the mobile channel (see Chapter 8) mean, however, that the data protocol must allow for redundant transmissions (repeating the same message several times) or error detection and the retransmission of erroneous blocks. A usable throughput of perhaps 1200 b/s is achieved within current designs [65].

This is a very low ceiling. When the cellular idea was promulgated almost twenty years ago, data transmission in the public telephone network was in its infancy. At that time 1200 b/s may have seemed adequate, if, indeed, data transmission was given much thought at all. Only a very small percentage of the wire circuits in the landline network would have been able to support much higher rates at that time, and it was not clear how many customers really wanted, or would pay for, the much more expensive facilities necessary to achieve those higher rates.

Today the situation is entirely different. As described in Chapter 7, the telephone system is being transformed into a *digital network.* All long-range telephone planning today is based on the premise that in the future the network must be able to make available to the end user fully digital circuits, at rates up to 64,000 b/s. Indeed, there is an evolving set of standards (known as ISDN standards — see Chapter 7) based around even higher data rates, up to 144 kb/s (for two voice channels and a data channel, simultaneously) to the telephone customer.

Viewed in the light of these emerging requirements, the analog cellular system seems constipated. The data user who might wish to generate and exchange data from mobile entry points finds cellular radio unable to accommodate even relatively modest requirements. It is generally agreed that rates similar to those being evolved in the wireline network — notably the 64-kb/s standard — will *never* come about within the current analog cellular framework. Although it may not be a burning issue for today's cellular customers, almost all of whom are interested primarily (if not exclusively) in voice communication, the incompatibility of cellular radio with newer digital network services is a source of considerable consternation among the telephone network planners. As long as the radio-link technology remains FM, the evolution of the telephone network toward high-rate digital capabilities must stop short at the cellular doorstep.

Some cellular operators are also aware that the absence of a true data capability in cellular radio has probably subtracted a slice of the potential market for mobile communication, even before the pie reached the table. Indeed there are some in the industry who believe that quite a large portion of the pie was eliminated; the potential for mobile data communication is seen by some as a larger market than the conventional mobile telephony segment. No doubt this reflects a view of the relative saleability of data *versus* voice services at the current price levels.

These visionaries may be right: in one respect, data transmission can be much more efficient than voice transmission. In the time it would take to read this sentence aloud, for example, it is possible to transmit several pages of text at moderate data rates, e.g., 9600 b/s of usable throughput. A cellular data service capable of offering users such rates could handle a much higher number of subscribers per channel, bringing down the system cost per subscriber, allowing more reasonable pricing, expanding the market. The ability to move relatively large quantities of data to and from mobile units under interactive control of the user could form the basis for a highly saleable service. A few of the applications envisioned include: repair instructions for mobile maintenance activities or technical information for craftsmen of all kinds, product catalogs for salesmen, police information for patrol cars, insurance claims adjustment in the field. We shall not soon learn whether these visionaries are correct: these applications are simply beyond the capability of today's analog cellular systems.

In itself, the absence of "modern" data capabilities may not seem as significant as some of the problems enumerated in the previous sections. Yet it could represent a missed market opportunity of major proportions, perhaps, ironically, the one opportunity that could have sustained cellular radio in its current top-heavy, single channel per carrier architecture.

4.3 THE SPECTRUM EFFICIENCY SYNDROME

In the eyes of the original proponents of cellular radio in the late 1960s and early 1970s, the primary shortcoming of precellular mobile telephony (IMTS) was not the lack of privacy, nor the inability to transmit data, nor even the cost of service. It was chiefly the problem of chronic spectrum congestion, high blocking rates, and severely limited capacity (especially in major urban centers) which inspired the cellular proposal. The cellular idea was built around the crucial concepts of frequency reuse and cell-splitting which promised a virtually unlimited potential for generating new mobile telephone circuits to meet future growth.

The failure of analog cellular radio to realize that promise of true spectrum efficiency has thus come as a shock and as a fundamental disillusionment to many. Yet the spectre of saturation haunts the industry once again. Almost as soon as the first large urban cellular systems began to reach full operational status, the old problems reappeared. By the end of 1985, the Los Angeles cellular system was reported to be virtually locked up during peak periods — after only two years of operation and with less than 50,000 users (under one-half of 1% penetration). "Between 3 p.m. and 5 p.m. it can take as many as 10 attempts to complete a cellular call,"

Communications Week reported [66]. Clearly the 2% blocking objective had fallen by the side. An executive at one of the resellers for Pacific Telesis admitted that it had become impossible to reach even his own cellular salesmen: "I put all my salespeople on beepers. I just couldn't get them on their car phones in the afternoon [67]." The reporter concluded, "In short, the main problems of conventional mobile service that cellular was supposed to solve — poor quality and long waiting times — have returned with a devilish vengeance [68]."

Apparently these problems were cropping up in other large cities. In 1984 — after less than one year of operation! — the Chicago wireline operator petitioned the FCC for an additional 12 MHz of spectrum. The Los Angeles operator asked for 20 MHz. The petition of NYNEX's New York City operator was equally urgent:

> There is a pressing need for allocation of an additional 12 MHz of spectrum to the top 10 cellular markets . . . However . . . a 12 MHz allocation would be insufficient to meet demand in the largest market — New York. The demand in this market has already exceeded initial expectations. [At this time the cellular penetration in New York was approximately three-tenths of 1%.] It would be stimulated further by the cost savings and resultant price decreases that would be possible with additional spectrum. Thus, demand would exceed capacity unless a full 20 MHz were authorized for cellular use in New York [69].

The chief reason given for this startling turn of events was that *cell-splitting was proving to be unworkable.* The problem is cost. Cell-splitting is frightfully expensive. Additional spectrum allocations would increase system capacity without cell-splitting, and would be much cheaper. NYNEX even prepared a quantitative estimate of the savings from an expanded allocation. "Capital expenditures can be reduced by almost 50% if an additional 6 MHz of spectrum is allocated to each cellular licensee [70]." In other words, cell-splitting would cost twice as much as simply growing through additional spectrum allocations. (See Figure 4.17.) The Commission was dismayed. What had happened to the cellular "spectrum engine"? "I am struck by your lack of attention to spectrum efficiency," said one Commissioner. "You don't focus on the efficiencies to be gained by cell-splitting — you talk about your need for more spectrum [71]."

The thought is chilling: the very heart of the cellular idea — spectrum efficiency through frequency reuse and cell-splitting — may not be viable. If this is so, then there is little likelihood that today's cellular architecture can ever propel mobile telephony out of its small, troubled niche into the sought-for status as a mainstream telecommunication service.

A Simplified View of Frequency Reuse and Cell-Splitting

Let us review the way in which the cellular idea addresses the question of spectrum efficiency. Consider again what we may call the "naive version" of cellular frequency reuse: a large area is divided into numerous small zones each served by a low-power transmitter, and every frequency can be reused in every cell. If indeed it were possible to reuse the same frequencies in adjacent cells, then the *reuse factor* (the capacity gained by reuse compared to a single-cell, IMTS-type system) would be very large, equal to the number of cells in the system.

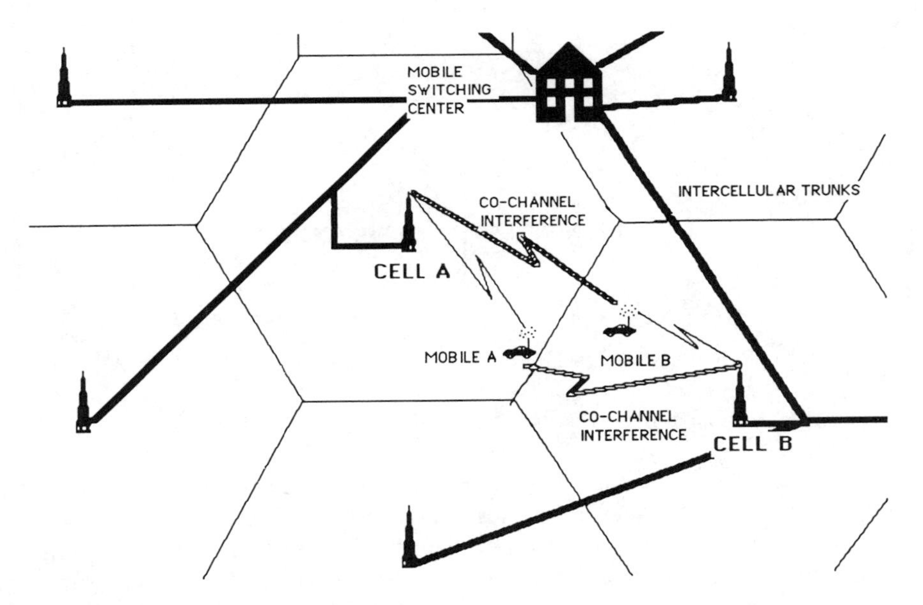

Figure 4.18 Obstacle to Frequency "Naive" Reuse.

Unfortunately it is not possible to use the same frequency in adjacent cells. This is because a signal on Channel 1 that is strong enough to reach a mobile near the edge of cell A will also be a strong source of cochannel interference for another mobile near the common edge of cell B. (See Figure 4.18.) The problem of cochannel interference — that is, the interference of two mobiles operating on the same frequency — forces the cellular designer to allow for considerable distance between cells that use the same channels (see Chapter 9). According to William C. Y. Lee, the reuse separation distance ranges from about four to six times the radius of the cells. In other words, in a system of 5-mile radius cells, two cells that reuse the same frequencies must be separated by from 20 to 30 miles. (See Figure 4.19.) Depending upon the assumptions regarding the required reuse separation and the type of antennas used (whether directional or omnidirectional), a reuse pattern is generated: a given set of frequencies may be reused every so many cells. Two common reuse patterns in cellular designs have been a 7-cell pattern and a 12-cell pattern. (See Chapter 9.)

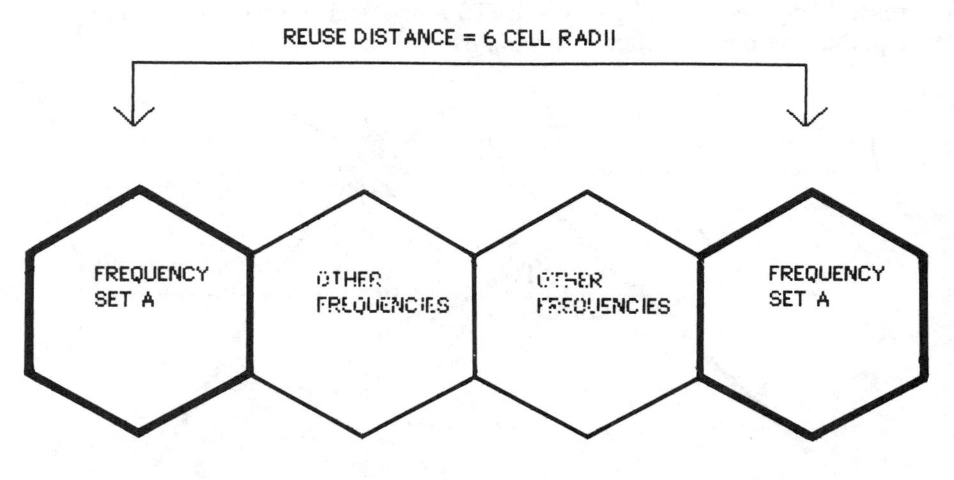

Figure 4.19 Concept of Reuse Distance.

Given a reuse pattern, the reuse factor for the system as a whole is dependent upon the size of the cells in relation to the size of the total area to be served. For example, assume we are dealing with a metropolitan zone that can be covered by a circle 60 miles wide. (Let us also assume the terrain is flat.) This region can be reasonably covered by 7 cells each with a radius of approximately 11⅓ miles. With a 7-cell reuse pattern, there will in fact be *no reuse at all* within this system, no gain in spectrum efficiency. The cellular system will provide exactly the same capacity, the same number of circuits for a given amount of spectrum allocated, as an IMTS-style, single-transmitter system.

As cell size shrinks, the benefits (capacity gain) of reuse begin to emerge. The initial gains, however, are small. The large-cell configurations that are built initially by cellular operators actually do not yield much of an increase in spectrum efficiency within a metropolitan area. In a 60-mile wide urban system, each frequency may initially be used only two or three times within the whole system. For example, if cells are initially implemented with a 6-mile radius, there will be approximately 25 cells in an urban region 60 miles across. If a reuse plan of 12 cells is used, the reuse factor will be about two. If a 7-cell configuration is used, the reuse factor will be a little more than three. In other words, as long as the cells are relatively large, the cellular system with all its massive overhead of cell-sites, hand-off, and interconnection and switching infrastructure is only modestly more spectrum efficient than an old-style IMTS system broadcasting from a single antenna would have been, had it been allocated the same 333 (or 666) channels.

In fact, for a capacity gain of only two or three times, a large cell system with a couple of dozen transmitters is probably much less efficient than a noncellular IMTS-type system, *if* economics are factored in — that is, if we look at the *cost per mobile telephone circuit.* How do cellular and IMTS (single transmitter) architectures compare? It is difficult to be precise. We may, however, assume that the fixed costs of an IMTS-type of cell-site are similar to the fixed costs of a cellular cell-site. The lower per channel costs per cellular cell-site (because in a multicellular system only a portion of the channels are used in each cell) are offset by the higher interconnection costs and the cost of the mobile switching center (not needed in the single-transmitter architecture). I believe that given today's spectrum allocations, *it is probably necessary to achieve a cell radius capable of yielding a reuse factor at least equal to the number of cells in the reuse pattern, before the expense of the multicell architecture will justify itself.* In a 60-mile wide system with a 7-cell reuse pattern, to achieve a reuse factor of seven would require cells with a radius of about 4.3 miles. With a 12-cell reuse pattern, the cell radius would have to be 2.5 miles to achieve a reuse factor of 12.

Fortunately, the reuse separation factor is constant regardless of cell radius, at least in theory. Thus, if the cell radii throughout the entire system are reduced to 1 mile, the reuse separation is now only 4 to 6 miles. In the same metropolitan system of 60 miles diameter, there would be about 900 1-mile radius cells. In a 7-cell reuse pattern, this would mean that each frequency would be reused 128 times. The capacity gain compared to a single-cell, IMTS-type system would be the same: 128 times more capacity. (See Figure 4.20.) For small cell sizes, the power of cell-splitting is enormous. Going to ½-mile radius cells would boost the reuse factor to 514. A ¼-mile radius cell would increase capacity by more than 2000 times over the single transmitter design.

This is why cell-splitting is vital to the realization of the promise of cellular reuse. The payoff in capacity only comes with relatively small cell

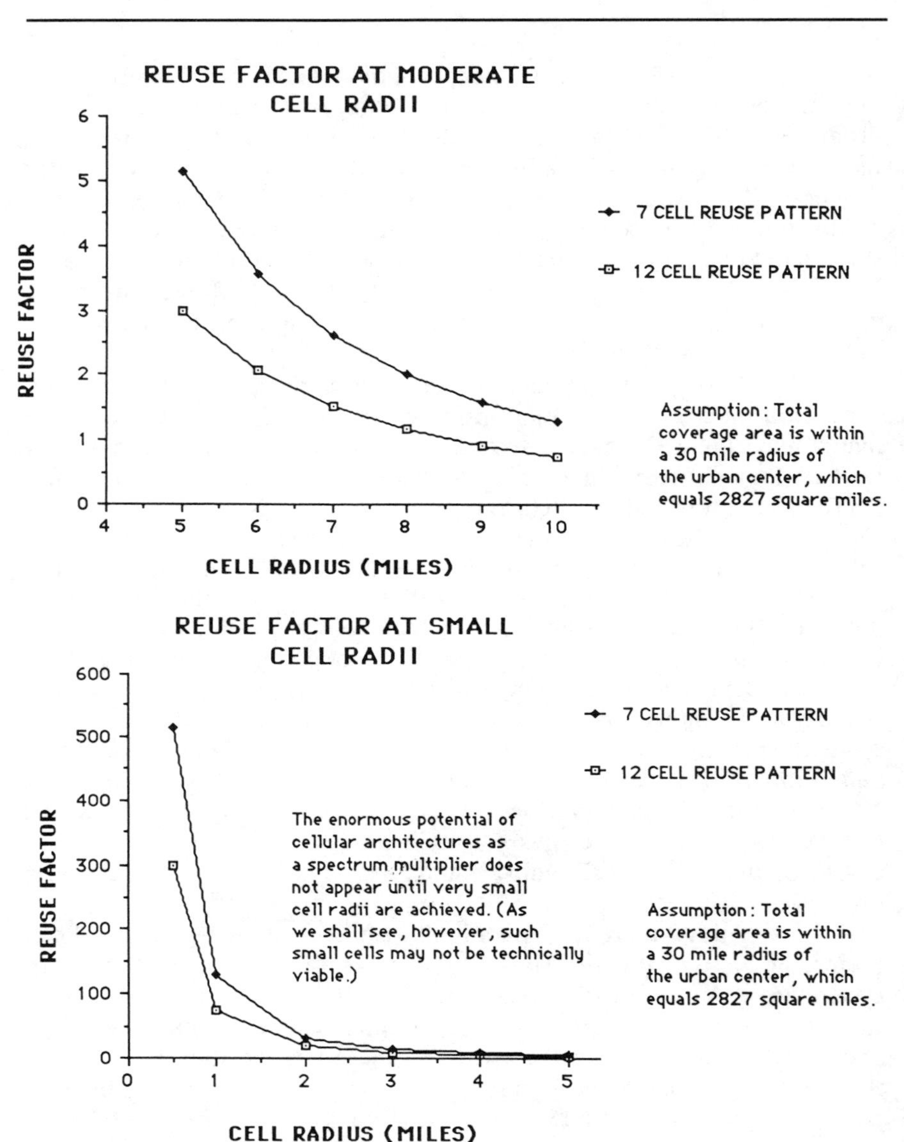

Figure 4.20 Relationsip between Cell Size and Reuse Factor.

sizes. Only by going to very small cells can the current technology yield the large capacities it needs to avoid congestion.

The problem, however, is that cell-splitting does not generate the same economic leverage. Whether the urban area is covered by 25 cells or 900 cells, the cost per cell remains basically the same. A system built of 1-mile radius cells provides 36 times more capacity than a system built of 6-mile cells — *but it costs 36 times as much to build.* In fact, the cost per cell for smaller cells tends, if anything, to be *higher* than for large cells. As Ameritech (the wireline operator in Chicago) stated in its petition to the FCC, "because cell-splitting normally occurs at sites located in the heart of the urban business districts where real estate is at a premium, these costs are increasing and will continue to increase sharply over the next few years [72]." Moreover, the tolerance in site location for a small cell is much smaller in absolute terms than for large cells. The transmitter for a 10-mile cell may be located much less precisely; the operator who positively must have the use of a very specific location for a 1-mile cell site antenna may find himself held hostage by the owners of the crucial plot of real estate. There is no principle of eminent domain in gaining optimum antenna sites for cellular operators.

In practice, cell-splitting probably *increases* the cost per circuit. In addition, the processing overload problems addressed in Section 4.2.2 — smaller cells produce geometrically higher rates of hand-offs, requiring more switching and interconnecting trunks — mean that the network costs of a small-cell system will be higher than for a large-cell system.

Quantifying these costs is not easy because of the lack of published cost data from cellular operators. To get some idea of the dynamics of these costs, however, let us consider a simplified hypothetical example (the numbers used are only illustrative and are chosen to simplify the arithmetic). Initially we assume a large-cell system (6-mile radius cells, covering a 60-mile diameter city) in which the system cost of each cell-site is made up initially of $500,000 of RF channel costs ($10,000 per channel) and $500,000 site costs (building, tower, antenna, interconnecting trunks, allocated switching, *et cetera).* Each cell provides 50 circuits; initially there are 25 cells; there are 1250 circuits in the entire system. The system cost is $25 million, or $20,000 per circuit. (If a loading factor of 20:1 is used, the system will be able to support 25,000 users and the system cost per user will be $1000.)

If the cells are split to a 3-mile radius, there will be 100 cells. The cost per cell is assumed to remain the same (an optimistic assumption). Therefore the cost for the entire system is $100 million. There are 5000 circuits, still at $20,000 per circuit, capable of serving 100,000 users.

Let us assume, however, that an alternative approach is also evaluated: instead of splitting cells, the operator wants to add more radio channels to each cell-site. To quadruple capacity without adding cells, he must add 150 new channels to each site. The cost of the new RF channels, at $10,000 each, is $1.5 million per site. This may require additional building floor space at each site, but otherwise he can use the same towers and other site facilities. The total system cost for 5000 circuits without cell-splitting is only $62.5 million — $50 million for the RF channels and $12.5 million for the cell-site fixed costs. He can serve the same 100,000 customers. The operator's cost per circuit under this scenario is only $12,500. In this simplified example, it costs 37.5% less to add capacity by acquiring more spectrum than by cell-splitting.

It is no wonder that the operators would rather receive additional spectrum than split the cells. Indeed, one wonders how far cell-splitting will ever be carried through. Given that cellular operators already have a problem with high system costs, cell-splitting only threatens to exacerbate this situation.

Thus the most unpleasant surprise of cellular reality is that the very logic of frequency reuse and cell-splitting may not be viable. The original analyses were obviously incomplete; they did not factor in the economics of cell-site proliferation. Furthermore, if cell-splitting beyond a certain point becomes increasingly unattractive economically, are cellular reuse factors destined to remain low? So low, in fact, that the purported advantage of the cumbersome and top-heavy cellular architecture over the simpler single-transmitter approach may never be realized? Would it have been easier and perhaps equally effective (or equally ineffective) to have simply granted the additional 40 MHz of spectrum to conventional single-transmitter, IMTS-type service and have done with it?

It seems that very small cells will be a particularly expensive approach to adding capacity, and yet it also seems unlikely that vast new blocks of spectrum will be handed over to cellular operators, at least in the United States. This leaves only one way out: cellular telephony must make a shift to a more spectrum-efficient radio technology, one that is capable of generating more circuits using the same spectrum, without forcing the uneconomic splitting of cells. Once again, we have reached the refrain: today's analog FM technology cannot solve the spectrum efficiency puzzle.

4.4 THE LICENSING SYNDROME

A final factor adding to the cost and uncertainty of the deployment of cellular radio in the United States has been the virtual breakdown of the licensing process. The FCC is commonly blamed, especially by the cellular operators, for maladministration of cellular licensing. The fact is, however, that cellular became a victim of its own "success" in stimulating interest in the mobile telephony business. Recall that in the beginning the

FCC had tacitly assumed that there would be only one operator — AT&T — and thus no need for licensing hearings at all. By the 1980s, the old AT&T was on its way out, and some competition was expected, even welcomed. A carefully tailored framework had been set up to allow for competitive operations in each market. But cellular generated a huge interest, especially against the backdrop of the mid-1980s stock market boom. It was vastly oversold as an investment opportunity, and the Commission was simply incapable of dealing with the tsunami of applications that were actually filed.

The licensing process has actually evolved (or devolved) through three successive phases, each incorporating further concessions to overwhelming market pressures:

1. A phase of comparative hearings;
2. A lottery procedure with controls on speculation;
3. A lottery procedure without controls.

4.4.1 The Comparative-Hearings Phase

We may recall that by the end of the 1970s, driven by external industry forces more than by its own agenda, the FCC had come to the conclusion that the cellular market could and should be developed in a competitive fashion. The standard arguments in favor of competition — lower prices to the consumer, more responsive, higher quality service, more rapid extension of the service to new markets — were bolstered by a growing feeling that nonwireline-radio common carriers were becoming capable, technically and financially, of developing and operating complex cellular systems. On the other hand, there was considerable concern at the Commission that the cellular service market was still marked by the tendencies of a "natural monopoly." The very large system investment required for cellular helped to convince many observers that a completely free market with unimpeded entry for new carriers would result in wasteful duplication, depressed rates of return for operators, and higher costs to the final customer. In short, the cellular regulators accepted the traditional arguments on *both* sides of the competition question — the free-market view which stressed the virtues of competition *and* the public utility-natural monopoly view which stressed its dangers. The goal of the licensing process was to make a marriage of the two immiscible principles.

The solution, as we have seen in Section 3.4, was the cellular duopoly — a device almost unique in the annals of regulated industry in the United States. The FCC embraced the notion that the cellular business could support two, but not more than two, operators in each market. (The duopoly was diluted by the rapid emergence of cellular resellers — essentially wholesale buyers who developed marketing expertise to resell the service "at retail," but who owned no cellular facilities themselves.)

As the Commission originally conceived the licensing process, there were to be two licenses for each city: a wireline license open to the telephone companies and a nonwireline license open to the radio common carriers or other credible applicants. The licensing procedure would involve the submission by each applicant of a detailed technical and business plan. The Commission would evaluate any competing proposals (hoping there would not be very many), select the best one, and award the license. The process would be similar in a general way to the licensing for radio or television broadcasters. It was a familiar, comfortable procedure. It was anticipated that cellular licenses would be awarded in several rounds, beginning with the largest cities, and concluding in a relatively short period of time to enable the winning licensees to begin construction and commence operations as quickly as possible. After so much delay, the Commission felt some urgency in bringing the service to the public.

Applications for the 30 largest markets were accepted in June 1982. It was immediately apparent that there was considerable interest in the cellular business. There were more than two hundred applications for the top 30 cities, and most had been prepared with scrupulous, even exhaustive, detail. Some of the applications comprised several thousand pages of engineering, financial, and market information, for the most part the result of a highly professional and thorough effort. The estimated cost of preparing each application ranged from $50,000 to $250,000. The applicants were almost all large and capable companies, including the major telephone operating companies, as well as impressive nonwireline applicants such as MCI, Metromedia, LIN Broadcasting, and Graphic Scanning — all experienced in communication services. All in all, the first round of applications represented an impressive effort by an impressive group of applicants. It was heralded as an auspicious beginning for the cellular industry, on both the wireline and nonwireline sides.

Unfortunately, the Commission was learning that it was possible to have too much of a good thing, even competition. Two hundred applications were far more than the administrative structure of the FCC was prepared to evaluate and process within any reasonable period. The FCC had publicly hoped for a much smaller number of actual applications and began urging the applicants to form cartels and to consolidate their applications to avoid the need for comparative hearings. Although the wireline applicants did fairly rapidly consolidate their applications, the major nonwireline applicants were veterans of other competitive processes (paging, for example). They had worked hard and tended to believe that they could win the comparative hearings on merit — this was after all the attitude typical of truly competitive sectors of the economy — and they resisted the blandishments of the FCC to merge their filings.

The very strength of the nonwireline proposals became a problem. Confronted with half a dozen or more serious, substantial, extremely detailed proposals, the Commission began to realize just how far its reach had exceeded its grasp. The applicants were prepared to fight their cases in whatever venue they might find themselves. The spectre of tough comparative hearings, with a virtual certainty of appeal by the losing parties, followed by an extended judicial review, threatened to delay the commencement of operations for years. The likelihood of appeal of the Commission's decisions was heightened by the complexity of the evaluation process, the myriad factors to be considered in judging one 1000-page application against another. In fact, the Commission proved prone to some questionable decisions, as in its judgment to award the very valuable Los Angeles license on the basis of a difference between two nonwireline applications in proposed coverage areas of less than 1% (overruling the administrative law judge in the process) [73]. Awarding a property of such value on the basis of such a slender margin (absurdly slender from the standpoint of the fuzzy realities of mobile radio propagation) left such verdicts open to challenge.

Moreover, the imbalance between the wireline applicants, who had divided up their markets among themselves to avoid this delay, and the nonwireline applicants, who had not, created a new problem, the so-called "wireline headstart" issue over which so much ink (and no doubt a few tears) have been spilled. It was argued that if the wireline licenses were awarded and the licensees permitted to enter the cellular business ahead of the nonwireline licensees (who would still be tied up in litigation of one sort or another), the wireline operator would quickly grab a dominant share of each cellular market. This would in turn threaten the viability of the competitive process. In the end, most cities did see the wireline licensee in operation well before the nonwireline licensee, and the available market-share data tend to indicate that the headstart advantage has been both significant and lasting.

The juggernaut of investor interest in cellular was only just getting started. The second round of license applications, for markets 31 through 60, were received in November 1982. There were more than 500 applications, about 16 per market. In March 1983 the third round, for cities 61 through 90, saw another 700 applications — 23 per market. Something had to give, and it turned out to be the FCC's licensing procedures.

4.4.2 The Lottery Phase

The Commission was neither organized nor staffed to evaluate hundreds of technical system proposals rapidly. Delays were inevitable unless some way to streamline the decision process could be found. In the 1970s, faced with similar problems in other areas, the FCC had experimented with spectrum "auctioning": licenses were to be sold to the high bidder, much as offshore oil leases are sold to the oil companies. There is much to recommend this approach; the courts, however, have held that spectrum auctioning exceeds the authority of the FCC's charter.

Another approach was simply to hold a "lottery" among qualified applicants for a particular license. The Commission had previously considered the lottery alternative for cellular, but had decided that the technical sophistication required to operate the new technology argued against it.

> We have carefully examined our decision not to select cellular licensees through a lottery procedure [they wrote] and we conclude again that a lottery would be inappropriate for cellular service.
>
> First, random selection appears to be inappropriate for determining which competing cellular applicant, from a group whose proposals may differ in significant ways, shall be granted a license . . . A cellular system requires a high capital investment and great technical expertise to realize the maximum benefits of cellular technology. Because of the limited number of licenses that can be granted in this service, it is especially important that the selection process be designed to award a license to an applicant proposing a high-quality system.
>
> In addition, we have set forth only limited standards . . . for determining whether applications "are acceptable for filing," [leaving] coverage and quality to be determined by the comparative process. If we were to resort to a lottery, it would be necessary to establish threshold standards for applications, in order to avoid a rush of superficially "acceptable" lottery entries that would be [incapable] of fully implementing the promise of cellular radio [74].

Now, under the crush of applications, the Commission decided to overlook its former reservations. A new assessment affirmed the pluses of the procedure. Applications would be screened to weed out spurious or duplicate filings, but no comparative assessment would be made. The winner would then be selected at random. After having demanded detailed engineering data, expensive market studies, solid business plans, the FCC was now deciding, in effect, that none of this would count. The licenses would be awarded by — a coin toss. The FCC saw this as a way of ensuring that a decision would be made, a license granted, within a reasonable period. There would be no comparative hearings, no delays for review, and fewer lawsuits because there would be no firm basis for the users to allege unfair treatment. The Commission would not have to search for some slim difference upon which to base its decision among essentially identical proposals. The public would benefit from rapid initiation of cellular service. The Commission also believed that the "threat" of a lottery might force the contentious and strong-willed nonwireline applicants to consolidate their applications ahead of time, as the wireline applicants had done. This would eliminate the headstart problem. On the second look, the lottery appeared to have much to recommend it.

The Outcome

The lottery decision threw the industry into chaos and permanently damaged the structure of the nonwireline market. The immediate consequences of this decision were almost exactly the opposite of what the FCC intended. The nature of the applications changed drastically. Cellular began to be perceived as an opportunity for the ordinary investor. It was touted by financial newsletters as "the greatest investment opportunity of the century [75]," a "speculator's dream [76]," an investment opportunity of "equal or greater potential than the railroads in the 1860s, telephone at the turn of the century, or TV in the 1940s [77]."

> I'm more convinced than ever [wrote one investment advisor] that cellular telephone could very well be the greatest investment opportunity of the century — *particularly for small and medium-sized investors (i.e., those with $5000 to $200,000 to invest)* [78].

Small fry were lured by outlandish, oversimplified market analyses:

> Here are some of the main reasons why cellular is such a unique opportunity [wrote one investment advisory]:
>
> It's not capital- nor labor-intensive.
> It can be built very quickly.
> Based on AT&T data and assuming subscription by less than 3% of the automobile market during the first eight years, it should earn nearly double the return per dollar invested of any communications technology in history . . .

> The numbers I have seen show that investors in the early stage of bidding on cellular licenses could reap a reward of 50 to 1 on projects they win during the first eight years alone, and as much as 10 to 1 each year after that [79].

Who would not be stirred by opportunities on such a scale? The public came eagerly to the trough. Companies began to spring up to help small investors with prefabricated applications and prepackaged engineering studies, for as little as $1000 to $5000. The lottery process was the great equalizer.

> The ramifications of this are mind boggling. *What it boils down to is that the FCC's decision to switch to a lottery method of selection has handed small and medium-sized investors virtually the same opportunity to win cellular licenses as the most well-heeled major corporations.* Unfair? Maybe. But this isn't an article on philosophy. What I'm talking about here is reality. And, in the real world, four Realities beat three Philosophies and two Unfairs every time [80].

The reaction of most of the "well-heeled major corporations" among the nonwireline applicants was exactly the opposite of this speculator's dream.

> The sudden injection of chance into their dealings made the nonwireline firms, including the likes of Metromedia, Graphic Scanning, MCI, plus lots of smaller players, very unhappy . . . Most applicants had filed applications that cost $50,000 or more per market, giving engineering and marketing details in an effort to impress the FCC and win a contested hearing. Suddenly, these were wastepaper, equated with mass-produced applications . . . [81].

In fact, many of the newer, small-time applicants saw their advantage precisely in the opportunity it offered them to leverage the big players. One cellular booster advised:

> If one of the larger wirelines — such as a Bell subsidiary or GTE Mobilnet — has also filed in your market, your job may be considerably easier. You may be able to "settle" with the larger telco for a piece of the action [82].

This is, of course, nothing more than the time-honored tradition of creating marketable value out of a nuisance. The small applicant really brings nothing to the table — he knows nothing about building or operating a cellular system and can probably contribute little money of his own — but he may force the major licensee to buy him out, either for cash or for a small percentage. The victims fully understood the situation: "It is obvious that most of the people in the filings are in it to make a quick deal," said the president of a major nonwireline operator. "This is the modern-day FCC story of Jesse James [83]."

The largest and best-prepared nonwireline players soon began to exit the market. MCI, which had won a number of lucrative "top-30" franchises during the comparative hearings (including the nonwireline license for Los Angeles, certainly one of the most attractive of cellular properties), decided to sell out in 1985. Metromedia, a paging and broadcasting giant which had sold its television stations to invest in cellular radio (including Chicago, New York, and Philadelphia), unloaded its licenses to Southwestern Bell the following year. Graphic Scanning began disassembling its cellular properties. Others followed. The "new crowd" crowed at their prospects:

> I believe that many of the heavier players will have second thoughts about competing in the upcoming smaller markets, given that the millions of dollars they had intended to invest would be pretty much neutralized by the lottery process. As just one example, the Knight-Ridder Newspapers recently announced that it has decided to pull out of cellular competition altogether, citing that "the odds were stacked against us" [84].

Cellular was vastly oversold as an investment opportunity. Some cellular "experts" bandied estimates that the value of franchises in markets 91–120 would reach $200 million each [85]! The result was chaos: more than 5000 applications were filed for markets 91–120, around 170 applicants per city. For example, there were 160 applications for cellular service for the city of Augusta, Georgia, population around 350,000 for the metropolitan area (including parts of neighboring South Carolina). A 1% penetration of the Augusta market would be only about 3500 cellular users; if this many users could be found and sold on cellular service, the Augusta market would generate several million dollars a year in revenue and perhaps a few hundred thousand dollars in operating profit (ignoring the capital costs). These revenues would be split between two operators, of course, so that one system might be capable of generating $200,000 a year in operating profits (before depreciation and return on investment). Considering that the cost of preparing and filing all those Augusta applications probably exceeded $1.5 million (about ten years of likely net cash flow from one system), the interest level was certainly extraordinary.

Inevitably, the process bogged down. The FCC was still committed to a prescreening to weed out incomplete, inadequate, or fraudulent applications, such as attempts by one party to file multiple applications for the same market. There were so many applicants that instead of speeding things up, the lottery slowed them down. It took twenty months — from

July 1984 until February 1986 — until things had been sorted out enough that the lottery for markets 91–120 could even be held. Moreover, the spectre of litigation had not been banished. Many lottery losers still challenged the qualifications of the winners. One observer wrote:

> By instituting a lottery for licenses, the FCC has made "competition on the basis of merit" irrelevant. Essentially, it now costs about $5000 to buy a lottery ticket. If you win, you have the right to be sued on the basis of your incompetence [86].

It got worse. For the next tier of markets, 121–150, there were an average of over 500 applications per city. Daytona Beach (770 applications) and Atlantic City (763 applications, 762 of them from nonwireline applicants) topped the list [87]. The smaller the market, the greater the investor interest. In December 1986, Midland, Texas, market number 295, broke the record with 1,058 applicants. That was shortly superseded by Aurora-Elgin, Illinois, which drew 1,105 applications [88].

Of course the FCC was inundated once again, administratively and even physically. "This place was like a fortress," an FCC staff attorney was quoted as saying after one filing deadline. "It was absolutely surrounded by cartons and cartons of applications." It took five days just to log them in [89]. The Commission began instructing applicants to send their filings directly to its Gettysburg, Pennsylvania, licensing center instead of the FCC headquarters in Washington. Even so, the steel shelving in the Gettysburg facility reportedly collapsed under the crush of paper — an event certainly symbolic of the entire situation [90].

In the meantime, many applicants had reached postfiling agreements to pool their chances (prefiling arrangements had been banned, but the Commission had allowed, even encouraged, postfiling agreements). If all parties could merge their applications, a lottery could be avoided and everyone would be guaranteed a piece of the pie. Of course there were no ground rules on how the pies should be divided, and even one holdout meant that a lottery would still be necessary. It was often hard to find a vehicle for integrating dozens or hundreds of disparate applicants under one structure. Lawyers, consultants, and brokers stepped forward to help develop these consortia. The logic of "a small slice of something is worth more than 100% of nothing" sometimes prevailed. One group reportedly represented more than 1200 applications, ranging from 30 to 50 applications per market [91]. In later rounds such umbrella structures were prearranged with "memberships" available at from $100 to $150 per market. One typical case offered an investor a share in an alliance for the Santa Rosa, California, market. When (if) his "alliance" won, he and the other 174 members of the group shared a 49% interest in the license (the other 51% had been traded by his alliance firm itself with another company).

The investor would have an individual share of around 0.3%, and analysts estimated that if he sold his interest he stood to make "a profit of ten times what he put into the application [92]."

The numbers of applications kept mounting. Soon it was clear that prescreening would no longer be possible. The Commission announced that it would cease prescreening and rely upon a postselection examination. Finally, in September 1987, the "antitrafficking" rules (which had forbidden transfer of licenses prior to construction) were eliminated; the lottery was now truly wide open. "Champagne corks were popping in law firms all around Washington," said one cellular attorney [93]. And with good reason: the removal of virtually all barriers to application for cellular licenses caused still another upsurge in the rate of new applications. By mid-1986, some observers were forecasting that by the end of the licensing process as many as 300,000 applications would have been filed [94].

4.4.3 Speculation

The cellular industry was coming under the spell of a different kind of economics. The "real" value of a cellular license — that is, the discounted future earnings stream from operating a cellular business — began to be overshadowed by its *speculative* value. Investors were now attracted to cellular licenses as a kind of tradable security, like a share of common stock, something that could be bought low and sold high. Just as commodities traders may invest in wheat futures without ever having the remotest intention of taking delivery of the wheat they "own," so cellular licenses began to be sought for their trading value alone.

The FCC has as a rule always attempted to discourage speculation in spectrum properties. In the case of cellular franchises, the FCC attempted at first to screen out duplicate filings (a natural tendency in lotteries is for participants to seek to improve their chances by purchasing extra tickets) and to eliminate applicants who appeared to be interested only in reselling the license, who had no intention of actually constructing and operating cellular systems. Yet since the FCC had neither the ability nor the desire to ban the sales of licenses completely, it was forced into an assessment of the motives behind individual transactions. "If licensees want to sell a license, we have to determine that they are not just speculating," said one Commission staffer [95]. This imposed yet another burden of judgment upon the overworked Commission staff. In the end, the antitrafficking rules were eliminated and the Commission fell back on its last resort: the losers would have to police the winners. One industry newsletter reported:

> In practice, the FCC may be faced with the problem that it can neither find these abuses upon examining applications nor investigate and enforce them with its limited resources. In fact . . . several Common Carrier Bureau staff members . . . said the commission has few means of detecting fraud unless petitions to deny are filed against a lottery tentative selectee . . . Since duplicate "turnkey" applications — that is, the same engineering information from different applicants — are allowed, there is no real way of determining whether similar applications filed under separate names actually represent legitimate applicants . . . It is mostly up to lottery challengers who file petitions to deny license awards to police the process, the FCC officials said [96].

On the other hand, from the applicants' perspective, the haphazardness of the lottery process created a need for some way to rationalize the random license allocations. Since even legitimate operating companies could no longer count on winning in particular markets, broad business strategies based on assembling a regional group of cellular systems could only be implemented by swapping licenses and shares of licenses with other winners. Entire strategies hinged now on whether companies were successful in obtaining a foothold in the markets they wanted. One paging operator, for example, filed in 23 smaller cities, but later became a partner in the Detroit license. This changed its priorities.

> Markets that once seemed important weren't important anymore. "We sought Flint, Grand Rapids, Toledo, and Lansing," [said the operator] [97].

It was as though the country had been carved into a huge jigsaw puzzle with each market divided into many small pieces and these pieces had been handed out at random to several thousand players. It was up to the players themselves to sort things out.

In September 1984 the first cellular "swap meet" was held in New York City to allow the hundreds of cellular players to trade their disparate pieces, each seeking to arrive at some reasonably coherent pattern suitable for actual operations. The deals became quite Byzantine. A typical reported trade, for example, involved the exchange of one-tenth of Oklahoma City, one-eleventh of Tulsa, one-twelfth of San Antonio, and two-twelfths of Tacoma for two portions of Honolulu, one share of West Palm Beach, and one of Toledo (*Barrons,* 5 November 1984). By announcing that no lottery would be held for markets where all applicants agreed ahead of time on how to share the pie, i.e., all parties to the agreement were guaranteed a share of the license, the FCC created a game of "chicken" (in game-theory terms) for all the participants.

> Traders huddled around the conference table, bluffed about selling their stakes, threatened to storm out if they didn't get their way, and built paper empires. "It was a cross between Monty Hall's 'Let's Make A Deal' and the baseball trading sessions," [said one participant] [98].

As this process went on, the distinctions between different markets with different demographics began to blur. A universal measure of the value of a cellular license emerged — the "pop," defined as the total population of the market multiplied by the percentage ownership of each player. As sophisticated market analyses dissolved into the universal language of "pops," it began to seem as though these were the units of value. Companies began to speak of buying "pops," and went out for financing with their "pops" listed on their balance sheets as though they were tangible assets.

And in a sense, through the self-fulfilling logic of speculation, the "pops" did become real assets. The value of a "pop" began to be quoted like a stock price, and almost immediately began rising. In 1984, an incredulous *Barrons* report cited average prices of $7 to $10 per "pop," but added (in arch journalistic style) that "two sources, one first-hand, one second-hand, said they knew of offers topping $20 for shares in fast-growing markets (*Barrons,* 5 November 1984)." By 1985, the average price per "pop" was about $15. Then Southwestern Bell purchased Metromedia's cellular properties for $41 per "pop." Pacific Telesis bought Detroit for $54 per "pop" [99]. Then in July 1987 McCaw Communications paid an astounding $70 per "pop" for Miami. What does this mean? The Miami area has a population of approximately 2.6 million. The "pops" valuation says that each of the two cellular licenses for that area is worth between $100 and $200 million before the system is even built. Some were predicting $100 a "pop" would soon be reached [100].

Do these numbers make any sense on the basis of likely business prospects? Assume that it takes another $100 million (over several years) to build the full-scale, small-radius cellular system in Miami, in addition to the, say, $100 million already spent for the license. A 10% return on this investment would require an after-tax $20 million per year from the system. If we assume that *each* operator could sign up 2% of the Miami population — 50,000 customers each, a 4% total penetration — then each of those customers must still generate at least $333 a month in revenue (excluding roaming and toll charges). In other words, there must be at least 100,000 potential customers in Miami, each willing to pay over $300 per month for a car phone, to sustain two cellular operators and provide them each with an unspectacular 10% return. (Of course this quick calculation does not include marketing costs or the costs to sustain operating losses during start-up.) Given cellular performance so far, as well as the available forecasts (Herschel Shosteck forecasts a *total* market for Miami in 1990 of about 26,000 customers), these figures would appear to be out of reach [101]. It seems clear that the valuation on these "pops" is in large measure speculative, rather than based on the realities of the business.

Nevertheless, the "greater fool" theory holds that a thing is worth what someone will pay for it. The speculative value of cellular licenses is, for the moment, quite real, as long as the "greater fool" is still to be found. One observer calculated that the expected value of an application for a small urban market would be "several hundred percent greater than the one-time cost of filing a single application [102]!" A more sober analyst pointed up the speculative value of even small percentages of winning licenses:

> Within days of the lottery, brokers had obtained bids from a low of $3.75 per capita [per "pop"] to a high of $11.50 per capita for 51% or more of the equity interest of the winning applications in almost all 10 markets . . . Eventually each consortium member's ownership interest of 2% to 3%, if sold (Why would one keep such a low equity interest?), will reap an average of $86,000 for each of the eight markets for which bids have already been received.
>
> Since the members paid from perhaps $2000 to $6000 per application per market, it is difficult to determine a rate of return on their Round 4 application preparation and filing costs. However, most members of this winning consortium will realize from five to 10 times the amount invested during a two-year period. Not bad! [103]

Note that in this passage the markets are not even identified; the licenses and their "pops" have become a commodity, completely fungible. Whatever one may think of the prospects of cellular radio as a business, it is clear that investing in cellular applications and trading cellular licenses has become very lucrative, at least temporarily. Success stories are common enough to fuel continued speculation. For example, one somewhat bigger-time speculator, a Dallas businessman, spent $5 million in 1984 to acquire stakes in seven franchises, including San Francisco, Nashville, and Memphis. He sold them a year and a half later for $34 million "even though he hadn't constructed a single antenna [104]."

The attractiveness of cellular licenses as a speculative vehicle has drawn the public into the game. "Applications mills" have appeared; for a few thousand dollars these companies will prepare and file a minimally adequate application. They will often sell the same engineering study to dozens of different investors. Inevitably, fraudulent practices have crept in. One notorious case involved a company called American National Cellular, Inc. (ANC), which may have been responsible for more than 5000 cellular applications.

> Promoters marketed applications for the lottery by promising novice investors they would make a mint. American National Cellular Inc. in Los Angeles hired television personality Mike Douglas to make video tapes and advertisements extolling telephones as "the investment of the decade . . . possibly of the century." And the company signed up 1000 agents to push applications for $5000 apiece [105].

ANC was later prosecuted by the Federal Trade Commission (FTC) for fraudulent representations and tax claims. The company was shut down and put into receivership in November 1985, after having taken in nearly $24 million from cellular applicants. A judgment of $23 million was later obtained against the company; by that time, however, only $800,000 was available for distribution to the plaintiffs. Several of the company's principals were also indicted; two, however, were already in prison, one for mail fraud and the other for grand-theft charges arising out of a fake diamond scam. The founder of the company was reportedly "last seen heading for Hong Kong with some $2.2 million of ANC's money [106]."

The ANC case might be dismissed as an aberration, but it was not. The ANC applications probably constituted about 25% of all the cellular applications filed for markets 121–150. The FTC was forced to petition the FCC to delay the lotteries for those markets, since the FTC suit had caused the suspension of the preparation of nearly 1200 applications that had already been paid for. The FTC argued that the clients who had paid for their applications should not be penalized for ANC's wrongdoings. "Caught in the crossfire between the fraudulent behavior of ANC and the efforts of the FTC to bring ANC to justice, these potential applicants are innocent victims," wrote the FTC [107].

The incursion by operators like ANC is one of the ancient side effects of speculative "something-for-nothing" markets. The Commission itself must bear substantial responsibility for creating an environment within which such an operation could flourish.

> American National vastly underplayed the risks, clients complain. Trudy Bullard, a Mansfield Center, Connecticut, widow with two young children, says she invested $10,000 in two applications after an American National agent assured her she would become a partner in lucrative franchises in Modesto and Visalia, California. She lost in the lottery, however, and now she says she isn't sure how she will pay for her children's education [108].

The first question is: What was a Connecticut housewife doing applying for cellular licenses in Modesto? The next one is: What was the Federal Communications Commission doing allowing such people to apply for licenses to build and operate cellular radio-telephone systems?

Indeed, why were other similarly unprepared individuals actually *winning* cellular franchises? The Nashua, New Hampshire, franchise was won by "an octogenarian, a disabled truck driver, and a highway engineer" all from California. As the *Wall Street Journal* reporter observed, "the three Californians barely know each other, know even less about car phones and haven't ever driven through Nashua." The Little Rock, Arkansas, license was won by an Odenton, Maryland, housewife and her husband, a switchboard repairman. Melbourne, Florida, went to a San Mateo, California, dentist and his wife, an artist [109].

Is this how it was supposed to be? A parody of the "free market"? Thousands, perhaps hundreds of thousands, of applications? Dentists trading "pops" with truckdrivers? The major nonwireline companies driven out? Cellular franchises awarded to retirees and housewives? Virtually all questions of qualification and ability to perform set aside, awards based on pure chance? Speculation displacing rational market forces? Fraud and trickery pervading the cellular-licensing process?

The clarification of regulatory attitudes toward competition is one of the key hurdles on the road to the next generation of cellular radio. In a sense, the licensing procedure has nothing to do with technology. The licensing disaster — that is the only word for it — has compounded the technological shortcomings of analog cellular radio, however, by removing from the field many of the stronger potential operators (particularly the major nonwireline carriers) who might have had the staying power and the long-term vision to work through the transition. The inexperienced nonwireline licensees in many areas will be ill-prepared for the technological upheavals ahead.

4.5 THE END OF AN ERA

The recurring, almost tedious refrain through most of this chapter has been the inherent technological inadequacy of the FM radio link to support the grandeur of the cellular vision. Certainly there have been other problems as well; the licensing fiasco has exacerbated a bad situation, and cellular networks have experienced the usual problems in developing and deploying network-control software. But the FM link determines to a large extent the cost of the system and the price of the service. It determines the performance of the system under interference-limited conditions, sets the reuse parameters, and defines the true geometry of cellular coverage. It shuts out modern digital network services and limits the ability of operators to assure the equivalent privacy of the wireline network. The high cost, poor spectrum efficiency, and obsolete service standards of today's cellular systems are all fundamentally rooted in the choice of FM. There are only Band-Aids, there is no fundamental solution, for these problems within the current technological framework.

There is nevertheless cause for optimism. Nearly fifty years after Edwin Howard Armstrong astounded the world with his development of frequency modulation, the long Eocene of FM is coming to a close. The stable technological environment within which mobile radio has evolved up until now is being superseded by a new set of technologies which will lead to unsettling and exciting breakthroughs. We are entering an era of technical innovation that will have us all gasping for breath within a very few years. The problems which seem so intractable today will fall very quickly before the advance of new technology in the relatively near future (assuming that we do not obstruct it with regulation). The rest of this book is about that future. It is about how the problems addressed in this chapter will be attacked with new technology.

The coming era will be the era of digital mobile-radio technology. This should not be taken to mean, however, that the monolith of FM will be replaced by a new monolith called "digital cellular." Digital techniques are proliferating throughout the communication network, and they are evolving so rapidly that it is not possible to capture all the specific alternatives in a book like this. The text would be out-of-date between the galley proofs and the printing.

What is possible, I think, is to lay the basis for an understanding of how the brave new world of digital communication will differ from the old world of analog communication (including FM). We are talking here about a development that, meaning no disrespect for Armstrong and his awesome achievement, will be as pregnant and as unpredictable as the introduction of steel tools to a Stone-Age tribe. Tasks that were incredibly arduous, like cutting down a tree with a stone ax, will become so easy that in the future they will wonder how we ever suffered with such primitive implements. Things that are simply inconceivable, like a power saw, become conceivable, then achievable, and finally commonplace. Today we really are chipping away at the problem of radio communication with a stone hand ax; in the not-so-distant future digital techniques will let us cut through with unbelievable precision and power.

REFERENCES

[1] Philip L. Forbes, "'Public Utility' Label Stifles Cellular System Competition," *Telecarrier,* April 1985, p. 21.
[2] US census data, 1980.
[3] Federal Communications Communication, Notice of Proposed Rule-Making, Docket 87–390.
[4] Presentation by L. M. Ericsson at Cellular Conference sponsored by Donaldson, Lufkin & Jenrette, St. Regis Hotel, New York City, June 3, 1987.
[5] Ford S. Worthy, "A Phone War that Jolted Motorola," *Fortune,* January 20, 1986, p. 43.
[6] Donald Schlosser, "Holding Back the Consumer," *Cellular Resources,* December 1984, p. 39.
[7] C. A. Laughlin, "Market Forces, Not 'Dumping,' Depressed the Cellular Telephone Market," *Personal Communications Technology,* December 1985, p. 14.
[8] Erwin A. Blackstone and Harold Ware, "The Emerging Mobile Communications Industry: Structure & Regulation," in Raymond Bowers, Alfred M. Lee, and Cary Hershey, eds., *Communications for a Mobile Society: An Assessment of New Technology,* Beverly Hills: SAGE Publications, 1978, p. 359.
[9] Worthy, *op. cit.,* p. 44.
[10] Herschel Shosteck, *Cellular Subscribers & Brand Sales,* Vol. I, No. 3, cited in *Industrial Communications,* October 17, 1986, pp. 6–8.
[11] Laughlin, *op. cit.*
[12] *Industrial Communications,* August 21, 1987, p. 7.
[13] Laughlin, *op. cit.*
[14] Shosteck, *Cellular Subscribers, op. cit.*
[15] Herschel Shosteck, *Data Flash: Cellular Subscribers & Brand Sales, Quarterly Survey,* Silver Spring, Maryland: Herschel Shosteck Associates, June 1987.
[16] *Industrial Communications,* February 27, 1987, p. 5.
[17] William C. Y. Lee, *Mobile Communications Design Fundamentals,* Indianapolis: Howard W. Sams, 1986, p. 256.
[18] Peter Dworkin and Roberta Ostroff, "The 'Celling' of America," *US News & World Report,* August 31, 1987, pp. 45–46.
[19] Schlosser, *op. cit.*
[20] Herschel Shosteck, "Can Cellular Be Sold?" *Telephone Engineer & Management,* July 15, 1987.
[21] Gene Costin, "Cellular in the Two-Way Marketplace: Bonanza or Bust?," *Mobile Communications Business,* September 1986, pp. 33–37.

[22] Schlosser, *op. cit.*, p. 40.
[23] Shosteck, "Can Cellular Be Sold?" *op. cit.*
[24] *Ibid.*
[25] *Ibid.*
[26] Remarks by John DeFeo, President of NewVector Communications, at Cellular Conference sponsored by Donaldson, Lufkin & Jenrette, St. Regis Hotel, New York City, June 3, 1987.
[27] Schlosser, *op. cit.*, pp. 39–40.
[28] Herschel Shosteck, *The Demand for Cellular Telephone: 1985–1995,* Silver Spring, Maryland: Herschel Shosteck Associates, September 1986, p. 43.
[29] Remarks of John DeFeo, *op. cit.*
[30] *Industrial Communications,* September 11, 1987.
[31] Interview with Chairman of Bell Atlantic, *Telephony,* July 14, 1986.
[32] Remarks of John DeFeo, *op. cit.*
[33] Shosteck, "Can Cellular Be Sold?," *op. cit.*
[34] V. H. MacDonald, "Advanced Mobile Phone Service: The Cellular Concept," *Bell System Technical Journal,* Vol. 58, No. 1, July 1979, pp. 19–20.
[35] Nathan Ehrlich, "The Advanced Mobile Phone Service," *IEEE Communications Magazine,* March 1979, p. 10.
[36] *Mobile Radio Technology,* "Broadband Booster Amplifier Overcomes Cellular 'Dead Spots,'" October 1985, pp. 62–63.
[37] *Philadelphia Inquirer,* "A Rocky Phone Problem," July 13, 1985.
[38] *Cellular Radio News,* "Cellular One Announces Roaming Agreements, Warns of Dread 'Green Leaf' Attenuation," August 1984, p. 5.
[39] James J. Mikulski, "DynaT*A*C Cellular Portable Radiotelephone System Experience in the U.S. and UK," *IEEE Communications Magazine,* Vol. 24, No. 2, February 1986, pp. 40–46.
[40] James F. Whitehead, "Cellular System Design: An Emerging Engineering Discipline," *IEEE Communications Magazine,* Vol. 24, No. 2, February 1986, p. 10.
[41] *Ibid.,* p. 13.
[42] Steven Titch, "For PacTel Mobile, Bigger is not Better," *Communications Week,* January 27, 1986, p. 54.
[43] Martin Cooper, "Cellular Does Work — If the System is Designed Correctly," *Personal Communications,* June 1985, p. 41.
[44] Max Greiner, Karl Low, and Rudolf Werner Lorenz, "Cell Boundary Detection in the German Cellular Mobile Radio: System C," *IEEE Journal on Selected Areas in Communications,* Vol. SAC-5, No. 5, June 1987, p. 850.

[45] James J. Sterling, "The Losee-Shosteck Effect: Does it Increase Cellular Switch Workload?," *Telecarrier,* August 1986, pp. 16–18; see also *Industrial Communications,* July 3, 1986.
[46] Greiner *et al., op. cit.*
[47] Shosteck, "Can Cellular Be Sold?," *op. cit.,* p. 76.
[48] Gerald Labedz, Ken Felix, Valy Lev, and Dennis Schaeffer, "Handoff Control Issues in Very High Capacity Cellular Systems Using Small Cells," *Proceedings of the International Conference on Digital Land Mobile Radio Communications,* Venice, June 30–July 3, 1987, pp. 360–366.
[49] *Personal Communications Technology,* "Metroplex Charges Southwestern Bell with Interception of Cellular Transmissions," December 1986, p. 30.
[50] Philip J. Quigley, "The Privacy Loophole: How Technology Leapfrogs the Law," *Personal Communications,* October 1985, p. 29.
[51] *Ibid.*
[52] *Morning Advocate* (Baton Rouge, Louisiana), "Edwards Mobile Phone Recording Ruled Legal," April 12, 1986, p. 6A.
[53] *Morning Advocate* (Baton Rouge, Louisiana), "Judge Dismisses Lawsuit Filed by Marion Edwards," November 13, 1986, p. 2B.
[54] *Morning Advocate* (Baton Rouge, Louisiana), "Marion Edwards Loses Appeal of Car Call Ruling," December 16, 1986, p. 3B.
[55] *Morning Advocate* (Baton Rouge, Louisiana), November 13, 1986, *op. cit.*
[56] *Personal Communications Technology,* "Metroplex Charges Southwestern Bell with Interception of Cellular Transmissions," December 1986, p. 30.
[57] *Telephony,* "Electronic Privacy Bill Passes House Committee," May 26, 1986, p. 90.
[58] *Ibid.*
[59] *Personal Communications Technology,* June 1986, p. 27.
[60] *Telephony, op. cit.,* p. 90.
[61] *Personal Communications Technology,* December 1986, p.4.
[62] *Mobile Radio Technology,* May 1986.
[63] *Telecarrier,* September 1986.
[64] Z. C. Fluhr and P. T. Porter, "Advanced Mobile Phone Service: Control Architecture," *Bell System Technical Journal,* Vol. 58, No. 1, January 1979, p. 59.
[65] *Ibid.,* p. 57.
[66] Steven Titch, *op. cit.*
[67] *Ibid.*
[68] *Ibid.*

[69] Edward Wholl and Joseph Di Bella, "New York City 'Needs' 20 MHz More for Cellular Mobile System Operations," *Mobile Radio Technology,* January 1986, pp. 22–23.
[70] *Ibid.,* p. 23.
[71] *Industrial Communications,* "Dawson Pushes for 'Flexible' Spectrum Allocation & Strikes Blow to Cellular's Push for 12 MHz More," May 31, 1985, p. 9.
[72] *Telecommunications Review,* "Reply Commenters Debate Efficient Use of Cellular Spectrum vs. Consumer Cost Burdens," August 20, 1984, p. 28.
[73] *Cellular Radio News,* "LA Grant Leads to Blurring of Comparative Criteria," February 4, 1985, p. 1.
[74] Response of the Federal Communications Commission to the "1982 Amendments to the Communications Act."
[75] Robert Ringer, "The Greatest Investment Opportunity of the Century," *Tortoise Report,* Vol. 2, No. 10, November 1984.
[76] Douglas R. Casey, "Cellular: A Speculator's Dream," *Personal Finance,* May 1986.
[77] *Ibid.*
[78] Ringer, *Tortoise Report, op. cit.,* p. 1 (emphasis in the original).
[79] Casey, *op. cit.*
[80] Ringer, *Tortoise Report, op. cit.,* p. 4 (emphasis in the original).
[81] Thomas G. Donlan, "No More Crossed Wires: Cellular Radio Gets Straightened Out," *Barrons,* November 5, 1984, p. 22.
[82] Stuart Crump, "Cellular Radio and the Telephone Industry: A Natural Team," *Personal Communications,* May 1984, p. 10.
[83] *Mobile Phone News,* "What Happens when the Individual Investor Becomes a Tentative Cellular Licensee?," May 8, 1986, pp. 1–4.
[84] Ringer, *Tortoise Report, op. cit.,* p. 2.
[85] *Ibid.* p. 4.
[86] Charles J. Mathey, "Cellular: A Cautious View, A Bright Future," *Personal Communications,* January-February 1985, p. 13.
[87] *Mobile Phone News,* "Fifth Round Nets 8100 and Counting," February 12, 1986, p. 1; *Mobile Phone News,* "FCC Schedules 5th Round Lottery for April: 7300 Applications Received for Markets 135–150," March 12, 1986, p. 4; *Mobile Phone News,* "Nonwireline Applicants for Past Two Filings Average Over 500 per Market," March 26, 1986, p. 5.
[88] *Mobile Phone News,* "Wireline/Nonwireline Cellular Lottery Results for Last of MSA Markets 281–305," December 18, 1986, p. 7.
[89] Neil Borowski, "Mobile-Phone Licensing is a Gamble," *Philadelphia Inquirer,* August 5, 1984, pp. 1, 11.

[90] Bob Davis, "Dialing for Dollars: Lottery for Franchises for Cellular Phones Mars Debut of Systems," *Wall Street Journal,* July 22, 1986, pp. 1, 6.
[91] Dean George Hill, "Private Auctions Abound in Round 4 Cellular Show-Down," *Telecarrier,* April 1986, p. 8.
[92] *Mobile Phone News,* May 8, 1986, *op. cit.,* p. 3.
[93] *Industrial Communications,* "FCC Makes Bold Stroke in Removing Anti-Trafficking Rules for Cellular," September 4, 1987, p. 5.
[94] Davis, *Wall Street Journal, op. cit.*
[95] *Mobile Phone News,* May 8, 1986, *op. cit.,* p. 3.
[96] *Industrial Communications,* "Challenges by Cellular Applicants May Be Tip of Fraudulent Cellular Application Iceberg," April 17, 1987, p. 8.
[97] Donlan, *Barrons, op. cit.,* p. 22.
[98] Davis, *Wall Street Journal, op. cit.*
[99] *Industrial Communications,* "McCaw Stock Offering is a Bellwether of Cellular Companies Going Public," August 14, 1987, p. 2.
[100] *Ibid.*
[101] Estimates provided by Herschel Shosteck Associates.
[102] Ringer, *Tortoise Report, op. cit.,* p. 4.
[103] Hill, *op. cit.*
[104] Davis, *Wall Street Journal, op. cit.*
[105] *Ibid.*
[106] *Mobile Phone News,* "FTC Wins $23 Million Judgment against ANC, Clients to be Refunded $1 Million," December 18, 1986, p. 3.
[107] *Mobile Phone News,* February 12, 1986, *op. cit.*
[108] Davis, *Wall Street Journal, op. cit.*
[109] *Ibid.*

Part III

DIGITAL COMMUNICATION: THE BASICS

Digital communication techniques are not new; in fact they predate analog electrical techniques. It has only been in the past twenty-five years or so, however, that electronic components have enabled the development of modern digital telephony systems. These systems all share certain basic principles and a core vocabulary which forms the foundation for a more focused analysis of the application of these techniques to mobile telephony. Digital systems are increasingly favored because they enjoy certain general advantages over analog techniques, the most important of which have to do with the emerging plans for a digital network that will possess capabilities far beyond those of today's telephone systems.

Chapter 5
THE REEMERGENCE OF DIGITAL COMMUNICATION

5.1 THE ANALOG REVOLUTION

Until Alexander Graham Bell, the world of communication was a digital world. The preeminent digital device was the telegraph, which communicated by turning an electric current on and off. It was simple, robust, and effective. By pressing down the telegraph key, the operator completed the circuit. Holding it down for a short time created a "dot"; a longer contact constituted a "dash." The presence or absence of gross electrical energy defined a message by means of an accepted code (the Morse code was one of several). The chief constraint was the speed of transmission. A good operator could manage 25 or 30 words, occasionally 40 words per minute [1].

Before long it occurred to a number of inventors (including Thomas Alva Edison) that the keys could be rigged to transmit mechanically at much higher speeds than human operators could attain. The Morse letters could be punched out on a paper tape ahead of time, and the tape could then be drawn through a special automatic keying device to activate the telegraph transmitter at a much faster rate. By the 1870s, automatic telegraphs were transmitting at rates in excess of 500 words per minute over test lines, and rates of from 100 to 200 words per minute were routinely achieved on ordinary intercity telegraph circuits [2]. This was the wireline corollary of today's goal of "spectrum efficiency." By utilizing the same wire circuits but transmitting at a much higher rate, the carrying capacity of the telegraph was greatly increased.

Automatic telegraph receivers (so-called "printing telegraphs") were developed to handle the higher speed messages (which were beyond the capacity of human recorders to set down). Soon it occurred to someone to connect an automatic receiver in series with another automatic transmitter to retransmit the same message over another telegraph circuit without human intervention. A chain of transmitters and receivers could be linked together [3].

This was a tremendous breakthrough. Long telegraph circuits tended to accumulate electrical noise, mainly due in those early days to electrical storms along the route and to improper insulation of the bare wires on telegraph poles. The longer the circuit, the greater the accumulated noise, the more attenuated the signal, the more garbled the transmission. For many years, it was necessary to limit transmission distances to a few hundred miles. A telegram originating in St. Louis and destined for Boston would be sent as far as Cincinnati, transcribed by a human operator, retransmitted to another telegrapher in Pittsburgh who took it down and sent it on to New York from where it was once again manually retransmitted to Boston. The multiple short hops were necessary to ensure that the number of errors introduced in any one electrical link was not too great. Obviously, the process was expensive, as it required many operators to transcribe and resend the message over and over. It was also fraught with the inaccuracies of multiple transcriptions.

By connecting automatic transmitters and automatic recorders, the whole process could be automated. This speeded up transmission over long distances and cut down on the labor costs of the telegraph operation. Telegraphy had discovered the valuable principle of signal regeneration. By completely regenerating the message at short intervals, it was possible to cover greater distances and still limit the number of errors that tended to creep in.

Telegraphers encountered a special problem, however, when the first transatlantic cables were laid in the 1850s and 1860s. Here were true long-distance circuits, reaching some 1600 miles from Ireland to Newfoundland. The laying of these huge cables out across the open Atlantic in waters two miles deep was a feat of engineering audacity comparable to digging the Panama Canal or landing on the Moon, and it captured the public imagination and came to symbolize technological progress for the Victorian era. Yet the physical challenges of laying the cable were matched by the unforeseen electrical challenges involved in actually transmitting signals on these immensely long circuits. Of course regeneration was impossible. For reasons that were not fully understood at the time, the effective transmission rate on the transatlantic circuits was reduced to two or three letters per minute. Here, on the most expensive circuit of all, where wireline "transmission efficiency" was most needed, the transmission rates were so slow — three hours or so for the text contained on one page of the average book — that even after the cable was installed, it was still quicker to transport a typical journal across the ocean by ship than to transmit its contents by undersea telegraph.

To improve this situation somewhat the English scientist William Thomson devised an ingenious technique to increase the throughput of the cable, even without altering its electrical or physical characteristics [4]. Thomson (who later became the first Baron Kelvin) realized that the ordinary Morse code was itself rather inefficient. The symbols only recognized two levels of current ("on" and "off") and two intervals of time ("short" and "long") and did not even utilize these code parameters very efficiently. With such symbols it was necessary to transmit up to five symbols ("dots and dashes") to signify a single letter of the alphabet. Thomson constructed an apparatus that was capable of transmitting symbols based on up to five distinct voltage levels. In Thomson's code, two of these five-level symbols were sufficient for unique designation of 25 of the 26 letters in the alphabet. Thomson's multilevel code had increased the information density of each symbol. It was the symbol rate that was limited by the electrical and physical characteristics of the cable. If each symbol could be made to carry more information, the overall throughput of the cable could be increased substantially, to from 16 to 20 words per minute.

A number of people began to speculate that it might be possible to transmit many more levels, for example by transmitting different electrical frequencies. This opened the door to a revolutionary thought: it might be possible to transmit tones electrically. It was noticed that if an electromagnetic circuit attracting a small armature (similar to the hammer of a doorbell) was connected to a device which caused a circuit interruption at a high rate, say, 500 times per second, a pure tone of 500 cycles per second would be generated in the receiver. In effect, the circuit became an electric tuning fork. Then, in 1861, a German schoolteacher named Johann Philipp Reis invented a device in which the vibrations of human speech were made to make and break an electrical contact reed [5]. The intermittent circuit could be transmitted to another location where it caused another reed to vibrate, reproducing to an uncanny degree many of the tonal qualities of the original voice. Reis called his device a "telephone." Although it was not actually capable of transmitting recognizable speech, many people believed that with time and further refinement a circuit could be constructed that would fully recreate the complex character of the human voice by simple, very rapid "on" and "off" pulses.

Then, in 1875 and 1876, Alexander Graham Bell solved the problem of speech transmission by coming at it from a completely different direction. Instead of using intermittent current (the "on-off" circuit of Reis and others), Bell found a way of causing an electrical current to vary *continuously* with the variations of human speech and other sounds [6]. Bell worked on two methods of accomplishing this: magnetic induction (the first approach) and variable resistance circuits (the superior approach). Bell's original "magneto transmitter" was extremely simple: speech captured by a diaphragm caused a small magnet to vibrate physically, which induced a fluctuating electrical current in a wire circuit. At the receiver, the process was reversed: the varying electrical current induced a fluctuating magnetic field that physically vibrated another magnet attached to a diaphragm for amplification.

Bell called his technique "undulatory current" and claimed it to be the fundamental invention underlying the telephone in all its forms. In hundreds of infringement suits filed during the life of the Bell patents, the principle of undulatory current was cited repeatedly by the courts in upholding Bell's priority and originality. Other inventors tried to find a method of transmitting speech by means of intermittent "on-off" current — which Bell himself freely admitted would have fallen outside the scope of his patent — but none was successful.

In modern terminology, Bell's undulatory current is referred to as *analog* transmission. Instead of transmitting discrete pulses of energy, like the telegraph or Reis's "telephone," an analog system like Bell's telephone transmits a complex, unbroken electrical waveform, which corresponds closely to the waveforms produced by the original sounds of human speech. The fidelity of the transmission depends to a large extent upon the faithfulness with which the original speech-sound waveforms are reproduced by the microphone transmitter as electrical waveforms.

Analog transmission proved to be simple, reliable, and economical for large-scale application. It was heralded as a breakthrough from the older telegraphic thinking. The idea of waveform reproduction guided early radio researchers as well. All the radio systems developed through the first half of the twentieth century utilized the same basic analog signal-processing principles of wireline telephony (although FM did represent a very different, somewhat less straightforward method of implementing an analog transmission). In general, from 1876 until about 1950, analog transmission reigned supreme among all communication media (with only vestigial survivals of the older telegraphic techniques).

5.2 THE DIGITAL RESURGENCE

In the fall of 1962 (while the cellular idea was still gestating) an epochal event was taking place in the world of ordinary wire telephone service. The first T-carrier system was installed on a telephone trunk line between the Dearborn central office in Chicago and the town of Skokie, Illinois, somewhat over 13 miles away [7]. This mundane event — little noted or understood outside the telephone industry even today — was quite simply the greatest milestone in modern telephony, culminating more than twenty years of difficult research and development. It was the beginning of the end of the analog network.

AT&T had been working for many years to find techniques to increase the capacity of the expensive copper wire plant by combining multiple telephone calls on the same wire simultaneously. T-carrier, which followed N-carrier, P-carrier, and so forth, was the latest of these "carrier" or multiplexer systems. It combined 24 telephone circuits on a single pair of wires, providing a tremendous savings in copper-plant costs for inter-exchange trunks connecting between telephone switching centers. Once introduced, T-carrier was an immediate success. Within four years, more than 4.6 million voice-circuit miles had been installed. By 1981, there were over 100 million circuit miles. In 1983, more than half of all exchange trunks were T-carrier based [8]. The T-carrier format was translated into microwave radio systems by the early 1980s and has become effectively the transmission backbone of the modern telephone trunk network.

What was unique about T-carrier was that it was based on *digital* transmission of the human voice, the first digital communication system ever fielded in the telephone network. Instead of following the standard analog approach of converting acoustical waveforms into electrical waveforms, the T-carrier circuitry converted speech sounds into a series of coded pulses, similar in a general way to the pulses of the telegraph — except much faster. In fact, the rate on a T-1 carrier channel is more than 1.5 million pulses per second. Each of the 24 voice circuits is represented by a stream of pulses brimming at a speed of 64,000 per second. All 24 pulse streams are combined together at the transmitting end, sent as a single continuous chain, and then separated out at the receiving end. New processing devices had to be developed to handle data moving at such phenomenal speeds, higher than the fastest telegraph by a factor of more than 1000. The length of each pulse is less than one microsecond. The timing on such systems has to be very sharp.

The technological impetus for digital communication was caught up in the electronic revolution that produced the transistor and the computer. In fact, the stream of data being pumped at high speed by a T-carrier system is very much like the data in a computer; it can be stored, copied, manipulated, added, analyzed, with many of the same electronic tools. In a T-1 transmission, the waveform of the human voice has completely disappeared, replaced by a stream of 1s (pulses) and 0s (spaces). At the basic level, the T-carrier circuit is no more complex (conceptually) than

Morse's telegraph. The current is either *on* or *off;* it is the pattern of switching from *on* to *off* that carries the information.

The most striking aspect of T-carrier transmission is its *abstractness.* The stream of 1s and 0s bears no tangible relationship to the overt, physical characteristics of human speech. Unlike nineteenth century attempts at telephony using "intermittent circuits," which attempted to transmit the pulses in tune with the tones produced by speech, digital T-carrier works on a more abstract plane. Everything is in code, and the code is everything. The speed and the logical structure of the coding system can be manipulated, as Thomson modified the Morse code for the Atlantic cable, to produce digital systems capable of enormous capacity, carrying all kinds of signals. Today a single digital circuit can contain thousands, even tens of thousands, of telephone conversations, all encoded in the Spartan language of *on* and *off.* Digital coding has been applied to music and photographic or television images. Encoded and reduced to a stream of 1s and 0s, speech, music, and video, as well as most computer data, are all indistinguishable. Digitized speech can be mixed with, say, digitized music, stored like computer data, manipulated mathematically, transported, and separated at the receiver into the original components.

The processes of digital communication are awe-inspiring, in speed, in precision, in terms of sheer technical accomplishment. Yet these processes are clearly much more complex than Bell's simple telephone. Two questions therefore arise:

1. What does "digital" really mean? How are these transformations actually accomplished? How is the rich and subtle character of human speech reduced to a simple string of 1s and 0s, and then reconstituted with sufficient fidelity?
2. Why bother? Or, to put it another way, why does digital, with all its expensive overhead, pay off?

The first question will be addressed in Chapter 6. The second question is the subject of Chapter 7.

REFERENCES

[1] Robert Conot, *Thomas A. Edison: A Streak of Luck,* New York: Seaview Books, 1979, p. 19.

[2] M. D. Fagen, ed., *A History of Engineering and Science in the Bell System: The Early Years (1875–1925),* AT&T Bell Laboratories, 1975, Chapter 7.

[3] Conot, *op. cit.*

[4] Harold Issadore Sharlin, *Lord Kelvin: The Dynamic Victorian,* University Park, Pennsylvania: Pennsylvania State University, 1979.

[5] Robert V. Bruce, *Bell: Alexander Graham Bell & The Conquest of Solitude,* Boston: Little, Brown, 1973, p. 117.

[6] *Ibid.*

[7] E. F. O'Neill, ed., *A History of Engineering and Science in the Bell System: Transmission Technology (1925–1975),* AT&T Bell Laboratories, 1985, p. 562.

[8] *Ibid.*, p. 563.

Chapter 6
THE DIGITAL VOCABULARY

A comprehensive presentation of digital communication techniques is well beyond the scope of this work [1]. One of the great advantages we have today over the telegraph engineers of the nineteenth century, however, is a more organized language for talking and thinking about communication problems in digital terms. What is important, I believe, is to familiarize oneself with the essential vocabulary of digital communication, the handful of key concepts which provide the basic framework for understanding the potential of these new systems and the opportunities they are creating for mobile telephony.

A basic block diagram of a digital communication system is shown in Figure 6.1. There are three basic steps or processes. The first, the *analog-to-digital* or A/D conversion, as it is usually called, is the main distinctive feature of a digital system. This process reduces the original complex waveform to an easily manipulated numerical form. The input (for telephony systems) is audible sound produced by human speech; the output is *data* in a specific logical format. In this form, it can be stored or transmitted in a variety of media — it is *independent* of the medium of transmission.

The second basic step, usually called *modulation,* converts the logical data input into a transmittable form, typically pulses of electric current. The transmission, however, may also be based on light waves (in fiber-optical systems) or radio.

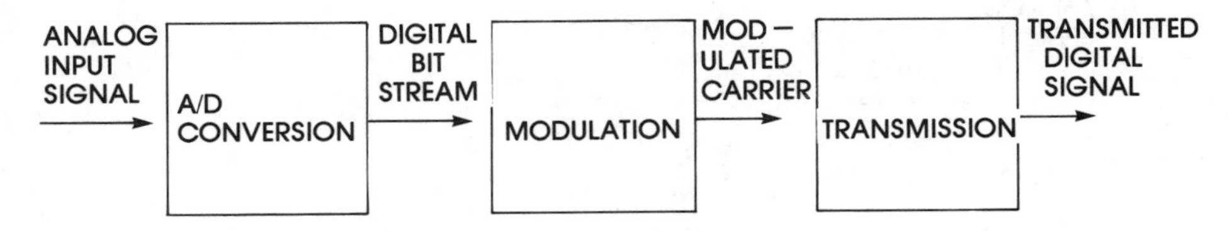

Figure 6.1 Block Diagram of the Digital Communications Process.

The third basic step is the *transmission* itself and, what is of much greater interest, the *signal processing* which takes place in conjunction with the transmission process. One of the most important types of signal processing is called *multiplexing* (see Figure 6.2), which involves the combination of several separate communication circuits in a single transmission channel (as we saw with T-carrier in Chapter 5).

The receive chain is usually a mirror image of the transmit chain: multiplexing is complemented by demultiplexing, modulation by demodulation, coding by decoding, A/D by D/A. Many devices are designed to work bidirectionally, to modulate as well as demodulate (whence the term *mo-dem,* for example).

The most basic of all digital concepts, however, which underlies both A/D conversion and modulation is *coding.* Coding may be defined as any process which produces a coded output: information that is represented in the form of a sequence of numbers (or numberlike symbols, such as Morse code's dots and dashes). Coding is in one sense very simple and straightforward. In other ways, the concept of coding is quite subtle, and the word is often used with somewhat different shades of meaning or emphasis to describe different stages in the digital process. For example, some types of coding work on analog inputs; others work on inputs that have already been previously coded. The problems facing the coder in each case are different.

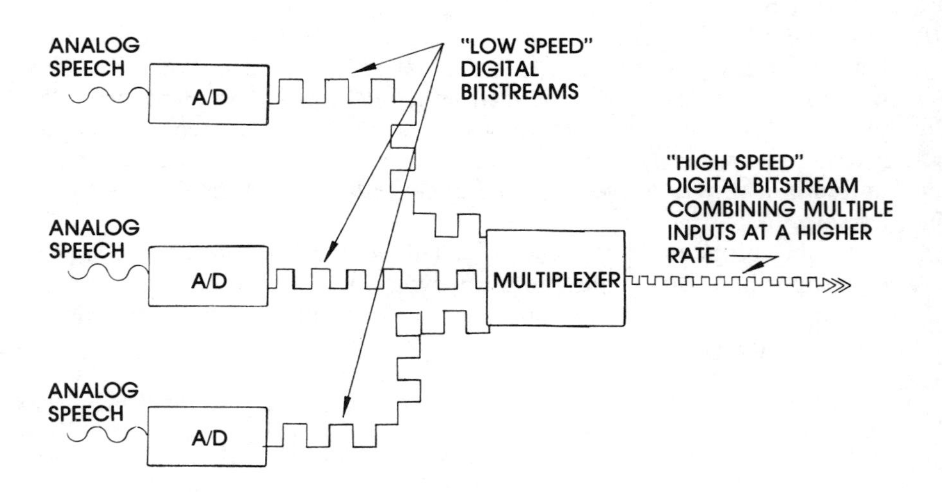

Figure 6.2 Multiplexing.

It is also important to recognize that a coded signal "exists" on two levels simultaneously. On one level, it can be described *logically*, or numerically. The normal description of a digital transmission as a string of 1s and 0s, for example, is a logical description. But the coded signal also exists *physically,* as a series of electrical or optical pulses, for example, or a pattern of variations of some characteristic of a radio wave (such as frequency or phase). The logical design of coding schemes is quite distinct from the physical design. The "same" coded signal (in logical terms) may be represented in many different physical forms. There can be confusion over whether a particular use of the word "coding" refers to logical or physical aspects of the coded signal.

Indeed, the distinction between logical and physical levels of coding is the basis for the most fundamental design partition in the engineering of digital systems: software *versus* hardware. It is quite often the case that a software engineer may have little knowledge of the actual physical characteristics of his code, the electrical form it takes. The hardware engineer, on the other hand, may troubleshoot his designs with logically meaningless (random) signals; he is more concerned with the physical characteristics of the signal.

It is useful, therefore, to relabel Figure 6.1 as shown in Figure 6.3. Every digital transmission actually employs *two* stages of coding. The first stage converts the analog input into a stream of numbers. The designers of this coding stage are primarily concerned with the relationship between the analog input and the digital output: has the *information* in the analog input been adequately captured in the code? This is a question of logic, or "meaning," not physics. Engineers who work on the first coding stage are usually well-versed in mathematics (especially algorithm design) and in the properties of the analog input signal (acoustics, in the case of telephonic signals). They accept the hardware characteristics as a constraint and may not concern themselves greatly with the physical characteristics of the system.

The second coding stage is mainly concerned with the physical characteristics of the transmission channel. The informational characteristics of the (coded) input are unimportant (and are usually simulated with random sequences). The logical aspects of the second coding stage are strictly determined by the physical parameters of the transmission channel.

Both coding stages "exist" on both levels, logical and physical, and each can be discussed on either level. But the main emphasis during the first stage is on the informational or logical characteristics of the code. In the second stage, the main emphasis is on the physical implementation of the code.

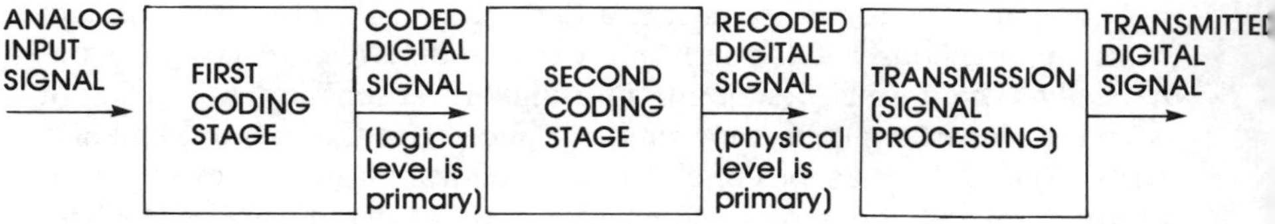

ANALOG INPUT SIGNAL → FIRST CODING STAGE → CODED DIGITAL SIGNAL (logical level is primary) → SECOND CODING STAGE → RECODED DIGITAL SIGNAL (physical level is primary) → TRANSMISSION (SIGNAL PROCESSING) → TRANSMITTED DIGITAL SIGNAL

Figure 6.3 Redefinition of the Digital Communications Process.

Another important consideration in the design of digital communication systems (as opposed to many data-processing systems) is the requirement that all of these processes must generally take place in *real time* — which means that they must be performed without introducing noticeable delays (no more than a small fraction of a second). All digital processing takes time; how much time depends upon the complexity of the processing algorithm and the speed of the processing hardware. Processing speed is a very important hardware constraint, especially in the first coding stage (which tends to be much more computationally intensive than the second coding stage). Indeed some of the first-stage coding algorithms proposed for digital mobile radio were devised many years ago, but could never be implemented in real-time systems until quite recently because of the unavailability of sufficiently fast processors.

6.1 THE FIRST CODING STAGE: A/D CONVERSION

The A/D or digitization process comprises three basic logical or mathematical concepts.

1. Sampling;
2. Quantizing;
3. Coding.

Today, all three steps are generally performed in what amounts to a single operation, but historically each represents an independent breakthrough in the understanding of the principles of digital communication. (Because the normal emphasis in the first coding stage is on the logical-level design, we shall not discuss the physical-level design issues.)

6.1.1 Sampling

The initial input to a digital communication system is a waveform of astounding complexity, representing the electrical analog of the sounds spoken into the telephone. (See Figure 6.4.) The speech input is rich in frequencies and overtones and rapid variations in amplitude and pitch which together make up not only the sounds of the spoken words themselves but also lend a characteristic tenor to each individual voice. No two speakers will produce the same waveform, even when speaking the same word. Each voice is as distinctive as a fingerprint. The electrical input waveform contains all of this information in addition to the bare meaning of the words. In fact, the words themselves could probably be transmitted in digital form with a bit rate of perhaps 50 to 100 bits per second. To reproduce a specific human voice faithfully today, however, requires more than 1000 times as much digital information to be transmitted. Another way of putting this is to say that if all human beings had voices with exactly the same characteristics and always said every word in exactly the same way, it would be necessary to transmit only about one-tenth of 1% of the information that must be transmitted to capture all the richness and emotion of real speech.

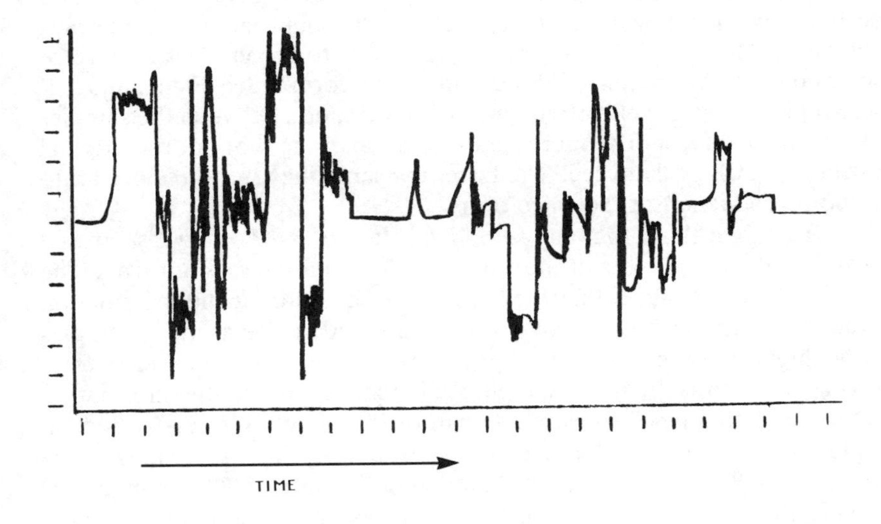
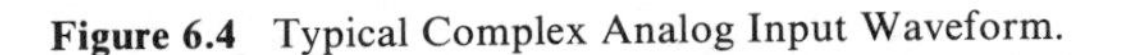

Figure 6.4 Typical Complex Analog Input Waveform.

The first challenge for a digital system is to reduce this complex waveform to a manageable numerical form, which normally means a string of 1s and 0s — without losing the ability to recreate the richness and the nuances.

There are two methods for attacking the problem of reducing a complex waveform into simpler elements. One approach is known as Fourier analysis. It is based upon the amazing proof, first offered by mathematicians in the nineteenth century, that any complex waveform can be viewed as the sum of a finite number of simple sine waves. In physical terms, the rich tones of the human voice can be conceived as a mixture of many pure tones, at distinct frequencies, all superimposed upon one another.

Fourier analysis, however, is computationally complex and for a long time lay beyond the limits of real-time processing. The alternative technique that forms the basis for digital communication is much more intuitive and is easily grasped by the layman, although it emerged later in the history of mathematics. Discovered, forgotten, and rediscovered several times, until it was finally recognized by Bell engineers in the 1930s, the *sampling theorem,* also called the *Nyquist* theorem after Harry Nyquist, the Bell engineer who identified its importance for telephony, is regarded by some as the most fundamental of all digital concepts. It is the idea, which can be proved mathematically, that any complex waveform can be reconstructed from an adequate number of discrete samples.

Sampling is the process of taking instantaneous measurements of a continuously changing parameter, such as the amplitude of a complex waveform. The sampling information allows us to reconstruct a more or less crude representation of the original waveform. (See Figure 6.5.) If the samples are relatively infrequent, the information between the samples will of course be lost. The more samples we take, the more of the original information will be captured. The faster the sampling rate, the more faithful the reproduction of the waveform.

One might think that any sampling rate, no matter how fast, would always result in *some* loss of information. This, however, is not true. The sampling theorem states that it is possible to capture *all* the information in the waveform itself *completely,* if we use exactly twice as many samples as the highest component frequency in the waveform. Another way of looking at it is to say that any sampling rate will capture all the information contained in frequencies *below* one-half of the sampling rate. So, for example, if it is decided that the highest frequency in a telephone voice circuit is 4000 cycles per second, then a sampling rate of 8000 samples per second will capture all the information necessary to describe the most complex waveform completely.

In the T-carrier architecture, the sampling rate was actually set at 8000 samples per second. In fact there are overtones above 4 kHz contained in the human voice which are effectively filtered out prior to coding. The absence of these higher overtones is part of the explanation for the noticeable reduction in quality of the telephonic voice (as compared to a high-fidelity medium such as broadcast radio or TV). The quality reduction was considered acceptable, however, for the following reasons:

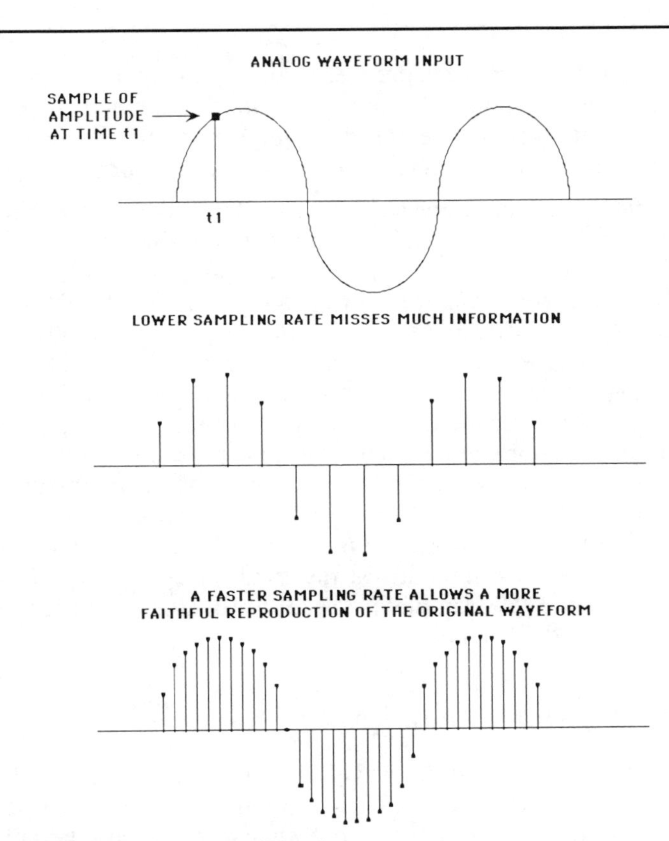

Figure 6.5 Sampling of an Analog Waveform.

1. Most of the voice energy, including speaker recognition cues, is contained in the frequencies below 4 kHz.
2. At the time T-carrier was introduced, the other components of the analog network, such as the analog switches (the universal form at the time T-carrier was introduced) and analog telephone instruments, traditionally passed only about 3 kHz of bandwidth. Any significant gain in digital quality (which was first applied only in intermediate transmission links, not throughout the entire circuit end-to-end) would not have been able to reach the end user. (This is less and less the case as digital systems begin to proliferate in the network.)
3. A higher sampling rate — say 30,000 samples per second to capture all frequencies below 15,000 cycles — would definitely have produced a higher quality voice transmission, but would have necessitated

much faster, and more expensive, processors. Again, such fast processors are commonplace today, but in the early 1960s a higher sampling rate would have posed a significant additional challenge to system developers for what was perceived as a small gain in quality.

The hardware limitations were critical. The chief advantage of sampling, especially as compared to Fourier analysis, is that the taking of each sample is an extremely simple operation. It is simply an amplitude measurement. The measurement need not concern itself directly with frequency or phase *per se,* nor with any other "complex" aspect of the input. It need not even be concerned with what the signal originally was. A waveform coder is a very efficient machine. Set the sampling rate and it will capture all the information contained in the waveform at frequencies below one-half that rate. To improve the resolution, to capture more information, it is necessary only to increase the sampling rate. Each individual sample still constitutes but a single measurement on a single dimension. The chief technical challenge arises from the need for speed. The sample must be captured in a very short interval. In the 1940s and 1950s when T-carrier was being developed 8000 samples per second was a formidable engineering challenge.

6.1.2 Quantization

Once the amplitude of the waveform has been sampled and a value obtained, the next step in reducing it to digital form is to *quantize* that value, to round it off to the nearest unit on a specially designed measurement scale. For example, assume that the difference between the maximum possible (or allowable) amplitude and the minimum (usually zero) is defined and this range is then divided into, say, ten equal steps. Each sample is then rounded off to the nearest of these ten steps, and a number from 1 to 10 is recorded. The series of samples has now been reduced to a series of numbers. It is now possible to speak of it as *digital* information. (See Figure 6.6.)

Several observations can be made concerning the quantization process.

1. Quantization introduces error, a deviation of the recorded value from the real value which represents "incorrect" information. The difference between actual sample amplitude and the recorded value that is rounded off cannot be recovered.
2. The average relative magnitude and, to a degree, the functional importance of this *quantization error* (also called *quantization noise*) is determined by the fineness of the scale employed, the number of steps within the overall range. A quantizer with a few steps will

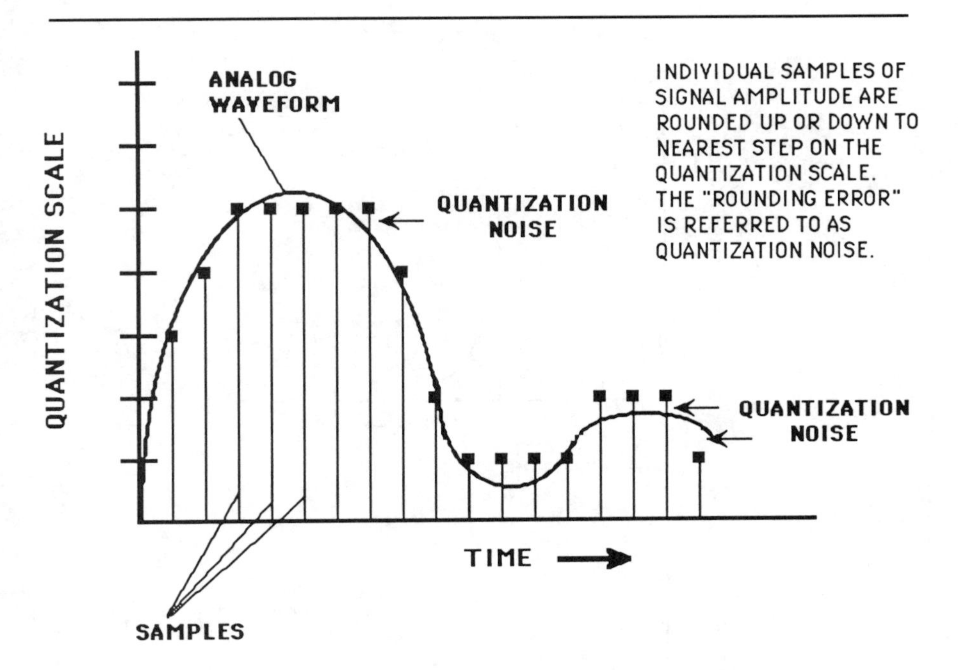

Figure 6.6 Quantization.

introduce more noise than a quantizer with many steps. (See Figure 6.7.)

3. The problem of quantization noise cannot be defeated by simply increasing the number of intervals on the scale, because the more intervals, the more information must be encoded for each sample (see Section 6.1.3, Coding), the greater the number of bits to be transmitted, the greater the capacity required in the channel. A very fine scale will reduce quantization noise but will increase the capacity requirements, and the expense, of transmission. It may also make the signal more vulnerable to transmission noise, as we shall see. A coarse scale will reduce the bandwidth needed but will degrade the transmission with more quantization noise. It is a design trade-off.
4. The quantization noise may be reduced by using a logarithmic scale instead of a linear scale. (See Figure 6.8.) Small changes are finely graded at low signal levels. At higher levels, the scaling is coarser. The result is to render the quantizing noise approximately a fixed percentage of the signal amplitude [2].

In the case of T-carrier, the system architects initially settled upon a quantization scale containing 128 steps logarithmically scaled. Each am-

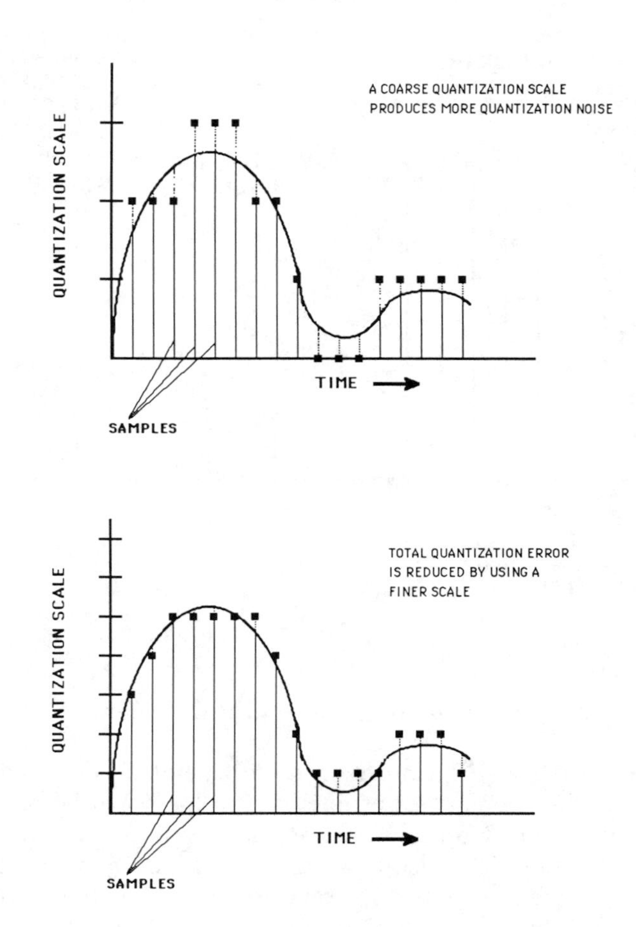

Figure 6.7 Effects of Quantization Scale Coarseness.

plitude sample was rounded to one of these 128 values. Subsequently, a 255-level scale was employed, using the full 8-bit word to carry voice information [3]. There are actually two slightly different versions of the "standard" quantization scale known as μ-law, used in North America, and A-law, used in Europe.

6.1.3 Coding

Once the sample has been quantized — converted to a discrete number — the next question is: What is the best way to encode that number?

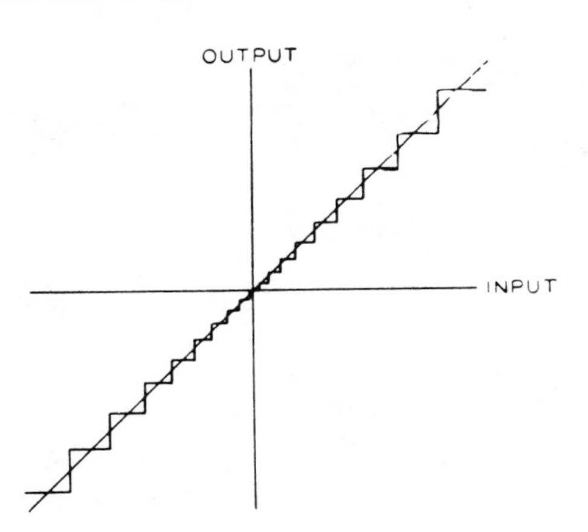

Figure 6.8 The Relation between Input Value and Output Value with Quantization Logarithmically Tapered.

The question may not make sense until it becomes clear that there *are* alternatives. As noted previously, coding involves both a logical and a physical, e.g., electrical, representation of the quantized sample. Both logical and physical considerations enter into the choice of coding. For example, the quantized sample in an original T-carrier system represented one of 128 levels in a logarithmic amplitude scale. It would be possible to represent each quantization level with a different voltage level, 128 voltages in all. Hypothetically, if the quantized sample happened to be a level 13 on the logarithmic scale, the coding for this could be a pulse of 13 volts. If the quantized sample happened to be a 99, a pulse of 99 volts could be generated, and so on. In fact, such a system would be highly efficient, because it would require only a single pulse of current to transmit each sample. The throughput of the system would be maximized. Such a system would be reminiscent of Lord Kelvin's multilevel telegraph which increased the capacity on the transatlantic cable.

The problem with such a superefficient coder is that transmission channels are noisy and subject to other problems such as attenuation and distortion (see Chapter 8). A 128-level coder requires a very fine discrimination on the receiving end. In our example, the receiver would have to be able to distinguish a 126-volt pulse from a 127-volt pulse, a difference of less than 1%. If enough noise has entered the channel, it will play havoc with such fine distinctions.

At the other logical extreme, a coder can be constructed to represent each quantized T-carrier sample as a string of seven separate pulses, each pulse representing either a 1 or a 0. This is called *binary coding* — the conversion of the sample into a base-two representation. (See Figure 6.9.) The disadvantage of this approach is that it increases the number of pulses

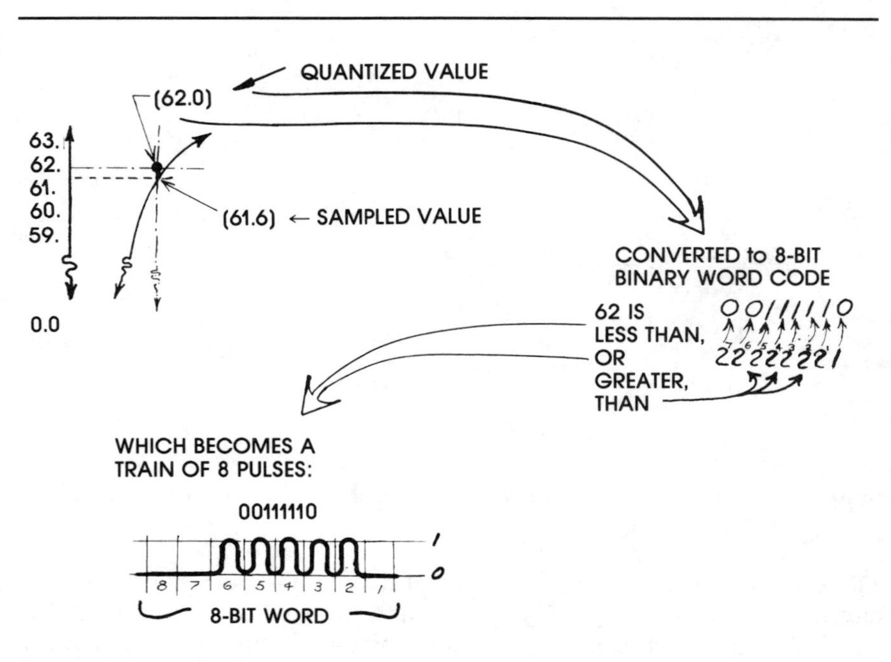

Figure 6.9 Binary Coding.

needed to transmit each sample and thus decreases the capacity of any transmission system. On the other hand, the discrimination problem for the receiver has been greatly simplified. Instead of looking for very small differences in signal level, the receiver has only to determine whether there is any current flowing or not. If the current is "on," it is registered as a 1. If not, then a 0 is recorded.

Historically, this issue of receiver sensitivity and ability to discriminate the incoming pulses dominated all other considerations. A 128-level discriminator would have been totally impractical. Even designing a receiver that could accurately distinguish 1s and 0s turned out to be a major challenge for the T-carrier architects for reasons that have to do with what happens in the transmission process. In fact, it was found in early T-carrier prototypes that simple on-off discrimination — that is, using a positive voltage for a 1 and a zero voltage for a 0 — were unreliable. A crucial decision was made to move to *bipolar* transmission, also called *alternate mark inversion,* in which 0 was still represented by zero voltage but 1s were represented alternately by either positive or negative voltage [4]. (See Figure 6.10.)

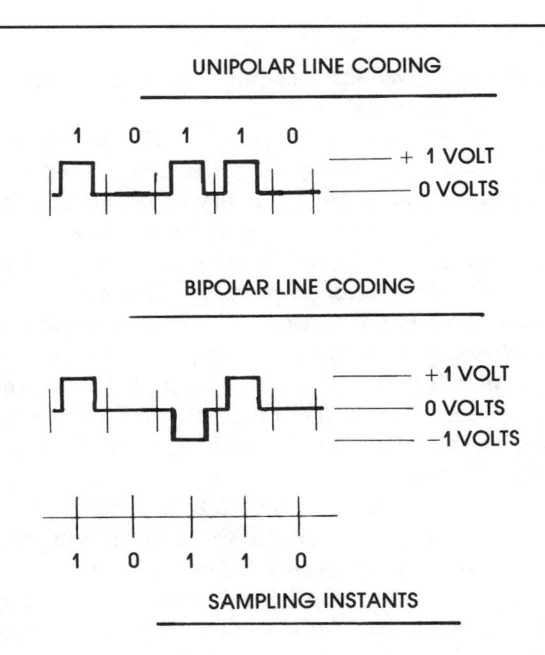

Figure 6.10 Bipolar Transmission.

Here we have actually crossed into the second stage of coding. The representation of binary code (logical) by unipolar or bipolar *line coding* clearly demonstrates the two levels in digital coding. The logical code is a two-level code (every symbol is either a 1 or a 0), while the transmission code is a three-level code (every symbol is either a + voltage, a 0 voltage, or a − voltage).

The assumption that digital systems would always operate, logically, in binary code was soon fixed. From that point coding schemes drove the selection of quantization schemes. For example, the chief question in the early development of T-carrier type systems was whether the samples could be quantized with a 32-level scale (corresponding to a "5-bit number" which could be coded as a string of five pulses, each pulse either a 1 or a 0), a 64-level scale (coded as a "6-bit number"), or a 128-level scale (coded as a "7-bit" number). One of the "inventors" of digital coding, A. H. Reeves of ITT, guessed that a 5-bit code word would do, but he did not have the apparatus to test it in the late 1930s [5]. One of the first demonstrations of digital coding applied to television signals, for example, utilized a 5-bit code (although the sampling rate was 10 million samples per second, much higher than for voice) [6]. The quantization noise is obviously less with a 128-level scale than with a 32-level scale, but encoded in binary form the 128-level scale requires two more pulses (two more "bits") to transmit each sample.

The originators of T-carrier decided to use a 128-level scale and encode each sample as a 7-bit number (also called a 7-bit "word"). An eighth bit was added to transmit supervisory signaling information. Thus, the T-carrier system was designed to generate 8000 samples per second, and 8 bits were transmitted for each sample. Later, the standard was modified to use the full eight bits for voice. The bit rate for the voice circuit was thus 64,000 bits per second [7]. The technique was designated *pulse code modulation,* or PCM. The standard form is 64-kilobit PCM (although 32-kilobit PCM systems are also being fielded in the 1980s). As indicated above, sometimes a further designation of *μ-law* or *A-Law PCM* is indicated; these variations refer to the technique by which the logarithmic quantizing scale is created.

One other result of binary coding is that some pulses, or bits, are more important than others (in terms of the information they carry). In a 7-bit number, the first bit is worth 64 steps in the quantization scale, while the seventh bit is worth only 1 step. (See Figure 6.11.) The importance of an error in the *most significant bit* (MSB) is very great. Instead of transmitting a sample level as, say, 100 on the PCM scale, an error in the most significant bit (causing it to be received as a 0 instead of a 1) would be decoded as a 36 on the PCM scale. An error in the *least significant bit* (LSB) would cause a change from 100 to 101, a much less noticeable difference. (In fact, some PCM systems occasionally "steal" the least significant bit from the PCM word to transmit additional signaling information [8].)

In standard PCM, no special measures are taken to protect the MSB, because it is generally assumed that very low transmission error rates can be maintained. As we shall see, however, in the case of mobile radio systems where transmission or channel errors are inherently much higher, the necessity of protecting the most important bits will become apparent.

Sampling, quantizing, and coding combine to produce a logically malleable digital signal out of a complex and difficult-to-manipulate speech waveform. Subsequent signal processing steps will be performed upon this digitized signal.

6.2 THE SECOND CODING STAGE: MODULATION

The digitized information must be coded a second time for actual transmission. The form of electromagnetic energy (radio, light, electricity) that is used in the circuit must be *modulated* — the logical 1s and 0s must

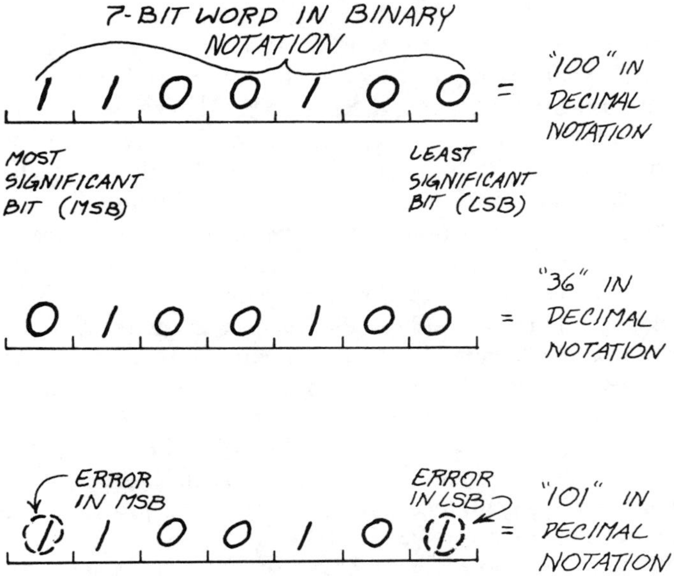

Figure 6.11 LSB and MSB Errors.

be translated into specific, discrete variations in certain characteristics of the transmitted signal.

For example, in some wire-based and optical-fiber systems, the bits are converted directly into energy pulses which carry the information in an on-off sequence similar to Morse code — electrical pulses in the case of standard T-carrier or light pulses in an optical-fiber system. This is the simplest form of modulation, sometimes called *on-off keying,* or *amplitude shift keying* (ASK).

Pure on-off keying, also called *unipolar transmission,* was found, however, to be susceptible to transmission errors. Designers of wireline digital systems have developed a wide variety of different line coding schemes, which are designated by esoteric acronyms like HDB3, 4B3T, or B6ZS [9].

In radio transmission, it is generally preferable to maintain a *continuous* transmission and to vary some characteristic of the transmitted radio wave. These variations can be coded to carry the binary information generated by the A/D converter. In the radio world, this is usually called *modulation.*

In most radio systems, a transmission is first established at a standard frequency (conceptually this frequency may be likened to a pure tone);

this is the *carrier* frequency. The carrier contains no information in itself. It is simply a vehicle, a baseline (like zero voltage in a wire circuit), from which the variations in amplitude or frequency which actually carry the information are referenced. Each carrier frequency becomes, in layman's terms, a radio channel (or TV channel, or microwave telephone channel). There are traditionally three dimensions of the radio carrier that are amenable to modulation: amplitude, frequency, and phase.

Amplitude modulation and frequency modulation are the familiar AM and FM of radio broadcasting. In *analog* AM or FM modulation, the analog waveform of the speech input signal is represented directly by *continuous* variations in either amplitude or frequency. In *digital* AM or FM systems, such changes are discrete, corresponding to distinct steps or levels much like the voltage levels for 1 and 0 in PCM systems.

For example, in a binary digital FM system, the frequency of the radio *carrier* wave is varied between two predefined frequencies, representing 1 and 0. (See Figure 6.12.) This is also called *frequency shift keying* (the word "keying" still recalls the telegraph key, the original digital modulator of an electrical wave).

Phase modulation is another approach, which codes the digital information in changes in the phase of the radio signal, that is, in the position in the sinewave cycle. Phase is often represented as a circle, and phase states are denominated in degrees. (See Figure 6.13.) In a binary phase modulated system, called binary phase shift keying, the transmission alternates between two different phase states, typically 0 degrees and 180 degrees. There are a great many radio modulation techniques (see Part V).

0 1 1 0 0 1

Figure 6.12 Binary Frequency Shift-Keying.

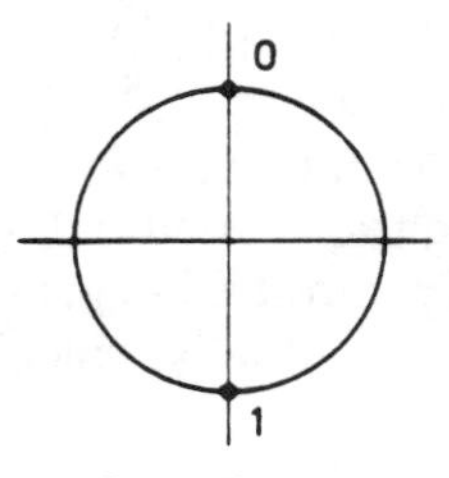

Phase diagram

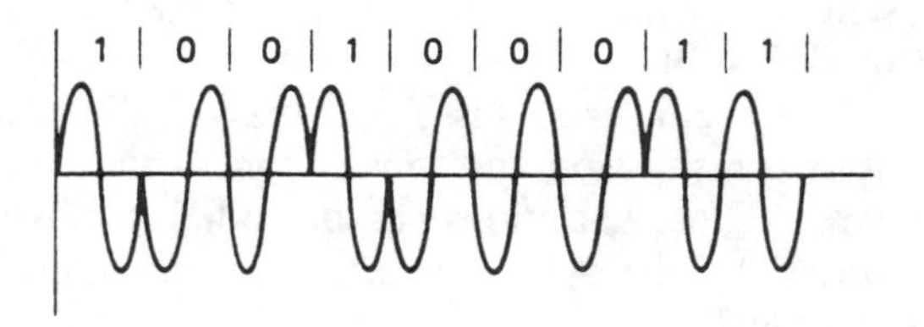

Time waveform

Figure 6.13 Binary Phase Shift-Keying Modulation.

Source: John Bellamy, *Digital Telephony,* p.283.

The Bandwidth Problem

Digital modulation usually expands the *bandwidth* needed to transmit the signal. Bandwidth is a subtle concept which equates roughly in the radio world to the amount of spectrum necessary to transmit a given signal. It is measured in terms of KHz or MHz of spectrum per channel. Bandwidth can also be defined for wireline systems, optical-fiber systems, or any other communication system (including even such unusual examples as human nerve cells or sign language for the deaf). For our purposes, we can think of the bandwidth of a *circuit* as the carrying capacity of that circuit and the bandwidth of a *signal* as the amount of that circuit capacity used up.

As noted earlier, there was a bandwidth penalty for PCM transmission. Each sample was converted into seven bits plus an extra supervisory bit. Encoded in binary form as eight separate pulses, the PCM signal needed a much wider bandwidth than the original analog signal [10]. It used up more capacity than its analog antecedent. Fortunately, the bandwidth of today's short-haul telephone wireline circuits is very large, hundreds of times greater than the bandwidth of the analog voice signal. As one text puts it, "wire pairs have no rigid bandwidth limitations [11]." The effective bandwidth of optical-fiber transmission systems can be truly astonishing — a single fiber not much thicker than a human hair can carry thousands of digitized voice circuits or hundreds of television channels. In both copper-based and fiber-based systems, the bandwidth penalty of PCM-type systems is usually not significant, or can be worked around, and the simplicity and low cost of the receivers dominates the transmission architecture.

In radio, by contrast, bandwidth is always limited. The radio spectrum is precious and must be utilized efficiently. Additional wireline circuits can always be obtained at a price. It is always possible to string another wire. Once the radio channels are used up in a given geographical area, however, there is no easy way to "manufacture" additional spectrum. Recognizing the scarcity of bandwidth in the radio spectrum, the FCC establishes strict bandwidth limits on radio emissions. This is partly to control adjacent channel interference and partly to ensure efficient use of the resource.

For this reason, digital radio has traditionally been at a serious disadvantage compared to analog radio, because it needed more bandwidth to transmit the same number of circuits. Radio engineers, therefore, began to follow the same approach used by Lord Kelvin in the nineteenth century to improve transmission on the transatlantic telegraph cable, a rare example of an extremely narrow-bandwidth wireline communication channel. Basically, instead of coding each transmission symbol to carry only 1 bit of information (binary coding) they used *multilevel-modulation* techniques. For example, a 4-level phase shift-keyed system, often called quaternary phase shift keying or QPSK, can transmit two bits of information in each symbol. (See Figure 6.14.) This reduces the number of symbols that need to be transmitted by 50% compared to a 2-level modulation system and cuts the bandwidth required by half. It, however, complicates the detection and discrimination task for the receiver (demodulator).

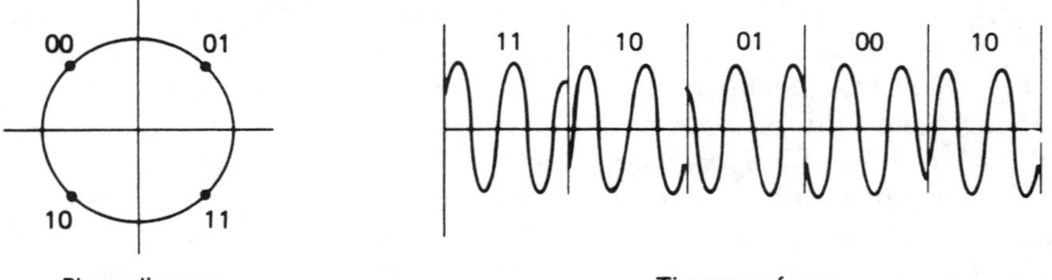

Figure 6.14 Quarternary Phase Shift-Keying Modulation.
Source: John Bellamy, *Digital Telephony*, p.283.

Multilevel modulation techniques have made steady progress in digital microwave applications in recent years. Today 16-level PSK systems, which transmit four bits per symbol, are quite common. They involve detection of phase differences of 22½ degrees. So is the 16-level quadrature amplitude modulation (QAM) system, a kind of hybrid phase and amplitude modulator. Both 16-level PSK and 16-level QAM are able to transmit a standard PCM 8-bit word in two symbols instead of eight, a 75% reduction in bandwidth. Even higher level systems are beginning to be deployed, including 64-level QAM (with 6-bit symbols), and developmental systems employing up to 256-level modulation with eight bits per symbol — enough to encode the entire PCM word in a single symbol — are being investigated. The detection and discrimination problems for such ambitious techniques are quite daunting, however, particularly in the mobile radio environment (as we shall see in Chapter 8), and it is likely that applications in the field of mobile radio will be limited for the foreseeable future to 16-level or perhaps to 32-level modulation (4-bit and 5-bit symbols, respectively).

6.3 TRANSMISSION AND SIGNAL PROCESSING

Once a signal enters the transmission channel — which may be a copper wire, an optical fiber, or the ethereal realm of radio — it begins to deteriorate. In a telephone wire, the pulsating electric current interacts with the metal itself. Some of the current is dissipated through resistance, converted to heat. Some leaks out through submicroscopic defects in the insulation. The purity of the copper and even the physical characteristics of the wire — old insulation, weak splices, manufacturing defects — greatly influence what happens to the signal *en route.* The transmission is also vulnerable to less tangible influences, such as the shimmering magnetic fields created by other nearby telephone wires, which can induce cross talk, or electromagnetic emissions from natural and man-made sources. The signal on a standard 19-gauge wire loses about 1.1 dB per mile at 1000 Hz [12]. A 10-mile line results in the loss of well over 90% of the transmitted signal. The signal deteriorates so rapidly, in fact, that reliable transmission over ordinary telephone lines is limited to relatively short distances (up to 50 miles or so), and current telephone standards require special conditioning for lines longer than 3 or 4 miles.

A radio signal is much more vulnerable. Rain, foliage, and other features of the natural environment can absorb huge amounts of the transmitted signal, depending upon the carrier frequency. Hills and mountains, as well as buildings, airplanes, and other man-made structures, can block, distort, and reflect the radio waves. Lightning, electric power lines, and even automobile ignitions generate radio noise which can obliterate the faint signals received by a distant site.

The received radio signal in a point-to-multipoint system (like mobile telephony) is typically very faint indeed. The "path loss" on a prospective 10-mile mobile radio link can range up to 150 dB or more. In layman's terms, the strength of the received signal may be less than *one trillionth of one percent* of the strength of the transmitted signal — which may have been only a few watts to start with. In fact, the total radio signal power received by all the mobile radios in a middle-sized city over an entire year is less than the energy generated by a grain of rice falling from the fork to the plate. This slim energy budget must be infinitesimally subdivided and modulated to carry tens of thousands of hours of human speech. The failure of a radio link for even a few milliseconds will produce a noticeable degradation of the communication channel. A handful of such failures and the total system will not be able to measure up to the standards of mobile telephony. From this perspective, the mobile radio link is an almost inconceivably fragile thing.

Here is where digital techniques pay off — in the transmission stage. Digital systems allow the communication architect to cope with the hazards of transmission *much* more effectively than is possible with analog techniques. This is because a digital signal can be manipulated or processed in a variety of ways to counteract different types of hazards and improve transmission quality and efficiency. Let us consider four types of digital signal processing:

1. Signal regeneration — the "conquest of distance";
2. Control of intersymbol interference — recovery of the physical code;
3. Error control — recovery of the logical code;
4. Multiplexing.

6.3.1 Regeneration: The "Conquest of Distance"

The effect of noise upon analog transmission is pernicious and pervasive. It becomes irretrievably mixed in with the signal itself. Like water running in a stream, the signal starts out pure. But along the way it dissolves out salts and minerals in the rocks, or is polluted with foreign contaminants, until, by the time it reaches the users downstream, it may be very dirty indeed. Filtering out the impurities is difficult because most of them have entered into solution and cannot be easily separated out. Fortunately, the downstream receivers of analog transmissions have one very powerful built-in filter — the human ear. The ear is a remarkable filter and integrator, probably because it operates digitally, by the way. The speech processing mechanisms of the brain are capable of extracting intelligible speech from an extremely noisy channel, as anyone who has ever followed a conversation in the middle of a loud cocktail party can attest. In some ways, the main challenge of the analog system is to make sure that whatever noise creeps into the system, the overall level stays below the threshold of the ear's processing capabilities.

The noise accumulates, however, as the analog signal proceeds along. The longer the circuit, the more noise enters the system. The farther downriver you are, the more polluted the stream. (Actually, the signal is also attenuated relative to the noise background; when the attenuated signal is amplified, the noise is also amplified and appears even stronger.) At some point, the noise overwhelms the ear, unless the signal-to-noise ratio can be improved. But this must be done at the source. Although analog repeaters, based originally on vacuum-tube amplifiers, were introduced as early as 1915 to boost the weakening signal and so extend the limits of transmission, the repeaters also boost and pass along the noise. Repeaters help with attenuation, but not with noise.

An analog transmission system has one basic strategy for coping with transmission noise. It must start with enough power to get enough signal through the uniform noise background. Also each repeater must develop an enormous signal-to-noise ratio over its section of the circuit to ensure that the overall quality of the circuit is maintained [13]. In a difficult environment — and mobile radio is one of the worst (see Chapter 8) where, moreover, repeaters are impractical — about the only analog counter-measure available is to increase the transmitter power.

A digital signal, by comparison, is extremely robust to begin with (see Chapter 7). The typical wireline digital receiver, e.g., in a T-carrier channel, looks for a pulse — the presence or absence of energy at some defined level. It is continuously sampling and quantizing the incoming signal, in much the same way that the transmitter originally sampled and quantized the original voice input — except that there are only two quantization levels (assuming a 2-level modulation of line code), so the decision is very easy. Within very wide limits, the receiver does not care what the exact amplitude or shape of the pulse is. It is operating with a very gross criterion: Is the signal on or off? Channel noise effects, which are typically small as a percentage of total signal amplitude over short circuits, will simply be rounded off. (See Figure 6.15.)

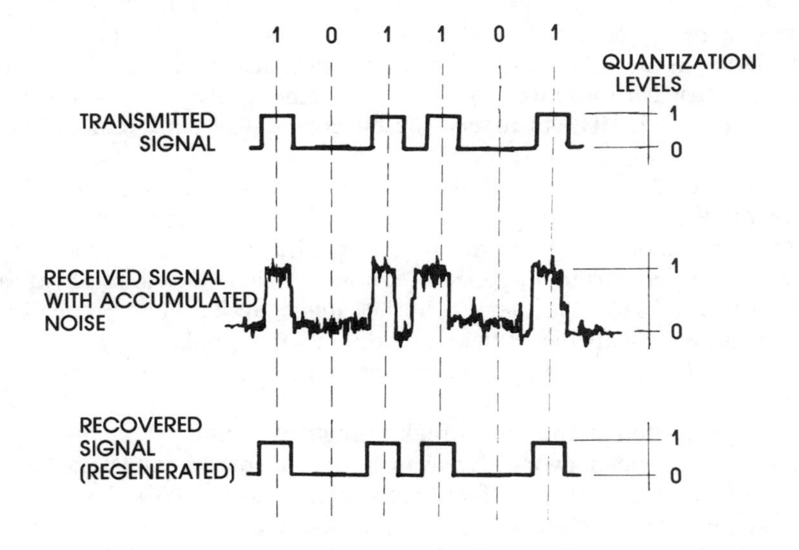

Figure 6.15 Control of Noise through Quantization in a Binary Signal.

Moreover, the perceptual effects of noise in a digital system are very different. If environmental disturbances or channel conditions do cause the noise to surpass the threshold, the result in a digital system is the creation of *bit errors.* The fundamental measure of digital signal quality is the *bit-error rate,* or BER. All types of noise are "homogenized" by being converted into bit errors; for example, the effect of cross talk in a digital wireline circuit is the same as the effect of, say, interference from a high-power electric line. The patterning of errors may create perceptual differences — a burst-error pattern resulting from impulse noise will sound different than a BER background created by evenly distributed white noise. Moreover the effects of bit errors may differ in different types of digital systems. But the perceptually troublesome peculiarities of analog signals — distortion (frequency-selective attenuation) and intelligible cross talk in particular, all of which tend to disturb the human listener far more than random background noise — are banished.

In an environment characterized by what is known as *Gaussian noise* or *white noise* — which is typical of most thermal and solid-state noise — the relationship between signal-to-noise ratio and the bit-error rate is described by Figure 6.16. As this figure indicates, the threshold signal-to-noise ratio required to generate a virtually error-free signal, one error in 100 million bits, say, is considerably less than that required to achieve satisfactory performance with analog systems — about 20 dB (or a ratio of 100 to 1), compared to 60 dB or more (1 million to 1) for a given link in a repeatered analog system [14]. Moreover, very small improvements in the signal-to-noise ratio will produce tremendous improvements in error performance, e.g., an improvement in signal-to-noise ratio from 20 dB to 21 dB will reduce the bit-error rate by a factor of about ten).

But the real magic of digital communication — "the payoff in PCM," as several of the inventors saw it back in 1949 [15] — lies in the opportunity for the application of *regenerative repeaters.* The principle of signal regeneration is the same as that applied by Thomas Edison to telegraphy transmission in the 1870s. At intervals in the circuit, an automatic repeater is inserted which is capable of detecting the incoming pulses, logically processing the digital information, and retransmitting a new pulse train. (See Figure 6.17.)

Regeneration is the core breakthrough of digital telephony in the wireline telephone network. By using properly spaced repeaters (about 6000 feet apart in the early T-carrier systems, up to 100 miles apart in some advanced fiber-optical designs), it is possible to "make performance essentially independent of distance," in the words of the authors of the engineering history of the Bell System [16]. The implications are truly far-reaching, as Bellamy writes:

> A fundamental attribute of a digital system is that the probability of transmission errors can be made arbitrarily small by inserting regenerative repeaters at intermediate points in the transmission link. If spaced close enough together, these intermediate nodes detect and regenerate the digital signals before channel-induced degradations become large enough to cause decision errors . . .

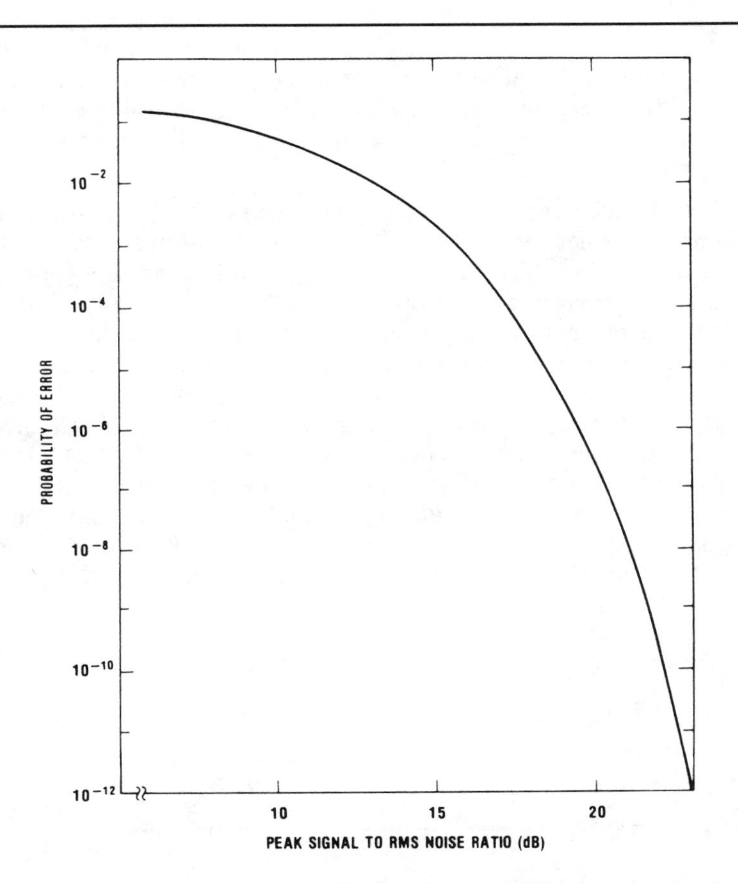

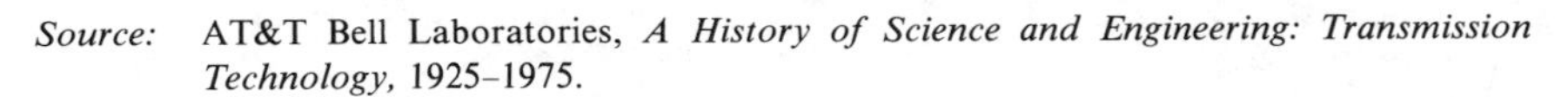

Figure 6.16 Bit-Error Rate in Binary PCM as a Function of Signal-to-Noise Ratio.

Source: AT&T Bell Laboratories, *A History of Science and Engineering: Transmission Technology,* 1925–1975.

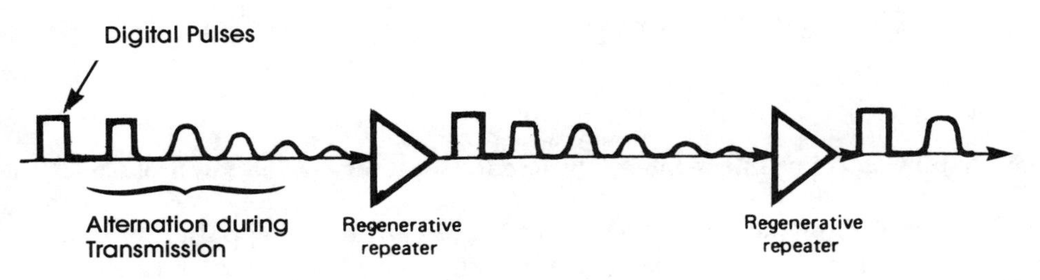

Figure 6.17 Digital Signal Regeneration.

Source: John Bellamy, *Digital Telephony*, p.73.

> When an all-digital network is designed with enough regeneration points to effectively eliminate channel errors, the overall transmission quality of the network is determined by the digitization process and not by the transmission systems [17].

In other words, the communication engineer can now control the quality of the signal in his network — regardless of how far it must travel or how severe the environment. Through digital regeneration, he has gained the upper hand over environmental noise.

Although regenerative repeaters cannot easily be incorporated into a mobile radio system, the principles of regeneration are still important for an understanding of what digital technology can mean for cellular radio. In fact, the main challenge facing the regenerative repeater is the same as that faced by the digital radio receiver: how to extract the digital bit stream from the degraded and noise-filled signal. Even where regeneration is not possible, there are related techniques available to recover and clean up the degraded signal. These recovery procedures can be utilized at both levels of the code, physical and logical. (See Figure 6.18.)

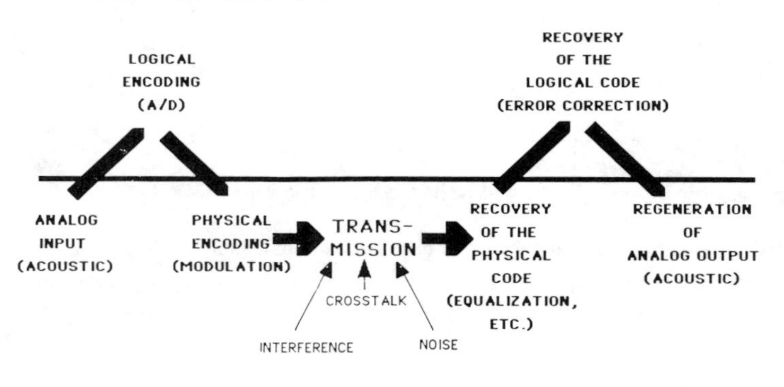

Figure 6.18 The Complete Digital Communication Process.

6.3.2 Control of Intersymbol Interference: Recovery of the Physical Code

Every digital system may be conceived as producing pulses in some form — discrete, rapid changes in some physical characteristic of the signal that are to be interpreted in an either-or, on-off fashion. Multilevel modulation schemes involve decisions by the receiver as to which of the defined set of values the received signal should be assigned at a given moment [18].

The detection of these pulses at the receiver is essentially similar to the sampling and quantization process that produced the transmitted digital signal in the first place. The receiver samples the amplitude (or frequency, or phase, depending upon what modulation scheme is involved) at the appropriate rate and makes a decision to classify the values according to the preestablished code levels.

The initial problem in pulse recovery is to acquire the proper timing. The sample should be taken in the middle of the pulse, not when the pulse is either starting or ending. (See Figure 6.19.) The receiver's sampling circuitry must be *synchronized* with the transmitter. The most common and straightforward way to accomplish this is by allowing the receiver to obtain its timing directly from the pulse stream. Some timing bits, or synchronization overhead, are usually built into the system, particularly for burst-transmission systems which must effectively turn their receivers on and off in a rapid, regular cycle (see Section 6.3.4, Multiplexing) and must reacquire the clock during each sequence.

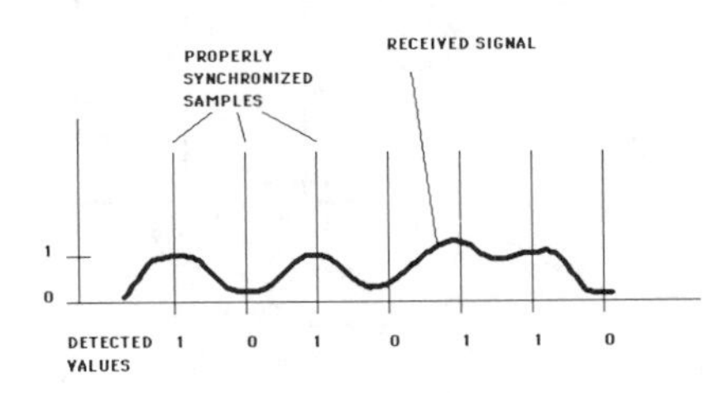

LOSS OF SYNCHRONIZATION CAUSES DETECTION ERRORS

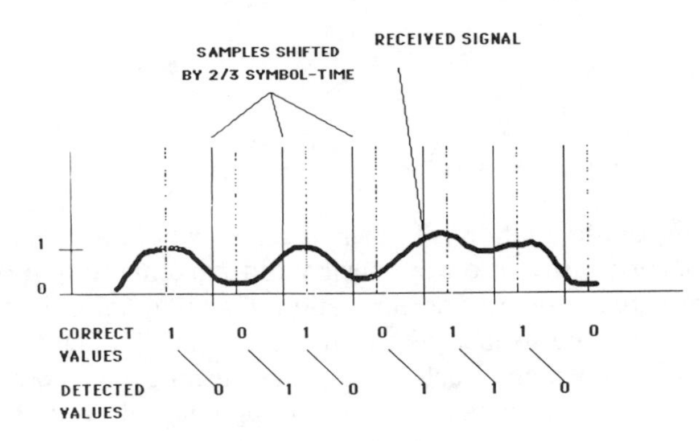

Figure 6.19 The Effect of Timing of Samples on Received Signal Quality.

The other, more fundamental problem to be overcome by the receiver is the *attenuation* of the pulses during transmission. This is an inherent property of all transmission processes. The transmitted pulses, which begin as ideally sharp and distinct shapes, tend to blur and smear together as they travel through the transmission medium. The original sharp, square pulse takes on a rounded, flattened shape, with preceding and trailing edges that overlap into the preceding and following pulses. (See Figure 6.20.) We can hear the process of attenuation in the sound of a thunderclap. A lightning strike nearby will produce a sharp, clear, very loud crack. A strike several miles away will be heard as a low, prolonged rumbling, building up over a few seconds in a slow crescendo, and slowly dying away. The effect on electrical pulses is similar. This is why the pulses on the transatlantic cable had to be transmitted so slowly. They were so attenuated that sufficient time had to be allowed for each pulse to die away before the next one could be detected. If the pulses were spaced too closely together, they simply slurred into one another in an unintelligible mess.

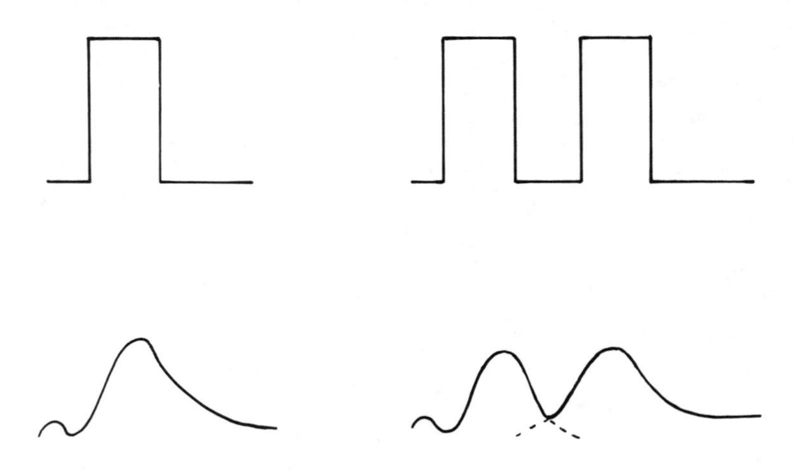

Figure 6.20 Attenuation (Spreading) of Pulses.

This effect of attenuation is known as *intersymbol interference* (ISI). ISI is a significant constraint on the design and implementation of digital systems. Since, like the relative accumulation of analog noise, intersymbol interference grows more serious with distance, it tends to dictate the spacing of regenerative repeaters. Each repeater retransmits the received pulses in a fresh, sharp form. It is also possible to apply signal processing techniques known as *equalization* to help recover the original pulse from the attenuated signal [19]. (See Chapter 8.)

The problem of pulse detection is critical for digital mobile radio, and limits the use of some high-level modulation schemes which otherwise could improve spectrum efficiency. (See Chapter 12.) The use of *adaptive equalization* and other techniques for pulse shaping is a virtual necessity for such systems and imposes a significant processing load on the demodulator, as we shall see.

6.3.3 Error Control: Recovery of the Logical Code

Depending upon the design parameters, some errors may creep into digital transmission, especially where regeneration cannot be fully exploited as in mobile radio. At this point, another set of digital signal processing techniques comes into play to control transmission errors after they occur.

There are basically two strategies to control errors in a digital system at the logical-coding level:

1. The designer can build robustness into the signal to reduce the BER probability for a given signal-to-noise ratio;
2. The designer can encode the signal such that errors can be detected and corrected prior to the final D/A output.

A digital signal may be made more robust in several ways. For example, the coding in either stage 1 (digitization) or stage 2 (modulation) may be stepped down from higher levels to lower levels of information density. A transmission with a low information density is inherently more robust. As an extreme example, if a 16-QAM modem, coding 4-bits per symbol, suffers a burst error that affects two adjacent symbols, an entire 8-bit PCM word will be lost. In standard wireline T-carrier transmission with one bit per symbol, the loss of two symbols could have far less effect, especially if they happen to be the two least significant bits. The more information each symbol carries, the more vulnerable the system is to bit errors. There is a bandwidth penalty, however, and the radio engineer is generally more sensitive to bandwidth than the wireline engineer.

Another approach is to encode the information redundantly. This is done in digital audio systems such as compact disks, where extremely high fidelity and error-free performance is desired. It is generally not employed in voice communication systems, again because of the significant bandwidth penalties. There are also techniques (such as *interleaving)* involving the way in which multiple digital voice channels are combined (see Section 6.3.4) which can improve signal robustness without necessarily expanding the bandwidth.

Once an error has occurred — for whatever reason the receiver has incorrectly decoded a particular bit — there are further steps that can be taken to detect and correct the error at the logical level. For example, a check-sum type of procedure (and there are many of these) can be used to identify whether an error has occurred in a given block of bits.

If an error is detected, a number of steps can be taken. The block containing the error can be thrown out, and the value of the previous block substituted. For data transmission this would never do, of course. But the inherent redundancies in voice signals (see Section 12.2.2) mean that it is often possible to utilize such gross expedients without noticeable degradation, at least as long as the BER is not too high.

Also, a value for the erroneous sample can be interpolated between the preceding value and the following value, assuming these are detected correctly. Depending upon the speed of the processor and the availability of prior and following samples, the interpolation techniques can work well with voice signals, because voice signals do not change very rapidly.

Again, however, error correction techniques consume bandwidth. A trade-off must be evaluated by the system designers between error correction and spectrum efficiency, particularly in bandwidth-limited systems such as mobile radio. Nevertheless, digital error correction holds particular promise for overcoming some of the performance problems of mobile telephone systems (see Chapter 8).

6.3.4 Multiplexing

The other main payoff of digital communication is the avenue that it opens to a different type of multiplexing (or combining of multiple circuits in a single transmission over a single channel, whether wire, fiber, or radio). The multiplexing advantage is central to the economic justification of digital techniques. For example, the driving force behind the development of T-carrier was the economic objective of saving copper-plant costs by combining 24 voice circuits on a standard wire pair.

The technique used is known as *time division multiplexing* (TDM). TDM works as follows. Imagine 24 individual PCM coders, each processing a single voice circuit, producing 24 bit streams each flowing at a rate of 64,000 bits per second. (See Figure 6.21.) Imagine further that we wish to transmit all of these bit streams over a single wire. A way to do this would be to construct a new multiplexed bit stream by taking the first, say, eight bits — the first whole PCM word — from the first channel, followed by the first eight bits from the second channel, followed by the first eight bits from the third channel, and so on, until a new string of 192

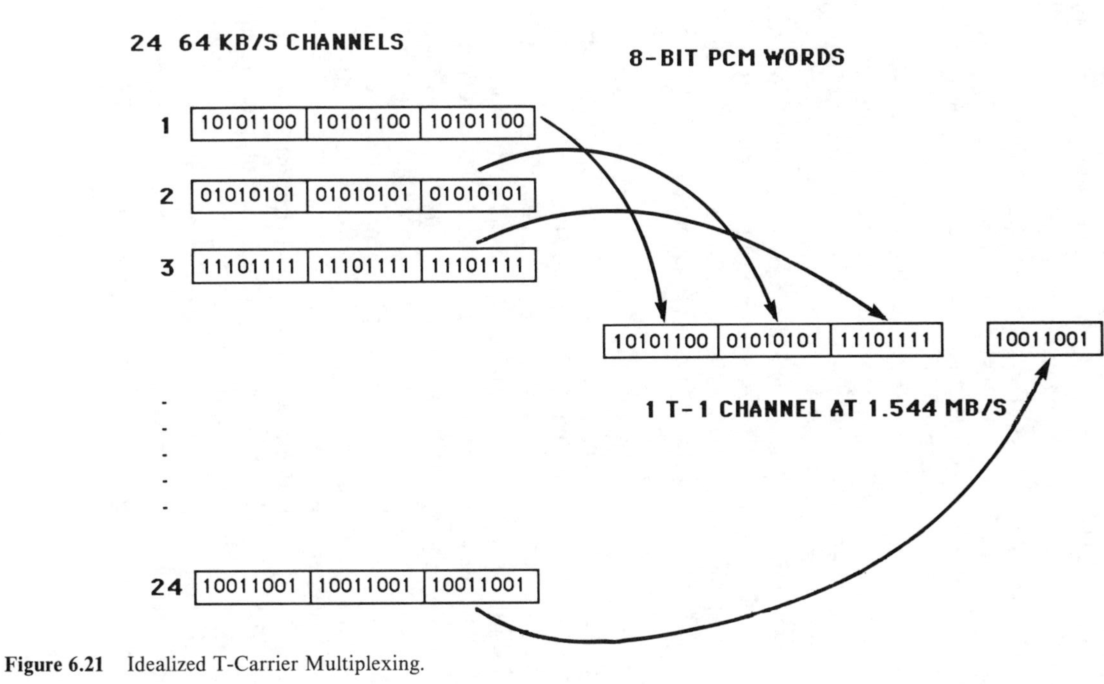

Figure 6.21 Idealized T-Carrier Multiplexing.

bits representing the first PCM word from each of the 24 channels has been created. Then the new string of 192 bits would be transmitted over the wire *at 24 times the speed of each individual channel.* In other words, the bit rate on the multiplexed channel would be more than 1.5 million bits per second. At the receiving end, the microprocessor would strip off the first eight bits and direct them to the first output channel, slowing them back down to the 64,000 bits per second rate. The entire string of 192 bits would be disassembled or *demultiplexed* into its 24 component parts. While the receiver was doing this, of course, a new string of 192 bits representing the second PCM word from each of the 24 channels is being constructed by the transmitter-multiplexer, so that it will be ready when the demultiplexer has finished the first string.

This is in fact more or less how a T-1 channel works. By operating the multiplexed channel at 24 times the data rate of each individual circuit, it is possible to combine 24 circuits for transmission — the format is called DS-1 (for Digital Signaling 1), and the facility is referred to as a *T-1 Carrier.* The individual channel components within the overall multiplexed channel are referred to as *time-slots.* The entire cycle of 24 time-slots is called a *frame.* Higher speeds allow the combination of multiple T-1 groups into T-2 Carrier, T-3 Carrier, and so forth, in what is called the TDM Hierarchy. (See Figure 6.22.)

There are a variety of multiplexing formats which differ mainly with respect to the manner in which the frames are constructed and particularly as to the method of transmitting signaling and framing information. The T-1 standard referred to here, with 24 voice channels (and incidentally using a 193rd bit for framing), is more or less standard in North America, Japan, and a few other countries. (See Figure 6.23.) An alternative standard, used in Europe and most of the rest of the world, is the CCITT standard, which utilizes 32 channels (each channel is still based on 64-kilobit PCM) of which 30 are available for voice traffic, one is reserved for framing, and one for signaling. The channel bit rate is 2,048,000 bits per second. (See Figure 6.24.)

As stated earlier, TDM is crucial for the economics of digital communication. There is a trade-off between the cost of the multiplexing equipment (along with the digital-processing equipment as a whole) and the cost (or, in the case of radio, the scarcity) of the channel facilities. As the costs of digital electronics have fallen, often spectacularly, digital communication systems are proving in over a wider and wider range of ap-

DIGITAL TDM SIGNALS OF NORTH AMERICA AND JAPAN

Digital Signal Number	Number of Voice Circuits	Multiplexer Designation	Bit Rate (Mbps)
DS-1	24	D channel bank (24 analog inputs)	1.544
DS-1C	48	M1C (2 DS-1 inputs)	3.152
DS-2	96	M12 (4 DS-1 inputs)	6.312
DS-3	672	M13 (28 DS-1 inputs)	44.736
DS-4	4032	M34 (6 DS-3 inputs)	274.176

DIGITAL TDM HIERARCHY OF CCITT

Level Number	Number of Voice Circuits	Multiplexer Designation	Bit Rate (Mbps)
1	30		2.048
2	120	M12	8.448
3	480	M23	34.368
4	1920	M34	139.264
5	7680	M45	565.148

Figure 6.22 Major TDM Hierarchies.

Source: John Bellamy, *Digital Telephony*, p.54.

plications. For example, when T-carrier was first introduced, it enjoyed only a very slim cost advantage over existing wireline analog systems. Yet within ten years, i.e., by the early 1970s, the cost of the multiplexers and repeaters had fallen by a factor of three in current dollars and probably by a factor of five in constant dollars [20]. During the same period, rising labor and copper costs continued to push up the cost of wire facilities. Today, digital systems are economically attractive in most segments of the telephone network because of facilities savings from the exploitation of TDM opportunities.

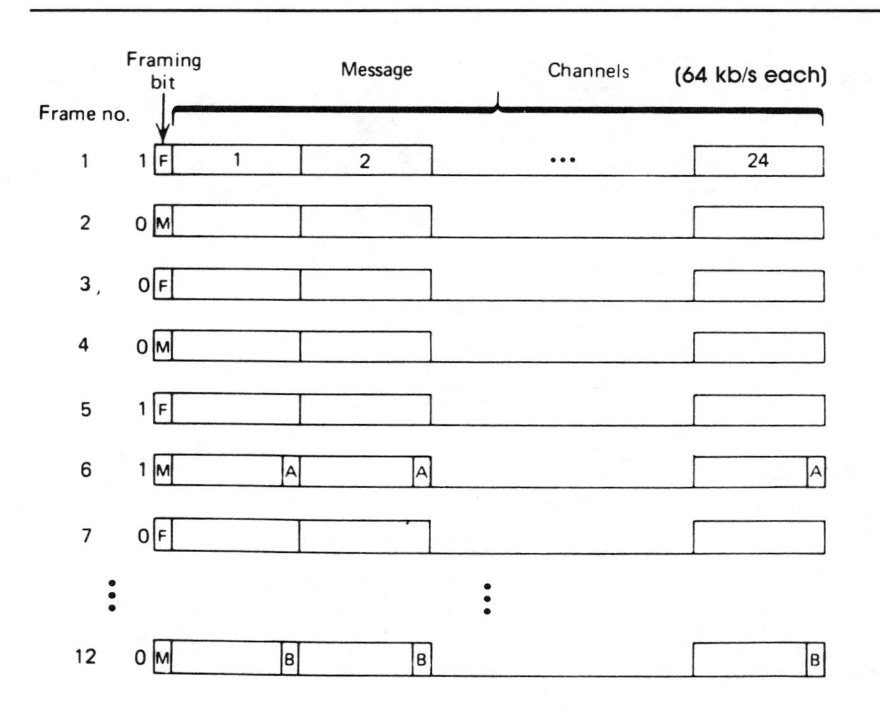

Figure 6.23 Framing Structure of the Primary T-1 Transmission.

Source: John Bellamy, *Digital Telephony*, p.208.

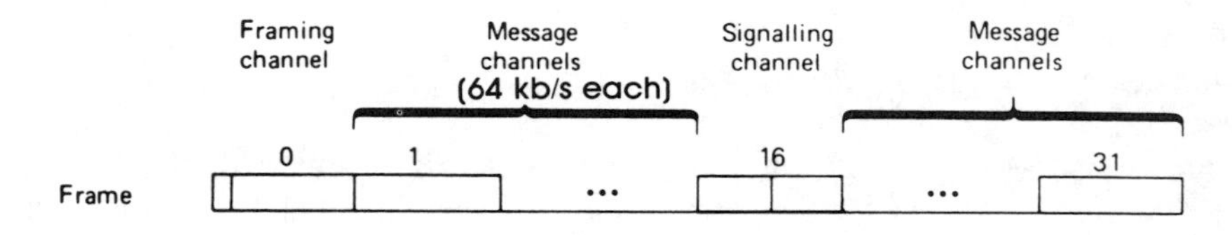

Figure 6.24 CCITT Primary Multiplex Transmission Standard.

Source: John Bellamy, *Digital Telephony,* p.210.

REFERENCES

[1] As good general texts on digital communication, see John C. Bellamy, *Digital Telephony,* New York: Wiley, 1982; and David R. Smith, *Digital Transmission Systems,* New York: Van Nostrand Reinhold, 1985. There are many others.

[2] E. F. O'Neill, ed., *A History of Engineering and Science in the Bell System: Transmission Technology (1925–1975),* AT&T Bell Laboratories, 1985, p. 530.

[3] *Ibid.,* p. 541; also Bellamy, *op. cit.,* pp. 101–102.

[4] *Ibid.,* p. 556.

[5] *Ibid.,* p. 534.

[6] S. Millman, ed., *A History of Engineering and Science in the Bell System: Communications Sciences (1925–1980),* AT&T Bell Laboratories, 1984, p. 412.

[7] E. F. O'Neill, *Transmission Technology, op. cit.,* p. 541.

[8] Bellamy, *op. cit.,* p. 209.

[9] *Ibid.,* pp. 171–190.

[10] E. F. O'Neill, *Transmission Technology, op. cit.,* p. 533.

[11] Bellamy, *op. cit.,* p. 272.

[12] Bell Laboratories, *Engineering and Operations in the Bell System,* AT&T Bell Laboratories, 1977, p. 125.

[13] E. F. O'Neill, *Transmission Technology, op. cit.,* p. 532.

[14] *Ibid.,* p. 532; also Bellamy, *op. cit.,* p. 71.

[15] E. F. O'Neill, *Transmission Technology, op. cit.,* p. 532.

[16] *Ibid.,* p. 536.

[17] Bellamy, *op. cit.,* pp. 72–73.

[18] *Ibid.,* p. 161.

[19] S. Millman, *Communications Sciences, op. cit.,* p. 422.

[20] E. F. O'Neill, *Transmission Technology, op. cit.,* pp. 561–563.

Chapter 7
THE ADVANTAGES OF DIGITAL COMMUNICATION

The charm of Alexander Graham Bell's telephone lay in its simplicity. A diaphragm to collect the sound, a small magnet, and a length of wire: Bell's original "magneto" design needed nothing more, not even a power source — the power was provided by the speaker's voice itself. The telephone was "so simple indeed," wrote the editor of the British journal *Engineering* in 1876, "that were it not for the high authority of Sir William Thomson [the hero of the transatlantic cable, later Lord Kelvin — who had attended Bell's first demonstration of the invention] one might be pardoned at entertaining some doubts . . ." that so primitive an apparatus could convey the complexities of the human voice [1]. Today any hobbyist who is familiar with the principle can build a workable analog telephone circuit without great difficulty or expense. It is not much more difficult to build an analog radio.

Digital transmission clearly involves another order of complexity. The conversion of the analog waveform into thousands of digital samples every second, the transmission and regeneration of these samples within a system that must be tightly synchronized end to end, and the faithful recreation of the voice signal from a stream of abstract 1s and 0s, all involve elaborate processing at both ends and extremely challenging hardware requirements. Although the layman thinks of the computer as the marvel of digital electronics, communication applications are *much* more demanding than most data-processing applications, because of the complexity of communication signals and the need to handle huge amounts of data in real time. Ten seconds of T-1 transmission encompasses almost two megabytes of data. A DS-4 digital transmission processes more than 300 megabytes in the same interval. Historically, the availability of fast electronic components was the limiting factor on the implementation of digital voice systems. The first PCM systems were experimentally developed during World War II — in fact the principles of PCM had been elaborated even before the war — but the inadequacy and high cost of digital-processing circuits held back the implementation of commercial digital systems until the mid-1960s.

The cost of digital circuitry still determines which applications are suitable for digital techniques. For example, digital microwave radio was unable to compete economically with analog systems until the late 1970s in some niches. Realizing the potential of digital techniques in applications like mobile radio will call for the development of much more sophisticated designs, more advanced materials, more complex software, faster processors, more ingenious mathematics — by comparison with which even T-carrier may look primitive.

Why bother? What are the benefits which offset the obvious costs and complexity of digital techniques?

In many texts on digital communication, it is conventional to construct a table of comparisons showing the advantages and disadvantages of digital *versus* analog communication. These can give the impression that comparing digital to analog is a bit like comparing one car to another: there are certain pluses and minuses for each alternative, and it is not surprising to learn that the more recently developed, more "modern" alternative is generally superior — "20% better, 30% cheaper."

Yet such point-by-point comparisons are misleading, because they miss *the vision*. A new way of looking at the communication network is emerging. Shared by a growing number of telephone engineers, though not as yet shared by very many radio engineers, the vision is inspired by a perceived technological discontinuity as distinct and historically significant as the development of the telephone itself. If the visionaries are correct, and I believe they are, digital communication is far more than a "20% better, 30% cheaper" alternative to conventional practices. It is a revolution.

The vision is called *the digital network*. It is a faith more than a current reality, a view of the future of communication. It is incompletely understood, even by those who profess it. It is subject to differing interpretations. Many attempts have been made to articulate the vision; most are slightly ridiculous, far too technology-driven to make sense to most nontechnologists. Part of the problem, too, is the inadequacy of our imagination. Digital capabilities have outstripped market needs, at least temporarily. Attempts to forecast the application of digital network services ring a bit hollow. We are somewhat like the early observers of the telephone who saw a variety of novelty applications of Bell's device — most notably to broadcast music and entertainment from the concert hall to the homes of remote listeners, much like radio — but who could not conceive that such a device might find its way into every home for two-way communication. Or, to draw a closer parallel, when the computer was first brought to the market, only a very small number of organizations had an identified need for manipulating large amounts of data. Market studies in the 1950s projected a demand for no more than a handful of these amazing, but apparently highly specialized, machines. The idea that within thirty years virtually *every* business would need a computer, or that millions of individuals would have computers in their homes, would have been incredible. Today the digital network, despite twenty-five years of field experience with isolated digital systems, is about at the same stage of development in the late 1980s as the computer was in the late 1950s. The hardware is becoming available; the applications are still somewhat unclear.

Nevertheless, to understand the advantages of digital communication, we must appreciate the totality of the vision. If the visionaries are correct, then the chief advantage of a specific digital communication system such as digital cellular over an analog alternative is the fact that it will be compatible with the emerging digital networks. The qualitatively new capabilities that will emerge with these networks will transfer to any digital system that connects to it, including a digital cellular telephone system. It is, therefore, important first to grasp what the communication engineers are really trying to say when they speak of digital networks and why more and more of the people close to the technology are convinced that these networks will represent as great an advance over the conventional analog telephone as the telephone itself represented over the telegraph.

7.1 THE DIGITAL NETWORK

An analog transmitter — such as a telephone microphone or a radio transmitter — is basically designed to launch a signal through a transmission medium — such as a copper wire or a radio channel. An analog receiver is designed to capture this signal from the medium. Particularly in radio, transmitter and receiver do not really function as a system after the initial call setup. They do not exercise control over one another. Both operate more or less independently; the transmitter blares away, and the receiver "gropes through the soughing regions of empty space," as Lessing poetically phrased it in his biography of Armstrong. The system is not really "intelligent"; it is opportunistic and hopefully well-engineered.

The enemy in this scenario is noise — the engineer's term for the sum of the effects of the environment upon the signal during transmission.

The signal begins to deteriorate, irrevocably, as soon as it is generated. The analog system cannot really do anything about that. It can only try to ensure that the level of the signal relative to the noise stays above a certain threshold. As long as the signal-to-noise ratio is maintained above this threshold, the quasi-digital circuitry of the human brain will be able to filter out the noise, and the communication will be satisfactory.

By contrast, the idea of a digital network involves the creation of an end-to-end *system,* which encompasses transmitter, receiver, *and channel facility* in a single, integrated whole. The elements of this system communicate among themselves and exercise control over one another. The system is capable of monitoring and controlling key signal characteristics during the transmission process. Transmitter and receiver, and indeed all elements of the network, are normally synchronized, or at least tightly coordinated. Both ends of the system possess sufficient intelligence to detect, diagnose, decide, and act upon various kinds of problems that may occur during transmission. In short, the system is not passive, but adaptive.

It is important to realize that the digital network is not a physical entity but a logical entity which is superimposed upon a highly heterogeneous set of physical transmission facilities. In the digital network, one telephone call may encompass a copper-wire local loop to the local central office, a fiber-optic interoffice trunk to the toll switch, a microwave or satellite link from the toll switch to the destination toll switch, and a mobile radio-telephone link to the called party traveling in his car. Yet the digital network possesses the ability to link these disparate facilities into a single digital circuit, the performance parameters of which can be precisely monitored and controlled. It is embodied in an interrelated system of processors which originate, transmit, detect, and decode the messages at different stages in the transmission path. Such a network offers a number of broad advantages over the analog system it is gradually replacing.

7.2 THE DIGITAL NETWORK IS ROBUST

A digital transmission system allows for communication to be established and maintained at uniform high quality levels in environments where analog techniques are costly and ineffective. The more difficult the communication link, the more digital communication will stand out over analog. One of the chief areas of digital application today is upgrading the quality of the long-distance segment of the network by the use of digital techniques, particularly optical fiber. Long-distance circuits have always been subject to an assumption of reduced quality. With optical-fiber systems, this is changing. Another particularly difficult communication environment is the field of mobile radio, where, until very recently, digital techniques have not been applied.

The robustness of digital communication is evident along several dimensions: resistance to noise; resistance to cross talk; error correction; signal regeneration.

Resistance to Noise. In part this advantage is due to the inherent resistance of the digital signal to degradation due to channel noise, which is in turn due to the *threshold effect* inherent in a binary coding system. For example, a noise source which is 5% as strong as the desired signal may alter the amplitude of an analog signal by up to 5%, enough to create a serious distortion. The same noise will probably have no impact on a PCM-type digital transmission where each bit is detected on an all-or-nothing basis, and a 5% deviation will simply be ignored by the detection circuitry (rounded in the sample).

The gain in digital signal robustness is dramatic. In analog systems, the design standard for signal-to-noise ratio on short-haul wire telephone systems is typically something on the order of 46 dB — the signal must be 40,000 times stronger than the noise [2]. Designing an analog transmission system to this standard is often extremely costly, especially as the length of the circuit grows. A digital system will produce virtually error-free transmission, i.e., with no channel noise and only the minimal quantization noise introduced by the digitization process itself, with a signal-to-noise ratio of as little as 15 dB — a signal only 30 to 40 times stronger than the noise. To state this in another way, a digital transmission system can withstand noise levels 1000 times higher and still deliver a good signal.

Resistance to Cross Talk. A very significant problem with analog wire systems, particularly analog carrier (frequency multiplexed) cable systems, is cross talk. Cross talk in cable pairs occurs when the electrical energy from one circuit is magnetically superimposed upon an adjacent circuit, and it has historically proved to be an intractable problem for wireline network designers. Cross talk can enter the circuit at almost any place in the analog network, and even after more than a century it has not been eradicated. Many analog systems are not noise-limited so much as cross talk-limited. In a digital system, cross talk is largely eliminated by the threshold effect, and even where it does obtrude with sufficient strength to cause detection errors, it appears not as an intelligible signal, which is highly objectionable, but as random, unintelligible background noise [3].

Error Correction. In a sense, a digital system does not care what the source of bit errors, the digital effect of noise, cross talk, interference, *et cetera,* may be. It treats all channel errors (as distinct from the irreducible quantization errors) arising from whatever source in the same manner. In early digital systems, little attempt was made to deal with errors other than through the inherent robustness of the coding and signaling format. In more modern digital systems, very powerful error-correction techniques have been developed which can be applied to scrub a signal and remove a very large percentage of errors from any source. The application of such advanced techniques becomes a matter of weighing costs (the additional digital circuitry, plus, in most cases, a bandwidth penalty) against the benefits (the ability to operate in even noisier channels or in the presence of even more severe cross talk or interference).

Thus the use of error correction highlights the remarkable interchangeability of processing power for raw signal power. Indeed, the effect of error coding can be measured as so many decibels added to the numerator of the signal-to-noise ratio. It is even possible to design digital communication systems that, by using advanced coding techniques, are able to function with a *negative* signal-to-noise ratio — i.e., where the noise may be many times stronger than the signal itself. (See Chapter 12.)

Signal Regeneration. As noted in the previous section, the most important design strategy for wireline digital communication lies in the ability to regenerate and retransmit a digital signal. By inserting regenerative repeaters at proper intervals, the effects of noise and other transmission degradations can be controlled to any level desired. To reiterate what was said in Chapter 6, digital regeneration techniques allow the system designer to control the performance of the circuit to almost any desired specification; the system is no longer at the mercy of the hazards of the transmission environment in the conventional sense.

7.3 THE INTELLIGENCE OF THE DIGITAL NETWORK

Many transmission problems involve time-variant or statistical sources of signal degradation against which an analog system is often largely helpless. Such problems can be controlled in a properly designed digital system. A digital system can be designed to monitor and adapt to changing channel conditions. This in turn allows the system to be limited not by the worst-case transmission conditions, but by the actual case.

Equalization. Real transmission channels, whether wire or radio, tend to introduce distortions into the digital signal. The distortion can affect amplitude, frequency, or phase of the transmitted waveform, producing intersymbol interference among the bit pulses. Moreover, the characteristics of the channel change over time, especially in a highly variable mobile radio channel [4]. The general solution to this problem is known as adaptive equalization. The distortion characteristics of the channel are measured, usually continuously, and the predicted distortions in the transmitted pulses are subtracted from the received waveform. Equalization is critical to allow a digital network to be superimposed upon a highly heterogeneous physical plant. It is essentially a method of allowing a coherent digital transmission channel to be established from end to end, without having foreknowledge of the precise physical parameters of the transmission facilities.

Echo Control. Another type of problem which arises because of the heterogeneity of the physical plant is the phenomenon of echo, which may be conceived of as a reflection of the transmitted signal back to the transmitter from some discontinuity in the transmission path, such as the improperly balanced interface between one segment of the physical facilities and another. Such echoes become disturbing to the talker for circuits longer than about 1800 miles [5].

The difference between the analog solution to echo and the digital approach illustrates the precision and power of digital techniques at the microlevel of the transmission process. Analog *echo suppressors* have a number of drawbacks. They are based upon fairly crude solutions, such as the total suppression of the path with lower volume or amplitude. This approach not only suppresses the echo, it effectively converts the long-distance circuit into a simplex channel, where only one party can speak at once. Digital *echo cancellers* utilize an approach similar in a general way to equalization. By storing the transmitted speech for a period of time equal to the round-trip delay of the circuit, attenuating the stored signal to the proper level of the returning echo, and then subtracting it from the incoming return signal, the echo is completely eliminated.

Channel Condition Monitoring and Decision Response Capabilities. In addition to automatic and continuous processes such as equalization and echo cancellation, a digital system is highly amenable to the implementation of decision rules whereby countermeasures can be implemented by the system if certain thresholds are reached in the deterioration of channel conditions. A digital system can easily monitor the quality of the channel measured as the bit-error rate. Other parameters such as signal strength and delay time, which in a radio system may be translated into propagation distance, can also be used as the basis for certain countermeasure decisions. In early digital systems, about the only type of decision response that was allowed for was the removal of a channel from service in the event that the bit-error rate rose too high. As we shall discuss in Parts IV and V, a digital radio system offers a wide variety of opportunities for countermeasures to be implemented, and such measures can greatly improve the theoretical performance of such systems compared to their analog predecessors.

It is also possible in a digital system for the transmitter to provide a great deal of information about its own functioning and condition on a more or less continuous basis. This becomes very useful when designing a large network, especially for diagnostic and maintenance purposes. Analog systems are generally less flexible in providing this type of information without disrupting the voice communication itself.

7.4 THE FLEXIBILITY OF THE DIGITAL NETWORK

One of the most important characteristics of digital systems is that they are usually under software control to a very significant degree, which greatly increases user and operator flexibility. An operation as simple as changing the telephone number associated with a particular telephone line is impossible for many analog central offices still in operation in the United States today. Even in early digital systems, much of the logic circuitry was customized and hard-wired. Today, more and more digital systems rely upon generic processors with outboard software in removable ROMs (read only memories) which can be easily switched out as improvements and new features are added. The developing microcomputer industry has accustomed many people to the idea that the basic capabilities of a computing device can be readily and dramatically altered by substitution of one ROM or printed circuit card for another.

In the evolution of digital systems, the trend is strongly toward generic mainframe systems which can be upgraded over a long lifetime through installation of new software releases to maintain the system close to the leading edge of current applications. Indeed, the trend is toward networks that can be reconfigured more or less on demand, even by the users themselves. One type of demand-configured architecture is known as *packet switching.* Another is based on the concept of a *virtual circuit,* or even a *virtual network* [6].

7.5 THE DIGITAL NETWORK IS GENERIC

The digital network transmits digits. It is the end user who determines what to do with the data he receives. Digital voice, digital images, digital music, digital data, all appear the same within the digital transmission and switching facilities. In principle, it should be possible to deliver any type of digital service over the same facility. This is precisely what the proponents of integrated services on digital networks have in mind: a single channel for digital television, telephone, data services, all delivered through a single integrated network.

This is perhaps the haziest area of thinking about the digital network. It is relatively easy to posit the capability; it is hard to say exactly what types of service could or should be offered in an integrated format. After all, television is available from broadcasters and coaxial cable systems which are not integrated with the telephone or data networks, and yet it is hard to see what the advantages of integration for the end user might be in this case. On the other hand, the advantages of sharing facilities between nodes in the network are very clear. The use of a single facility to transmit voice, video, and data from one collection point to another offers definite savings through the reduction in duplicative facilities.

7.6 THE EFFICIENCY OF THE DIGITAL NETWORK

The original driver behind the development of T-carrier was the desire to find some way of reducing the cost of interoffice trunking facilities by multiplexing many conversations on a single copper wire pair. T-carrier reduced the amount of copper by approximately 95% over ordinary wire-pair cable facilities.

The logic of digital multiplexing is compelling — particularly, as we shall see, in the world of radio. It is based on the idea of balancing the additional cost of digital circuitry against the more efficient use of the transmission channel and its associated equipment and facilities. In the case of wireline digital systems, it offers a reduction in copper costs, as well as in all the logistical and overhead costs associated with the construction and maintenance of copper plant. In the case of radio, digital multiplexing points the way to reducing the amount of transmitter-receiver radio equipment, which remains the most expensive element in any radio-based communication system. Of course, digital multiplexing also allows for more efficient use of the radio spectrum itself, which is a very important goal from the standpoint of the regulatory community and the radio communication industry as a whole.

7.7 SECURITY AND THE DIGITAL NETWORK

The first real work on PCM systems was stimulated during World War II by the need for a truly secure radio-telephone system. Digital communication systems were utilized in military and sensitive government applications long before they became economical in commercial settings, precisely because of the benefit of communication security. Outside of the government and defense industries, communication privacy and security is not normally a high concern among telephone users — partly because it is routinely assumed that the wireline telephone *is* secure. But the growth of network data services is bringing security more to the forefront in facilities planning. Banks, automatic teller machines, credit agencies, point-of-sale terminals for customer transactions, all may involve the transmission of sensitive information *via* the public switched network. Existing analog facilities provide a degree of privacy that is largely illusory, based mainly on the physical difficulties of tapping a wireline connection.

As discussed in Chapter 4, the problem of communication privacy is much more serious in mobile-radio systems, where the ease of interception greatly compromises even routine cellular radio-telephone calls. Analog scrambling systems do exist, but they are not accepted as having a high level of security, and they typically degrade voice quality.

By contrast, a digital radio system lends itself very readily to intensive encryption, which can be designed to provide almost any level of security desired and which has a relatively small impact on system performance or economics. Even without encryption, the digitization process itself provides a fair degree of privacy by making inexpensive interception devices harder for amateurs to construct [7].

7.8 THE DIGITAL NETWORK IS DYNAMIC

Digital technology is accelerating. For example, in the area of voice coding, the standard for many years was the 64-kilobit coding rate established by the PCM architects in the 1950s and 1960s. The 64-kb/s standard is still regarded by many as a kind of technological plateau. In the past few years, however, digital communication has entered the era of much lower bit rates. Now 32-kilobit coding is widely accepted as an emerging alternative standard. Most authorities today believe that 16-kilobit coders can achieve telephone-quality voice transmission. It is quite likely that within a very few years the standard for telephone quality will have dropped to 9.6 kb/s, or even 8 kb/s. By the mid-1990s, I believe that 4.8 kb/s will be available for toll quality. This trend is already producing a major rethinking of capacity and spectrum-efficiency assumptions, especially in bandwidth-limited radio systems. The same rapid, indeed breathtaking, advances are being seen in other areas of digital signal processing.

Digital technology is in an extremely fertile period, and many network capabilities that may seem out of reach today will actually arrive much quicker than many telephone planners (who are accustomed to the glacial pace of innovation prevalent under the old regulated monopoly) may expect.

7.9 THE INTEGRATED DIGITAL NETWORK

The installation of digital technologies in the telephone network has been piecemeal: a switch here, an interexchange trunk there, a digital patchwork as the older analog facilities are slowly being replaced. Even at the rate of one new switch per day in the United States, the installed base of central-office equipment will not be completely digital until sometime in the next century. For a long time, each digital facility was viewed as an isolated upgrade. It was only over decades that the digital elements in the network began to be numerous enough that end-to-end digital facilities could grow together.

As this began to happen, digital systems began to complement one another both in economics and in performance. At first, linking two analog central offices by means of T-carrier required the installation of expensive analog-to-digital channel banks at either end. Today, the installation of a T-carrier trunk between two digital central offices can eliminate channel banks. Direct integration of digital switching and digital transmission facilities is becoming a major economic driver of further digitization of the network.

The digital network is only now coming into being. The piecemeal implementation of digital systems over the past twenty-five years has established the advantages of digital transmission on the microlevel. The advantages of digital networks on the macrolevel have yet to be realized. And yet the chief advantage of any digital transmission system today is its compatibility with these emerging networks. The application of digital techniques to the design of mobile-radio systems will alleviate many of the most difficult performance and economic problems experienced today by analog cellular systems. In the end, however, the greatest benefits will come from the integration of mobile telephony into the technological mainstream of digital telecommunication.

REFERENCES

[1] Fred DeLand, "The Development of Telephone Service: XIII. The Parent Bell Companies," *Popular Science Monthly,* August 1907, p. 142.

[2] John C. Bellamy, *Digital Telephony,* New York: Wiley, 1982, p. 71.

[3] *Ibid.,* p. 72.

[4] David R. Smith, *Digital Transmission Systems,* New York: Van Nostrand Reinhold, 1985, pp. 223 ff., p. 552.

[5] Bellamy, *op. cit.,* p. 38.

[6] *Ibid.*, pp. 367 ff.
[7] *Ibid.*, p. 75.

Part IV
DESIGN CHALLENGES FOR MOBILE TELEPHONY

Mobile telephony is an extremely challenging application for digital technology. There are a number of broad performance and policy issues which must be addressed by the designers of any system intended for large-scale implementation:

1. *The Mobile Environment.* The system must be designed for high performance under the severe transmission conditions found in mobile radio channels.
2. *Frequency Reuse.* To achieve high capacity and spectrum efficiency within a finite band of the radio spectrum, the mobile telephone system must be designed to allow for efficient reuse of a common set of radio frequencies in different cells within a large region of continuous converage.
3. *Modularity and Geographical Flexibility.* The system should be capable of functioning in large and small configurations, for high-density urban areas as well as low-density rural areas.
4. *Low Cost.* As indicated in Chapter 4, the chief problem facing analog cellular systems today remains the high cost of system hardware and high operating costs. The system designer must keep in mind the central issue of cost in any evaluation of alternative techniques.
5. *Compatibility with Future Network Services.* The next generation mobile-telephone system should be compatible with the new services emerging in the public-telephone network, particularly those associated with digital network capabilities.
6. *Compatibility with a Competitive Marketplace and Continuing Technological Evolution.* It may not be desirable, or even possible, to

seek a single, worldwide, monolithic technical standard covering all aspects of the transmission technology for the next generation of cellular radio; instead system designers may develop a more generic architecture based on interface standards designed to allow for continuing technological evolution.

Chapter 8
DESIGNING FOR THE MOBILE ENVIRONMENT

Designing a mobile radio system is arguably the most demanding technical challenge in all of telephony. No communication channel is more variable, or more uncontrollable, than the radio link to and from a moving vehicle. For example, a representative signal-to-noise ratio objective on an analog wireline link (short-haul) is about 46 dB, according to a standard text: in other words, the speech signal should be 40,000 times stronger than the noise level [1]. Every effort is made through quality control on materials and careful facilities engineering to ensure a relatively stable electrical environment within the cable-transmission facility. Short-term fluctuations in the basic signal-to-noise ratio of more than 1 or 2 dB for a short-haul circuit are generally taken as an indicator of something abnormal. Over a longer period of time, measured in years, wear and tear may induce a gradual degradation by as much as 10–15 dB, at which point plans are generally made to rebuild the transmission facility [2].

By contrast, it is quite normal for a mobile radio channel to experience *fades* — sudden decreases in signal strength — of 40–50 dB (a reduction by a factor of 10,000 to 100,000) in a fraction of a second. In fact, a mobile telephone in a fast-moving automobile in an urban environment may expect to experience dozens of significant fades (20 dB down or more, a factor of at least 100) per second [3]. The receiver is fighting not only noise but interference from other transmitters on the same channel or nearby channels. During a fade, the desired signal may suddenly become vastly weaker than an undesired, interfering signal from a distant cell-site operating on the same channel, and the receiver may lock in on the undesired signal.

The impact on conventional FM is severe. Without fading, the capture effect of FM, its ability to reject a competing signal as strong as 1 or

2 dB below the desired signal, can essentially suppress cochannel interference. According to one of the most authoritative studies of mobile propagation, however, "this rapid fading [of the mobile channel] alters the signal-to-noise performance markedly, washing out the sharp threshold and capture properties of FM [4]." The impact on other analog modulation schemes is even greater:

> Rapid Rayleigh fading generally has a disastrous effect on single-sideband (SSB) and AM communication systems . . . The distortion introduced by fading is larger than the output signal, independent of how much the transmitter power is increased [5].

In a digital system, the effect of noise and interference in the mobile channel is a startling increase in the bit-error rate. The BER in the mobile channel may typically run up to 1,000,000 times higher than in nonmobile point-to-point digital radio. It is very different from the virtually error-free digital transmission which can be engineered in a wireline environment. Voice coders, modulators and demodulators, and synchronization schemes, which work well enough on wireline digital systems like T-1 trunks or point-to-point microwave radio systems, may simply disintegrate in the mobile channel. The challenge of maintaining acceptable transmission under such severe conditions is the continental divide between technological alternatives that may work for mobile telephony and those that will not.

The *mobile environment,* as it is often referred to, has been the subject of an enormous amount of theoretical analysis and field experimentation, and a great deal is known about its peculiar characteristics, some of which can be stated in seemingly precise mathematical or statistical terms [6]. The problems, however, are of sufficient complexity that mobile systems designers are still highly constrained by performance uncertainties in different types of setting. Indeed, it should be noted that there is no such thing as a single mobile environment, describable with uniform characteristics or statistics. The mobile environment in a large city with a great many tall, man-made structures is very different from the mobile environment in flat, open farm country. The important variables that define the mobile environment for a particular area include:

1. Physical terrain (mountainous *versus* hilly *versus* flat or land *versus* water);
2. Number, height, arrangement, and nature (construction materials) of man-made structures;
3. Foliage and vegetation characteristics;
4. Normal and abnormal weather conditions;
5. Man-made radio noise.

In addition, the importance of these factors is greatly influenced by the radio frequency at which the system is designed to operate. Finally, the effects are related to the behavior of the mobile subscriber — especially the speed of travel. Slow-moving or stationary automobiles face quite different problems in the mobile environment than those moving at highway speeds.

The complexity of the mobile environment can seem overwhelming to engineers who are used to the stable conditions of wireline telephone transmission. Electricians and physicists began modeling the electrical environment of metallic conductors as long ago as Lord Kelvin's mathematical depictions of the transatlantic telegraph cable in the 1850s. Today it is possible to describe, indeed to specify, the wireline transmission environment with a high degree of precision. Even conventional nonmobile radio engineering is relatively easy: most microwave links are engineered over a clear line-of-sight path where propagation takes place well above terrain irregularities. In contrast, the mobile propagation path quite often involves intervening terrain or structural blockages, which cannot be predicted with any precision; the mobile receiver is very close to ground level and is constantly moving into different settings. The mobile environment must be addressed through statistical treatments, which necessarily involve drastic simplifications of assumptions about user behavior and microvariations in terrain, foliage, and building patterns. Published studies can create an aura of certainty which is misleading. There is still disagreement over the mathematics for modeling a single transmitter-receiver propagation path in rough terrain [7]. Different models offer various predictions about the performance of specific systems. Most mobile engineers would agree that mobile system engineering is frustratingly imprecise and that to find out actually how a particular system is going to perform in a given physical setting it is necessary to build it, operate it, and modify it as shortcomings appear. The procedure is costly, and still may be unsatisfactory: Jakes *et al.* write "the testing of mobile radio transmission techniques in the field is time-consuming and often inconclusive, due to uncertainty in the statistical signal variations actually encountered [8]." The following conclusion of a recent study of propagation in nonurban areas, quoted virtually in its entirety, is typical:

> This investigation has revealed unexpected and widespread variations [of mobile signal characteristics] which cannot be simply related to the environment or the seasons. There is no simple explanation to give at this moment but it can be conjectured that one is dealing with the results of a complex scattering phenomenon occurring in the environment consisting chiefly of trees and other types of vegetation. To describe scattering by trees according to type, season, or temperature

one would certainly require a major experimental undertaking even for one frequency [9].

Conceptually, the mobile systems engineer must understand three things:

1. The way in which the radio signal may be altered or degraded in the mobile channel;
2. The effects that these alterations can have on link quality and system performance, given a specific radio-link technology;
3. The countermeasures that are available to combat these effects.

8.1 THE VOCABULARY OF RADIO

A radio wave is a form of electromagnetic energy, propagating at the speed of light, which may be visualized as a sine wave. (See Figure 8.1.) For our purposes, the wave has three important characteristics:

(Insert Figure 8.1 hereabouts!)

1. *Amplitude.* The magnitude or height of the sine wave crests and troughs;
2. *Frequency.* The number of crests that occur in one second; the basic measure of frequency is the *Hertz,* named for Heinrich Hertz, which is defined as one cycle, one crest-to-crest event. Most radio frequencies are measured in thousands or millions of cycles per second, *kilohertz* (KHz) or *megahertz* (MHz), respectively. (*Wavelength* is the inverse of frequency — the distance between two crests; it is measured in terms of meters, millimeters, or feet and inches, and so forth.)
3. *Phase.* The particular angle of inflection of the wave at a precise moment in time. It is normally measured in terms of degrees.

As we shall see, all three of these parameters may be used to carry information in a radio transmission.

Most radio transmission utilizes a continuous wave of a fixed frequency, called the *carrier.* The frequency of this carrier is stated as so many MHz (in the range of interest here). For example, the cellular carrier frequencies are generally about 800 MHz. Different carrier frequencies define different regions of the spectrum. For mobile communication, the range of carrier frequencies that are of interest is between about 1 MHz and, say, 3000 MHz (or 3 *gigahertz,* 3 GHz). Different carrier frequencies interact in different ways with the physical environment; thus the carrier frequency determines many of the most important characteristics of mobile propagation.

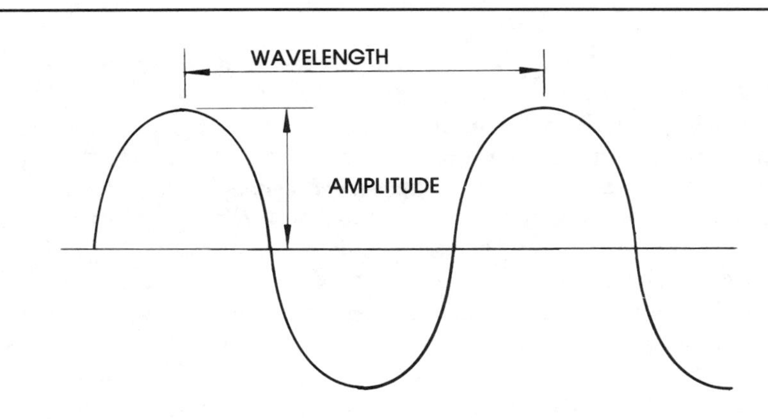

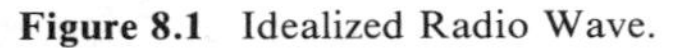
Figure 8.1 Idealized Radio Wave.

The modulated carrier — i.e., the carrier with the information, the voice signal, impressed upon it through slight changes in either its amplitude, frequency, or phase — actually occupies a narrow region of the spectrum around the nominal or unmodulated carrier frequency, sometimes called the *center* frequency in this context. (See Figure 8.2.) The width of this region — the *occupied bandwidth* — is also measured in KHz or MHz. This is what is commonly referred to as a radio *channel.* The *carrier spacing* is determined by the channel bandwidth, as well as the necessary guardbands.

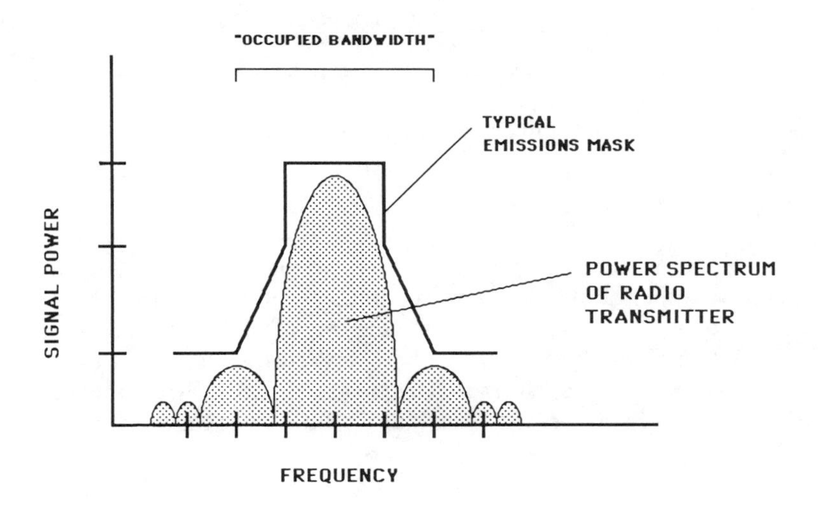

Figure 8.2 Emission Characteristics and Channel Definition.

The radio signal in a given channel is affected by *noise* and *interference.* Noise is considered to be the result of random environmental processes which produce radio energy such as lightning or automobile ignitions, or of thermal noise in the receiver itself. The ratio of the signal strength to the noise level is called the *signal-to-noise ratio* (SNR) and is the first fundamental measure of signal quality. Interference is the result of other man-made radio transmissions. There are two types of interference: *adjacent channel interference,* which occurs when energy from the carrier spills over into the adjacent channels (see Figure 8.3), and *cochannel interference,* which occurs when another transmission on the same carrier frequency impinges upon the receiver (see Figure 8.4). The ratio of the carrier to the interference (from both sources) is called the *carrier-to-interference ratio* (C/I).

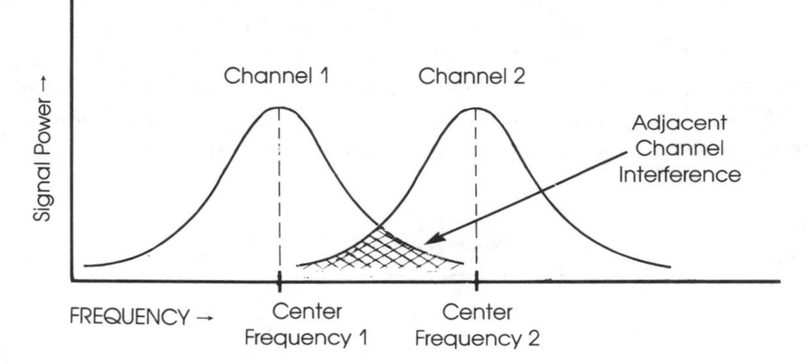

Figure 8.3 Adjacent Channel Interference.

The final basic propagation concept is the *link budget,* which defines the quality of the radio link, measured in terms of *decibels* or dB (or more precisely, dBm — where the "m" stands for milliwatts) of signal power. In a simple link-budget analysis, there are certain factors which "add decibels" to the link, such as the transmitter, the gain on the receiver antenna, and so forth, and others which "subtract decibels" from the link,

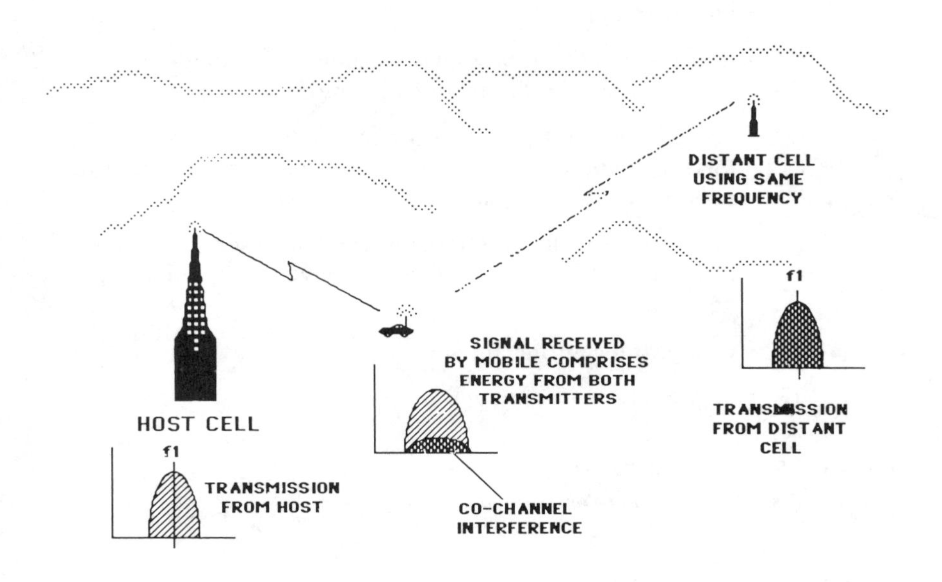

Figure 8.4 Co-Channel Interference.

like blocking terrain or fades. If enough dBs are left after the link budget has been constructed, the signal will get through with sufficient quality. A typical link budget is given in Figure 8.5. It should be noted that this is an extremely simplified propagation analysis.

$$G_s = L_p + F + L_t + L_m + L_b - G_t - G_r$$

where G_s = system gain in decibels

L_p = free space path loss in dB,

F = fade margin in dB

L_t = transmission line loss from waveguide or coaxials used to connect radio to antenna, in dB

L_m = miscellaneous losses such as minor antenna misalignment, waveguide corrosion, and increase in receiver noise figure due to aging, in dB

L_b = branching loss due to filter and circulator used to combine or split transmitter and receiver signals in a single antenna

G_t = gain of transmitting antenna

G_r = gain of receiving antenna

Figure 8.5 Simplified Link Budget.
Source: David Smith, *Digital Transmissions Systems.*

8.2 THE FATE OF THE RADIO WAVE

Several things happen to a radio wave transmitted to or from a moving vehicle.

8.2.1 Free Space Loss

First, as the signal radiates in all directions, assuming for the moment an omnidirectional antenna, the power of the signal at any given point steadily diminishes as a function of the distance of the receiver from the transmitter. In a vacuum, in free space, the signal strength would diminish as the inverse of the square of the distance — the famous *inverse square law.* In free space, if the received signal is 100 watts at 1 mile, it will be 25 watts at 2 miles (¼ the power), 4 watts at 5 miles (1/25 the power), and 1 watt at 10 miles (1/100 the power), assuming the same antenna. (See Figure 8.6.)

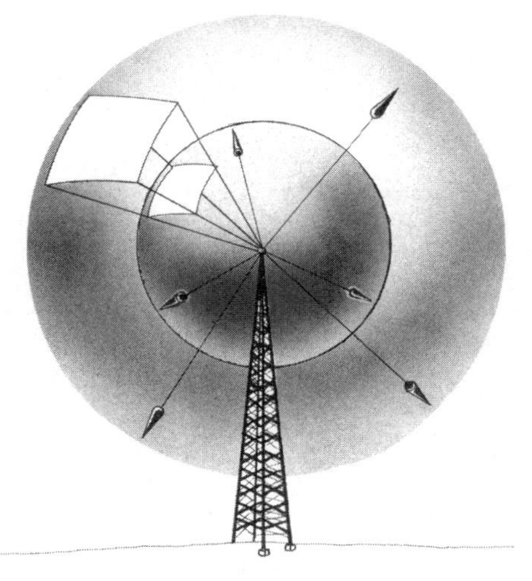

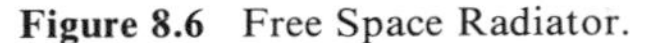

Figure 8.6 Free Space Radiator.

In practice, since mobile telephone systems are not set up in outer space, the path loss is *more severe* than the inverse square law would predict. Path loss for mobile systems may be evaluated as the inverse of the cube of the distance, or of some higher exponent up to the 5-th or 6-th power. Such statements reflect averaging of terrain, atmospheric, and other real-world effects. If the inverse of the 6-th power of distance is assumed, then the 100-watt signal level at 1 mile from the transmitter will decline to about 1.5 watts at 2 miles. (See Figure 8.7.)

The additional path loss is highly dependent upon frequency and upon assumptions about antenna design and other aspects of the physical locale. It is useful only in a very general way to indicate feasible transmission limits. It is, however, important to emphasize that even so fundamental a transmission parameter as the average rate of attenuation with distance must often be derived empirically for new frequency bands on the basis of field experience.

8.2.2 Blockage (Attenuation)

The second hazard facing the radio wave in the transmission path is the possibility that it may be partially blocked, or absorbed, by some feature of the environment. The degree of attenuation and the specific

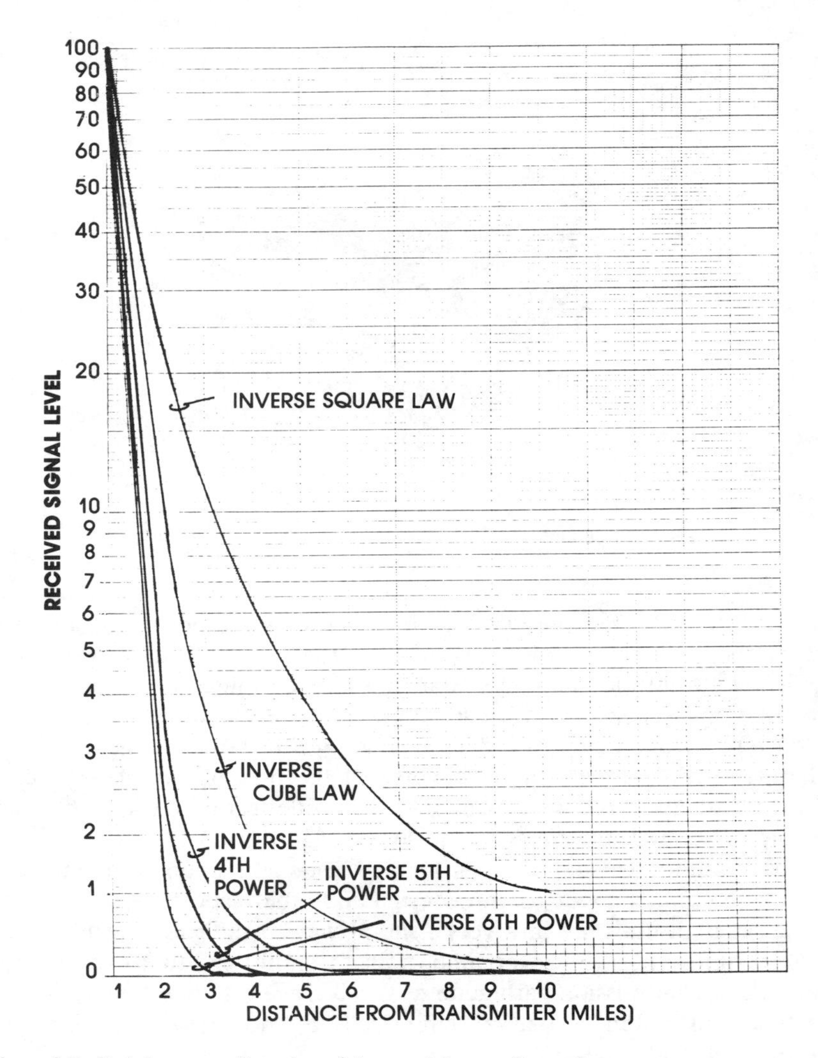

Figure 8.7 Path Loss as a Function of Assumed Inverse Power Law.

factors that may cause attenuation depend chiefly upon frequency. For example, frequencies below 1 GHz (1000 MHz) are essentially unaffected by rain or atmospheric moisture. Frequencies above 10 GHz are often severely affected, and frequencies above about 30 GHz are basically unusable over lengthy outdoor paths. (See Figure 8.8.) The general rule is that the lower frequencies have much greater penetrating power and will propagate farther. The higher the frequency, the greater the attenuation, the more power needed at the transmitter, and the shorter the radius of effective transmission. For example, for TV transmission at the lower VHF band (50–90 MHz) the Federal Communications Commission authorizes maximum transmitter power of 100 kilowatts. For UHF stations, operating at 500–800 MHz, the FCC authorizes station power of up to 5000 kilowatts to achieve approximately the same quality over the same range.

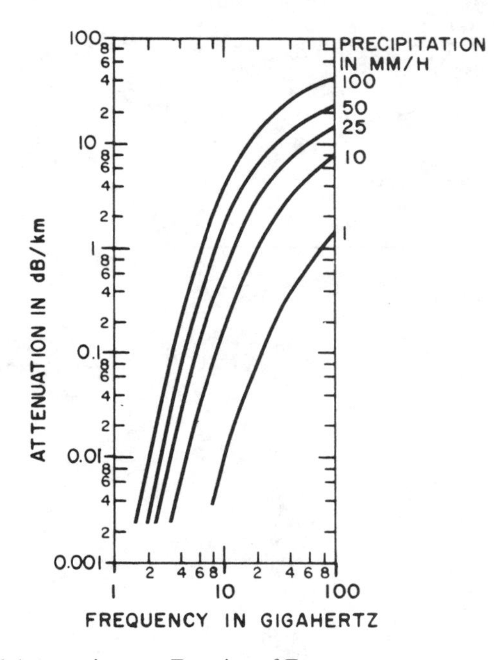

Figure 8.8 Rainfall Attenuation as a Function of Frequency.

Source: William C. Jakes, *Microwave Mobile Communications,* p.89.

Around 3–5 GHz, the attenuation becomes so great that the inverse power law results in impractically short transmission distances for omnidirectional transmitters. Highly directional antennas must be employed to concentrate the energy along a fixed point-to-point path. By focusing the radio waves in a narrow beam, analogous to a beam of light from a flashlight, the attenuation can be overcome for usable distances. These are called *point-to-point* systems and are characteristic of long-haul microwave telephone systems, which typically involve single hops of 10–50 miles. (See Figure 8.9.)

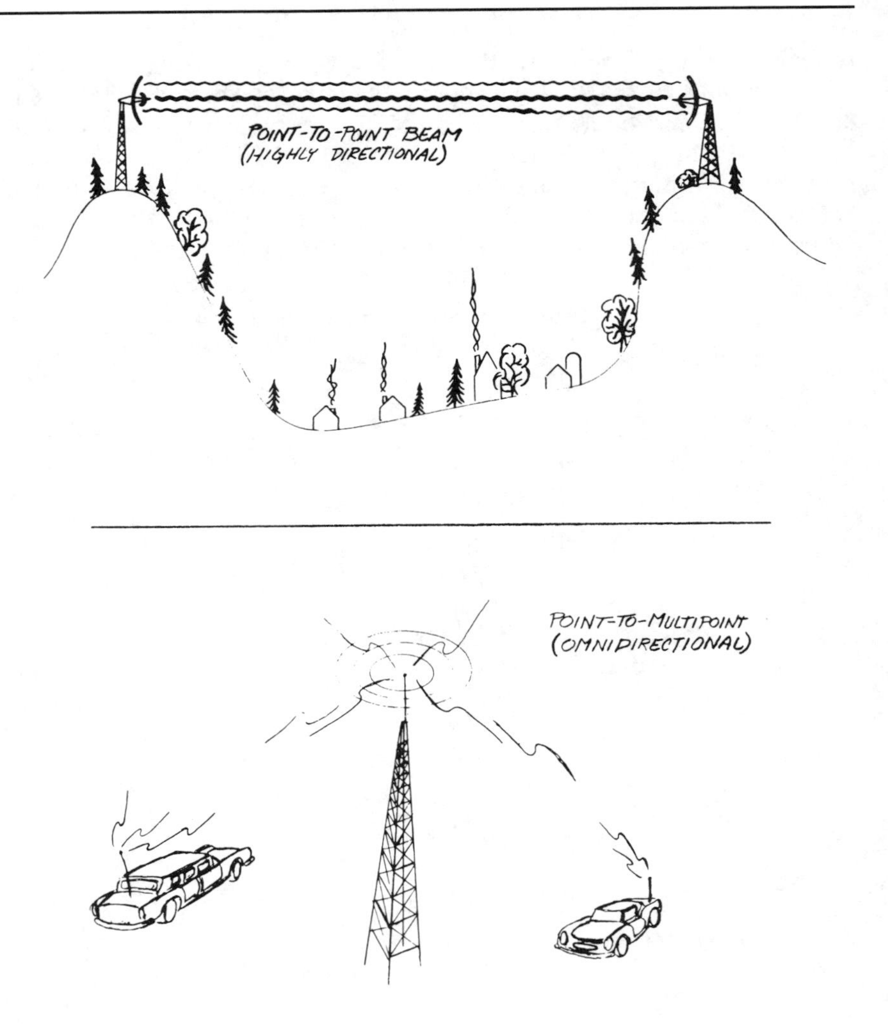

Figure 8.9 Point-to-Point *versus* Point-to-Multipoint Transmission.

It becomes more difficult to implement *point-to-multipoint* mobile systems at higher frequencies. Indeed, one of the principal concerns in the development of cellular radio was the fear that the radio link at 800–900 MHz would prove to be significantly less robust than the older IMTS systems at 150 MHz and 450 MHz. In practice, the transmission radius and effective coverage achieved by 800-MHz systems *is* significantly less than for 450-MHz or 150-MHz systems — but this is not necessarily a disadvantage (see Chapter 9).

At typical mobile-radio frequencies (150–900 MHz), the most important environmental attenuation effect is *shadowing,* where buildings or hills create radio shadows. The problems of shadowing are most severe in heavily built-up urban centers. Early mobile studies found shadows as deep as 20 dB over very short distances, often literally from one street to the next, depending upon orientation to the transmitter and local building patterns [10]. (See Figure 8.10.) According to some studies, average signal strength in suburban areas is around 10 dB better than in urban settings because of fewer large buildings. In turn, open rural areas show typical received signal strength up to 20 dB higher than suburban areas [11]. All things being equal, a mobile radio signal will propagate much better in rural areas than in built-up zones. Terrain irregularities also produce such effects, although the shadowed areas tend to be larger and the rate of change slower and somewhat more predictable. Another significant factor is the foliage in wooded suburban or rural areas. One study measured an attenuation of 2.4 dB per 100 feet at 210 MHz through a grove of live oaks [12]. Jakes also reports that for a suburban New Jersey setting using 836 MHz "the average received signal strength in the summer when the trees were in full leaf was roughly 10 dB lower than for the corresponding locations in later winter [13]." Man-made noise characteristics also differ from urban to nonurban environments [14]. (See Figure 8.11.)

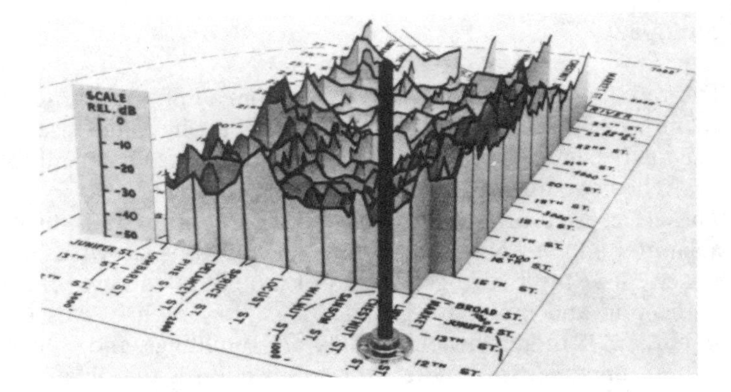

Figure 8.10 Signal Variations in a Segment of Downtown Philadelphia.

Source: William C. Jakes, *Microwave Mobile Communications,* p.95.

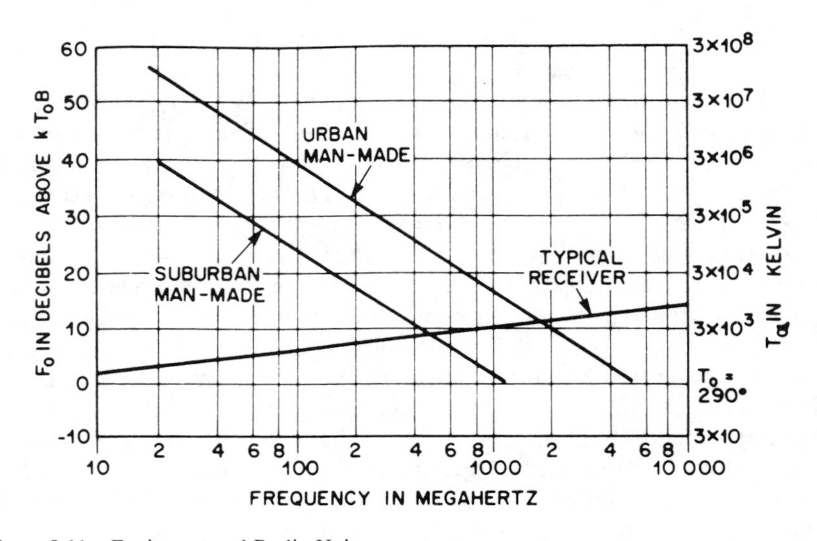

Figure 8.11 Environmental Radio Noise.
Source: William C. Jakes, *Microwave Mobile Communications,* p.298.

The fading effects produced by shadowing are often referred to as *slow fading,* because from the perspective of a moving automobile the entrance and exit to and from such a shadow takes a fair amount of time, since the area of the fade is large (many feet or hundreds of feet across).

8.2.3 Multipath

The radio wave may also be *reflected,* from a hill, a building, a truck, an airplane, or a discontinuity in the atmosphere; in some cases, the reflected signal is significantly attenuated, while in others almost all the radio energy is reflected and very little absorbed.

The effect is to produce not one but many different paths between the transmitter and receiver. (See Figure 8.12.) This is known as *multipath propagation;* it is the two-edged sword of mobile telephony. On the one hand, reflection and multipath propagation allow radio waves to "bend around corners," to reach behind hills and buildings and into parking garages and tunnels. (See Figure 8.13.) Reflection is the obverse of penetration. High frequencies reflect better, generally speaking, than lower frequencies, which tend to penetrate. For example, VHF frequencies (150 MHz) do not propagate well in long tunnels, but higher 800-MHz frequencies follow the tunnel like a waveguide and do considerably better [15].

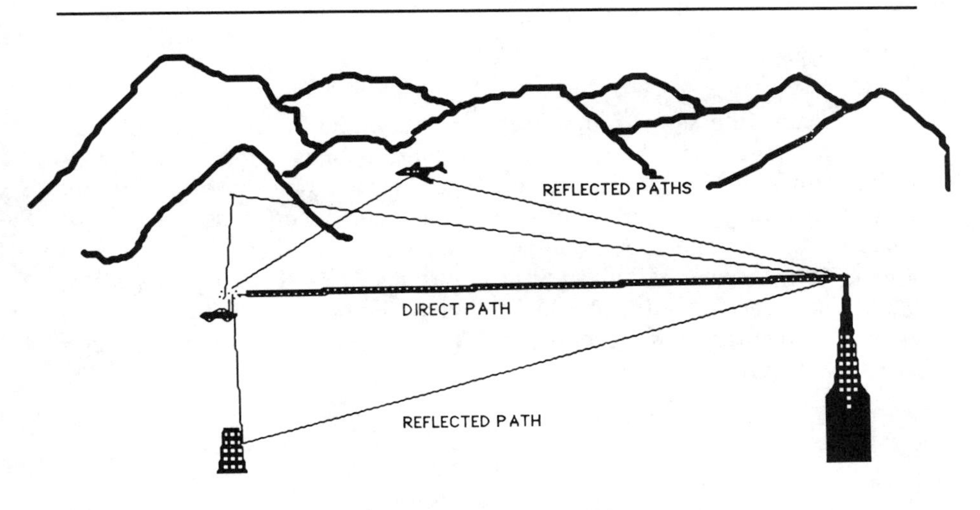

Figure 8.12 Multipath Propagation.

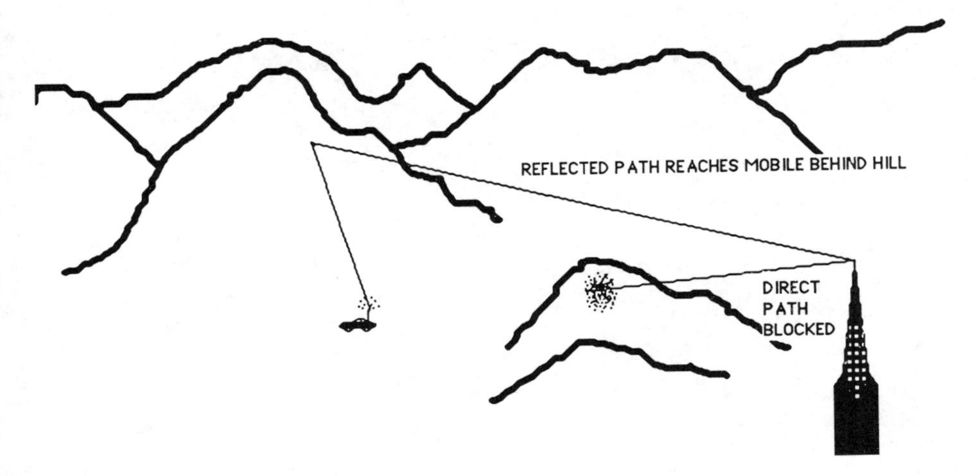

Figure 8.13 Multipath Allows Propagation around Obstacles.

On the other hand, multipath propagation creates some of the most difficult problems associated with the mobile environment. The three most important multipath issues for the digital designer are:

1. The delay spread of the received signal;
2. Random phase shift which creates rapid fluctuations in signal strength known as Rayleigh fading;
3. Random frequency modulation due to different Doppler shifts on different paths.

8.2.3.1 Delay Spread

Because the signal follows several paths, and because the reflected paths are longer than the direct path, if there is one, the multiple signals arrive with a slight additional delay. (See Figure 8.14.) Because different paths result in slightly different times of arrival, the effect is to smear or spread out the signal; for example, a single sharp transmitted pulse will appear to a receiver as indicated in Figure 8.15 [16]. In a digital system, particularly one operating at a high bit rate, the delay spread causes each symbol to overlap with preceding and following symbols, producing *intersymbol interference.*

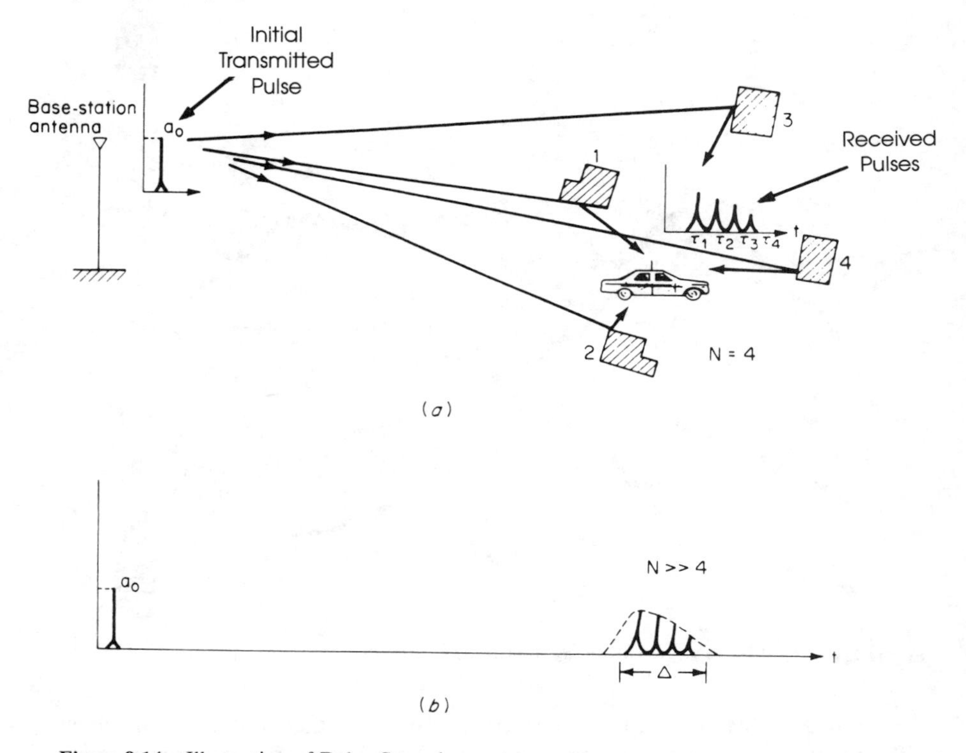

Figure 8.14 Illustration of Delay Spread.

Source: William C.Y. Lee, *Mobile Communications Engineering,* p.40.

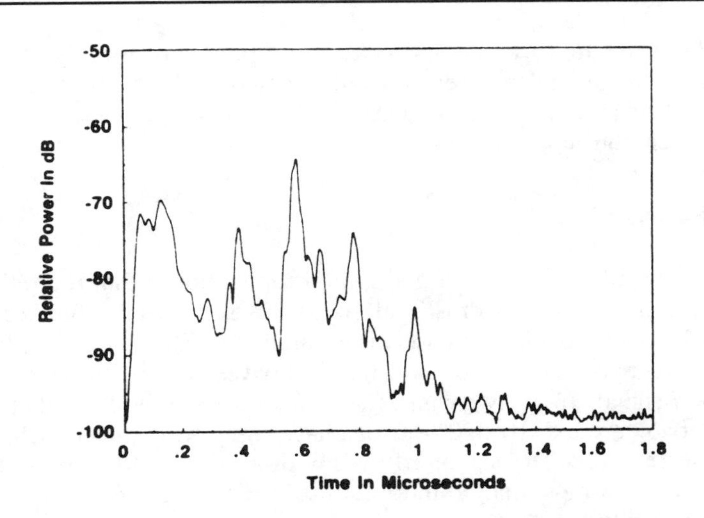

Figure 8.15 A Measured Profile of Average Received Power *versus* Time Delay at 850 MHz.
Source: Cox, et al., *IEEE Proc.,* 1987, p.767.

The delay spread is fixed, depending upon frequency. Therefore, the degree of intersymbol interference from this source is dependent upon the transmission bit rate and the modulation level, which sets the symbol time. If the bit rate is high and the symbol time is similar in magnitude to the average delay spread, the rate of intersymbol interference can be quite high. In effect, delay spread sets a limit on the transmission symbol rate in a digital mobile channel. One study found that a delay spread of 1 μs was typical for indoor office and work environments at 800 MHz [17]. Assuming 4-level modulation, or 2 bits per symbol, and a rate of 200 kilosymbols per second (equal to 400 kilobits per second), each symbol would occupy about 5 μs. This would imply an average symbol overlap of up to 20%. (Note that this is a statistical property; some percentage of delay spreads would be significantly greater than this.) The study concluded that for these frequencies in these environments, this was the maximum intersymbol interference that could be tolerated without adaptive equalization.

In outdoor environments in which a mobile telephone would have to operate, both direct paths and reflected paths may be significantly longer, and delay spreads may range up to several microseconds or more [18]. The longer delay spreads are typically found in more urban areas, where there are more reflectors. (Office buildings are usually excellent reflectors.) Longer delay spreads mean either that the symbol rate must be reduced or that adaptive equalization must be employed (see Section 8.5.3), which adds cost.

8.2.3.2 Rayleigh Fading

The second chief effect of multipath propagation is that the reflected radio wave may undergo drastic alteration in some of its fundamental characteristics, particularly phase and amplitude. The phase of the reflected wave may be rotated such that it arrives out of phase with the direct-path signal. If the two signals, assuming for the moment that there are only two, are exactly 180° out of phase, they will cancel each other out at the receiver. The signal effectively disappears. Other partial out-of-phase relationships among multiple received signals produce lesser reductions in measured signal strength.

The concept of a *fade* is therefore a *spatial* concept. Assuming the transmitter is stationary, at any given spot occupied by the receiver the sum of all direct and reflected paths from a transmitter to a receiver produces an alteration in signal strength related to the degree to which the multipath signals are in phase or out of phase. This signal strength may be somewhat more — or considerably less — than the expected signal strength, which can be defined as that which would be expected on the basis of the direct path alone, based solely on free-space loss and environmental attenuation. If the actual measured signal is significantly weaker — say 20 dB or 100 times weaker — than the expected signal level, we may conceive of that spot as a *20-dB fade,* for *that frequency* and for *that precise transmitter location and precise configuration of reflectors.* As long as we hold these factors constant, if we place our antenna in this spot we will lose 20 dB of signal strength.

What can we say about the number, the spacing, and the depth of these fades? There has developed a body of statistical knowledge which can be used with some success to characterize the incidence of fades in the environment.

The fades are said to fall within a statistical distribution known as Rayleigh distribution (after Lord Rayleigh, the great turn-of-the-century English physicist), and for this reason the phenomenon is often referred to as Rayleigh fading. The mobile environment is often called, from this perspective, the Rayleigh environment. The Rayleigh environment is peppered with fades of varying depths. The depth and spacing of the fades is related to the radio frequency: at microwave frequencies (800 MHz) the maximum fades occur every few inches. These maximum fades are very deep: the signal strength is reduced by 10,000 to 100,000 times down from its expected value. In between are thousands of shallower fades. It is as though, from a radio point of view, the static world we have created is full of countless holes of varying depths. (See Figure 8.16.) The radio signal cannot reach into the bottoms of these deep holes, which means that if our receiving antenna is inside one of these holes at any instant in time it will not receive the transmitted signal. The individual holes are small: moving the antenna a couple of inches will take it out of a deep fade and restore a good signal.

Now imagine an automobile antenna moving through this strange Swiss-cheese radio world at 60 miles per hour, 88 feet per second. The antenna passes through hundreds of holes of varying depths every second, causing the signal strength to fluctuate very rapidly between normal levels and fades ranging up to 40 dB or more. An amplitude monitor on a mobile receiver will draw a graph like Figure 8.17. This is the way Rayleigh fading is usually experienced and portrayed.

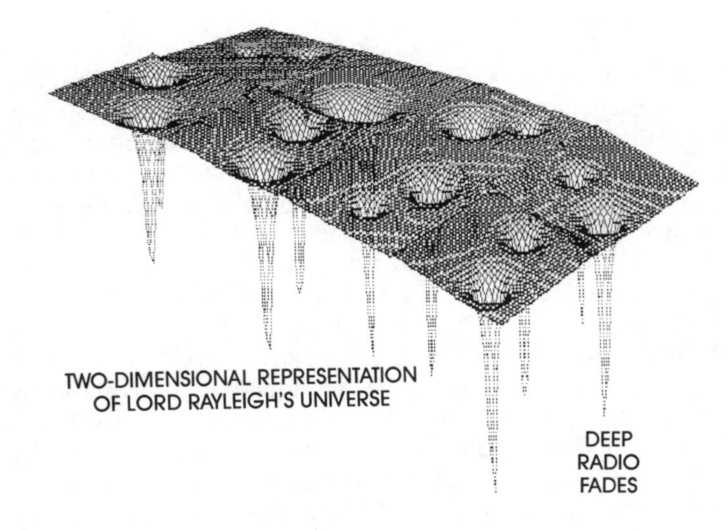

Figure 8.16 Lord Rayleigh's Universe.

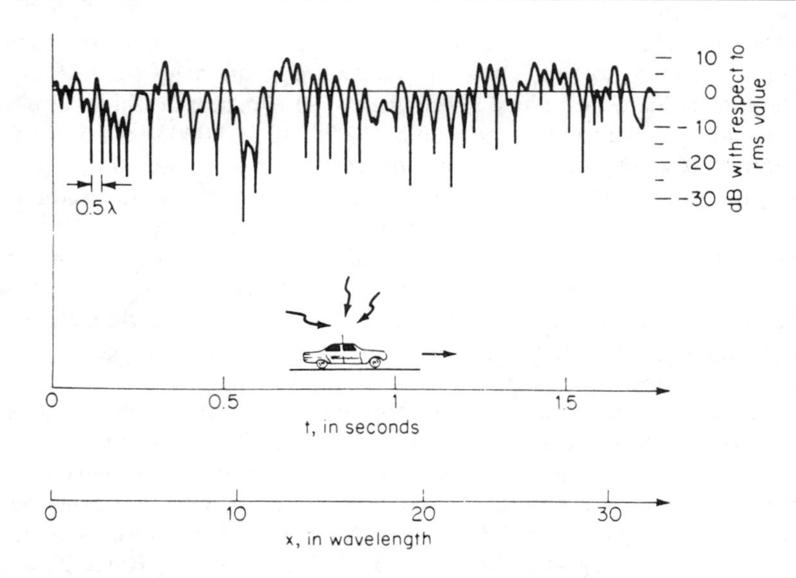

Figure 8.17 Typical Fading Signal Received While the Mobile Unit Is Moving.
Source: William C.Y. Lee, *Mobile Communications Engineering,* p.46.

These signal amplitude fluctuations constitute far and away the most difficult challenge of the mobile environment. Rayleigh fading is the dominant design challenge for any digital mobile-radio proposal. It is a characteristic of the environment and cannot be directly altered by the mobile systems engineer. Although there are several available countermeasures (see Sections 8.4 and 8.5), all involve additional cost and complexity.

Multipath fading is the great destroyer of mobile-radio signals. It overwhelms AM and single sideband systems. The great advantage of FM when it was applied to mobile radio in the 1930s was due to the fact that the FM receiver suppresses (ignores) amplitude modulation, and thus the degradation due to amplitude fading is greatly reduced [19].

Rayleigh fading, however, poses an additional problem for cellular architectures. A second transmitter in a different location will create an entirely different spatial pattern of fades, because the relative positions of the reflectors are different. The strengths of the received signals from two cell-site transmitters are not correlated. Chances are that, at a given instant in time, if a mobile is in a fade with respect to the correct base station transmitter, it is not in a fade from a distant "incorrect" base station transmitting on the same frequency in another cell. In an FM system, the capture effect is lost; actually, FM capture works against us: "When Rayleigh fading is included in the analysis, most of the interference is caused when the receiver captures on the interfering stations, even if it is for a very short time [20]." Consider a hypothetical chart of the signal levels received from two competing transmitters for a mobile moving through two superimposed Rayleigh environments, one for each transmitter. (See Figure 8.18.) The more transmitters there are in the general area operating on the same frequency, the more complex the Rayleigh patterning.

Another complicating factor is that some of the reflectors are also in motion, changing the Rayleigh characteristics over time irrespective of the movement of the mobile subscriber. Atmospheric reflectors, inversion layers, and so forth, are obviously transitory. Large trucks and buses can be significant reflectors for nearby mobile receivers. Indeed, one of the commonest demonstrations of multipath propagation occurs when an airplane — an excellent reflector — flies low over a housing area and creates a new and rapidly changing set of multiple paths for TV broadcast signals, causing the rhythmic flutter in the picture. (See Figure 8.19.)

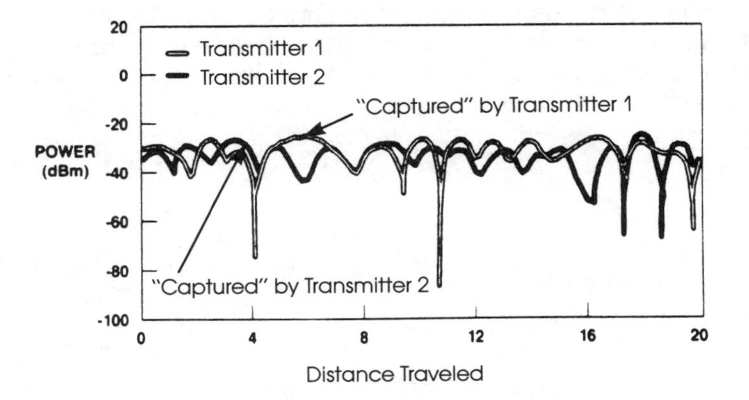

Figure 8.18 Rayleigh Patterns from Two Co-Channel Transmitters.

Source: dervied from Cox, et al., *IEEE Proc.*, p. 768.

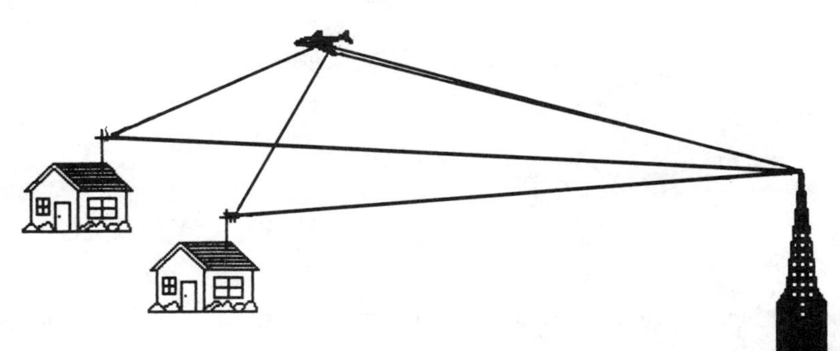

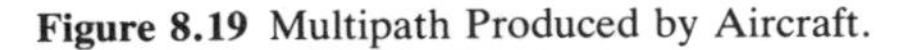

Figure 8.19 Multipath Produced by Aircraft.

8.2.3.3 Doppler Shift

Another aspect of the mobile channel is caused by the movement of the vehicle relative to the transmitter. This is the variation in the frequency of the received signal known as the *Doppler shift* (after Christian Johann Doppler, a nineteenth-century Austrian physicist who first called attention to frequency shifts caused by relative motion). Much as the sound of a horn on a moving car appears to the stationary observer to be slightly higher in pitch when the car is approaching rapidly and slightly lower when the car is receding, so radio transmissions are frequency shifted due to the relative motion of the vehicle. This frequency shift varies considerably as the mobile unit changes direction, speed, and is handed off from one cell to the next, and it introduces considerably random frequency modulation in the mobile signal. Moreover, the Doppler shift affects all multiple propagation paths, some of which may exhibit a positive shift, and some a negative shift, at the same instant. (See Figure 8.20.) Roughly speaking, Doppler-induced random FM is correlated with vehicle speed for a given frequency. At 22 miles per hour and 900 MHz, the median Doppler offset is about 30 hertz, which under certain conditions can be sufficient to "introduce distortion objectionable to the ear [21]."

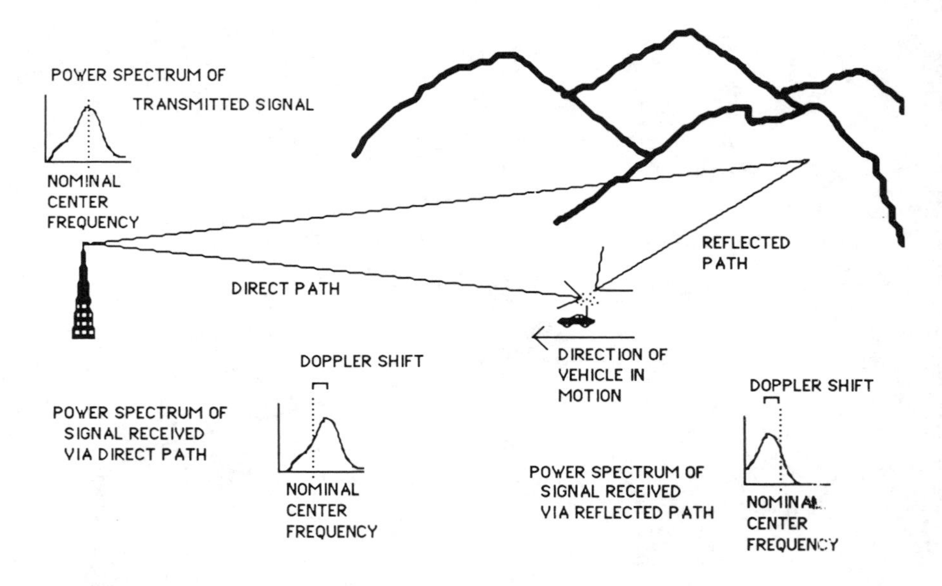

Figure 8.20 Idealization of the Doppler Effect on Received Signals in a Mobile Multipath Environment.

8.3 EFFECTS ON SYSTEM PERFORMANCE

The influence of the mobile environment on system performance has two aspects:

1. Effects on the quality of the individual radio link and the voice channel;
2. Effects on overall system performance.

8.3.1 Path Effects

The deleterious effects of the mobile environment on the radio signal have already been enumerated. Signal strength fluctuates rapidly and severely due to fast Rayleigh fading, and the time-averaged signal level also fluctuates due to shadowing or slow fading. In addition to multipath-induced variations in amplitude and phase, Doppler shifts induce random frequency variations. In short, every important parameter of the radio signal is subject to violent, sudden, and unpredictable alteration. Whatever modulation scheme is utilized — based on amplitude, frequency, or phase — the information-bearing parameter will be attacked.

In digital systems, all forms of signal degradation are converted into a common measure: the bit-error rate. The digital systems designer does not really care whether the source of a bit error was an amplitude fade produced by phase cancellation (Rayleigh fading), intersymbol interference due to delay spread, or signal attenuation caused by ice buildup on the antenna. The digital mobile-radio system must be able to tolerate a very high BER. In wireline digital systems, BERs of 10^{-8} or better are routine. In mobile applications where reuse is required, channel error rates of 10^{-2} or worse must be anticipated. In other words, from a digital point of view the mobile environment is more than a million times dirtier than the normal wire circuit. In fact, systems developers often simulate error rates of 5% to 25% to evaluate the bounds of system performance. Certain military digital mobile communication systems are designed to survive a 50% BER — one out of every two bits in error!

Strictly speaking, the BER is controllable by the system engineer. That is the liberating principle of digital communication, as we saw in Chapter 6. In the mobile radio circuit, however, it is generally not feasible to use multistage signal regeneration as is done on wireline T-carrier trunks, for example. Moreover, the bandwidth-limited character of radio transmission limits the engineer's ability to improve the BER by using less efficient but more robust coding and modulation techniques. As we shall see in Chapter 12, the digital mobile system designer is driven toward higher-level modulation schemes which are inherently more susceptible to errors at a given signal-to-noise ratio than lower-level modulation. For example, for a signal-to-noise ratio of 12 dB (actually the reference here is to the energy-per-bit to noise density ratio, which Bellamy suggests as the most appropriate measure for measuring signal-to-noise in a digital system) a 2-level phase modulation scheme will produce a BER of about 10^{-8}, while a more bandwidth-efficient 16-level phase modulation suffers a BER between 10^{-2} and 10^{-3} [22]. In practice, given the other design constraints of the mobile system, high error rates are inescapable.

A second important path-related effect, already touched upon, is the limit imposed by the average delay spread upon the rate of digital transmission in a digital mobile channel. As we shall see, there is a wide range of proposals as far as the channel bit rate and symbol rate are concerned, and for many architectures this becomes a very significant limit.

8.3.2 System Effects

The most serious effect of the mobile environment on cellular system performance in a cellular architecture is the creation of fuzzy cell boundaries. The signal strength contour — the idea upon which traditional mobile engineering is based — is actually only a convenient idealization of what turns out to be a very messy reality. Consider the actual signal strength received at a moving automobile as it moves away from the cell-site transmitter. Figure 8.21 is adapted from actual measurements, reproduced here with permission of William Lee [23]. Figure 8.22 presents a finer-grained version of a hypothetical small portion of this signal record. Under such conditions, *there is no signal contour:* any criterion signal level will actually be crossed many times over a wide fuzzy boundary zone. Moreover, two mobiles following the same path at different times will experience generally similar *but specifically distinct* signal-level patterns. The characteristics of the mobile environment are constantly changing.

The cell boundary becomes a statistical property, rather like the position of the quantum mechanical electron. It cannot really be known; we can only state a confidence level associated with a particular estimate of its value at a particular location. The instantaneous unpredictable actual value, however, may differ from our estimate by as much as a factor of 100,000 or more. (See Figure 8.23.)

In Chapter 4 we discussed the effect of fuzzy cell boundaries on the hand-off decision. The system controller in today's cellular architecture knows nothing except the strength of the signal it is receiving from the mobile, which is subject to the same multipath fluctuations. Called upon

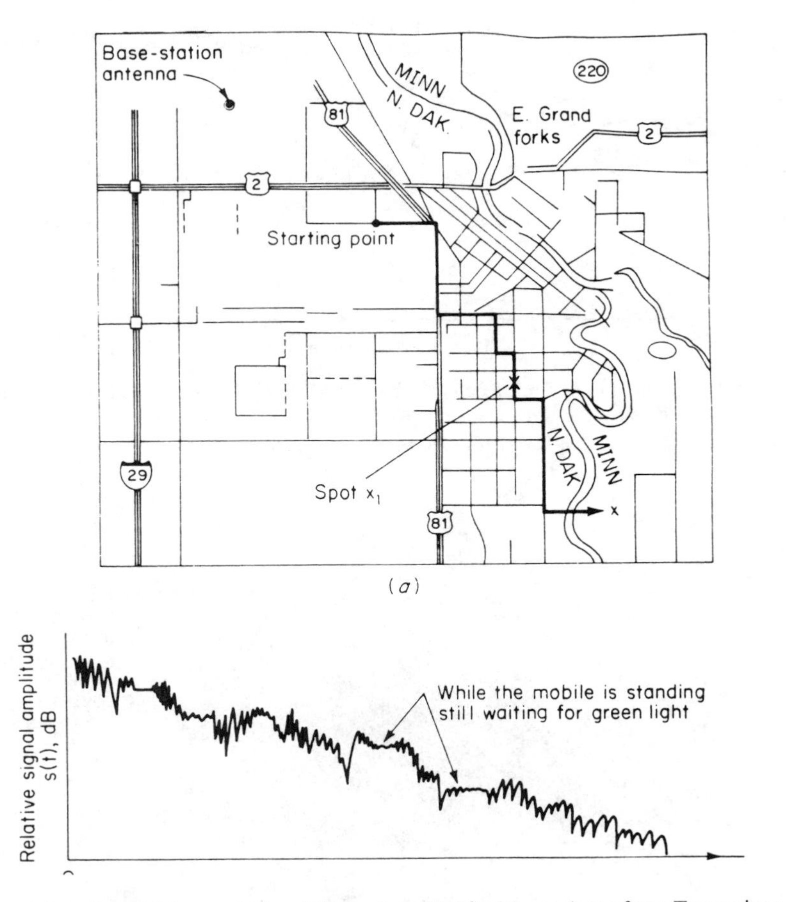

Figure 8.21 Actual Measured Signal Strength as Mobile Moves Away from Transmitter.
Source: William C.Y. Lee, *Mobile Communications Engineering,* p.17.

to make a decision as to whether a hand-off is necessary, a short-term average of the signal strength is used to determine whether the criterion level, the cell boundary, has been reached. Because of the inherently high variability in the signal levels due to the multipath phenomena, a high rate of false hand-offs may be obtained, especially where (1) the mobile unit is moving slowly or in a stop-and-go mode and (2) where it is moving parallel to a cell boundary.

The fuzziness of the cells poses a more fundamental problem for cellular operations, however. It tends to undermine the cellular architec-

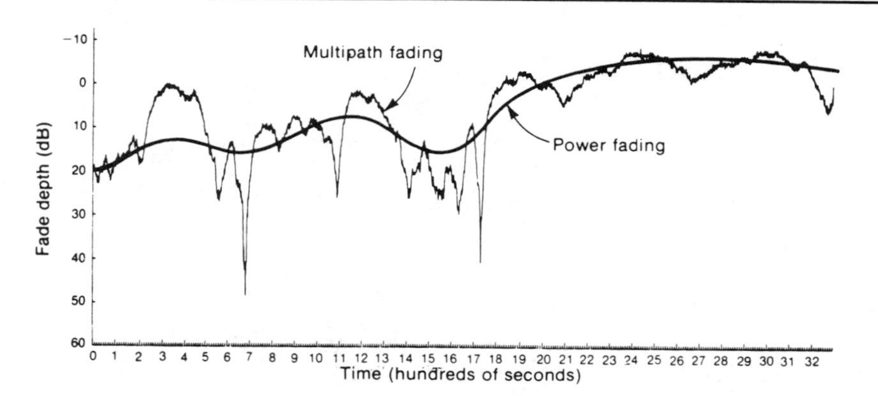

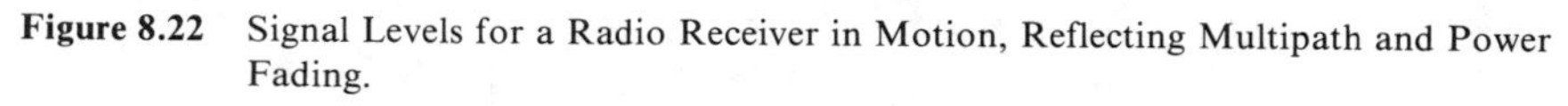

Figure 8.22 Signal Levels for a Radio Receiver in Motion, Reflecting Multipath and Power Fading.

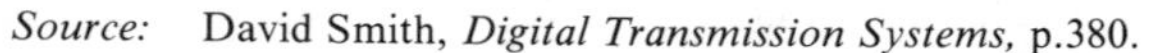

Source: David Smith, *Digital Transmission Systems,* p.380.

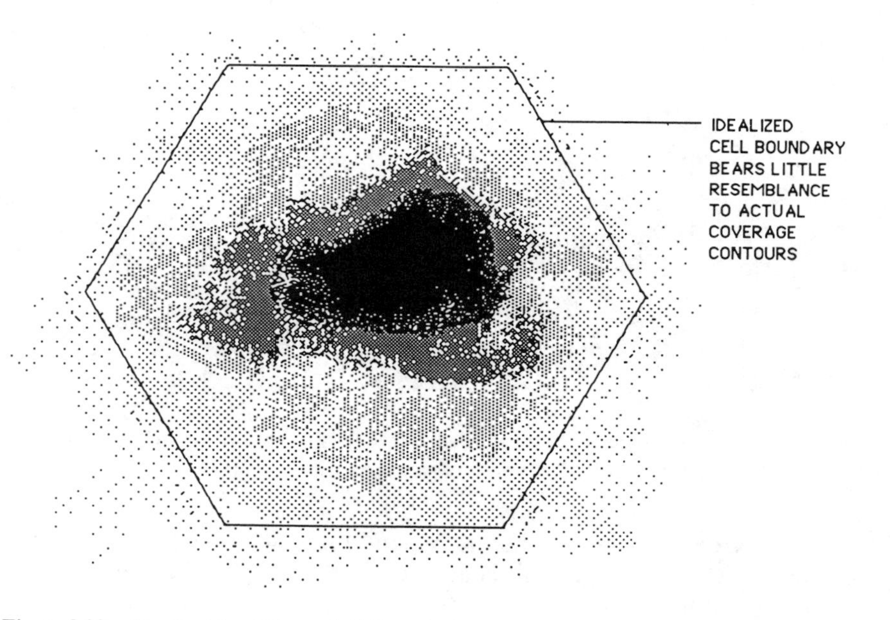

Figure 8.23 The Reality of Fuzzy Cell Boundaries.

ture at its foundation. Cellular engineers are beginning to realize that there is no such thing as "a cell." Instead, there may be one size cell for mobiles, another for portables, based on different antenna and power characteristics and different average antenna heights above the ground. (The fading characteristics are three-dimensional; the pattern of fades at four feet above street level will be very different from that at nine feet up.) According to informal industry reports, the cell boundary for trunk-mounted antennas is significantly different than the cell boundary for glass-mounted antennas. Optimizing the overall system — sometimes called balancing the cells — is tremendously complicated by these irreducible statistical uncertainties (see Chapter 9).

In short, the mobile environment is so severe that it may well defeat the cellular architecture itself, at least in some of the more ambitious extensions such as very small cells to gain truly high spectrum efficiency.

8.4 TRADITIONAL COUNTERMEASURES

The strategies for dealing with the problems of the mobile channel may be divided into two groups: those that are, so to speak, conventional — those which have been applied fairly widely in mobile systems to date, including analog cellular — and those that are based upon robust digital signal-processing techniques, which require a digital architecture and, therefore, have not been as widely applied. Among the conventional countermeasures, the most important are:

1. The use of fade margins;
2. Various types of diversity;
3. Supplementary base stations.

8.4.1 Fade Margins

The concept of a fade margin is straightforward: extra power is added to the transmission to overcome potential fading [24]. If a system needs a signal-to-noise ratio at the receiver of x dB, and the maximum likely fade is calculated to be y dB, then the way to ensure a received signal of x dB is always to transmit enough power to produce a normal received signal of $x + y$ dB. (See Figure 8.24.) The fade margin is normally equal to the maximum expected fade. Most propagation studies call for a fade margin of about 40 dB [25].

In an analog system, the fade margin is strictly a function of transmitter power. For example, large fade margins are built into all radio broadcast signals, which is why we generally do not hear the effects of multipath fading on car radios. (Such effects are, however, quite audible in fringe areas, often as a fluttering or picket-fence effect, analogous to the wavering of the TV picture when the airplane flies over.) Broadcasters are not concerned with frequency reuse *per se*. They can simply blast out their signal and let the curvature of the earth take care of isolation from the next town. In mobile radio systems, however, the use of large fade margins complicates the implementation of frequency reuse, since trans-

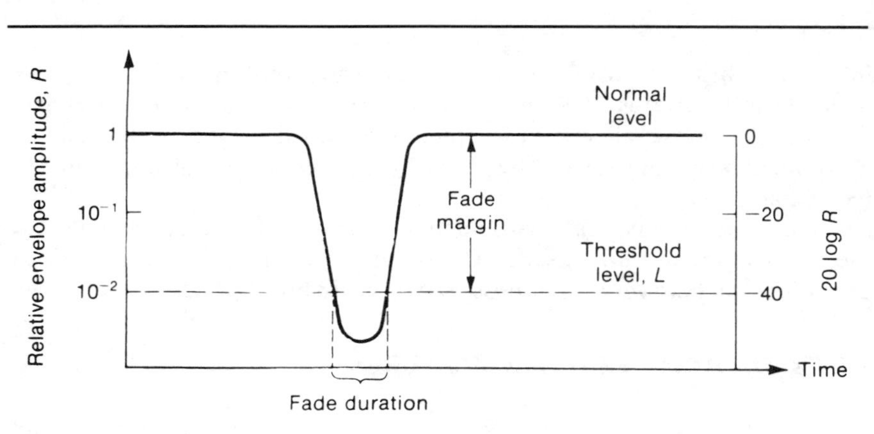

Figure 8.24 Fade Margin.

Source: David Smith, *Digital Transmission Systems,* p.382.

mitters are consistently running hotter than necessary for the average reception conditions and are transmitting farther, pushing out the distance at which the frequency is available for reuse. (Actually, reuse distance is not strictly related to transmitter power, but to the ratio between cell radii and the distance between transmitters. See Chapter 9.)

8.4.2 Diversity

Diversity refers to any of several techniques for sampling the received signal more than once and, by either combining these signals or selecting the best of them, improving the signal-to-noise ratio at the receiver. Of course, to counteract the fading problem the samples must be taken in such a way that the fading characteristics of the different samples are uncorrelated.

For example, the most common form is known as *space diversity,* which in layman's terms means simply having two or more antennas separated by a minimum of half a wavelength (several inches at 800–900 MHz) [26]. Two antennas so separated will show uncorrelated fading patterns; if one antenna is in a deep fade it is quite likely that the other antenna is not in a fade. (See Figure 8.25.) The radio circuitry can be programmed to select the antenna *branch* with the better signal. The more branches, the better the average signal. (See Figure 8.26.)

Many types of antenna-combiner systems, capable of producing space diversity, have been analyzed. Jakes *et al.* report that an eight-branch diversity system at the mobile unit — that is, eight separate antenna elements — is capable of ensuring that the received signal will be within 3

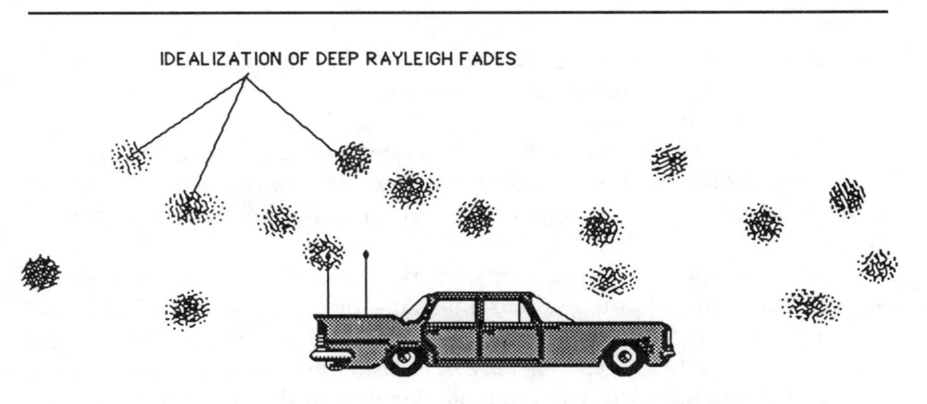

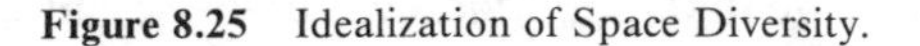

Figure 8.25 Idealization of Space Diversity.

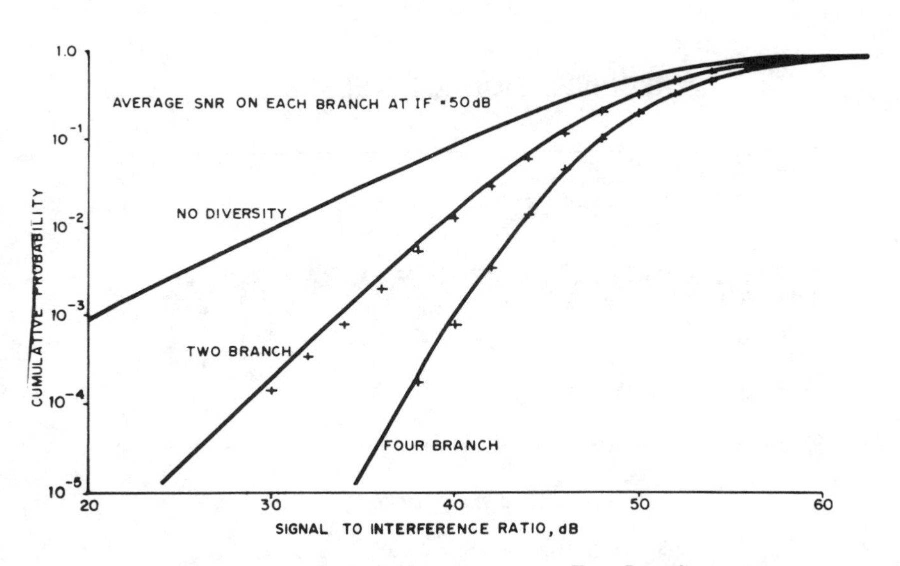

Figure 8.26 Effects of Spatial Diversity: Two-Branch and Four-Branch.
Source: William C. Jakes, *Microwave Mobile Communications,* p.363.

dB of its normal value 95% of the time [27]! This is clearly a tremendous improvement over the raw statistics of the Rayleigh channel. In other words, it is possible through multiple-antenna systems to reduce the effect of fast Rayleigh fadings below the magnitude of slow, shadow-induced, fading which multiple antennas do not help. Another important benefit of diversity may be that it significantly reduces cochannel interference by reducing FM capture of the interferer [28].

On the other hand, the cost of the equipment and the increased complexity of the circuitry necessary to deal with eight separate antenna inputs constitute additional overhead. Consumers have been unwilling to employ diversity techniques that might turn their automobiles into antenna farms.

Polarization diversity is another option, potentially an attractive one. It has been shown that the same signal transmitted with both a horizontal polarity and a vertical polarity exhibits uncorrelated fading statistics and can be used for diversity. This may be considered a special case of space diversity, since there are two antenna elements at the receiver. Cox *et al.* have shown the results of measurements taken on two orthogonally polarized signals at 800 MHz [29]. As the chart shows, if the receiver is designed to select the better of the two, the maximum fade would be about 10–15 dB, as opposed to almost 60 dB if the vertically polarized signal alone were considered. (See Figure 8.27.)

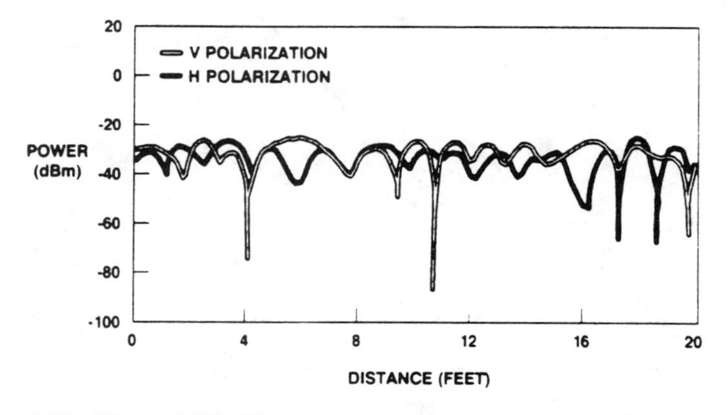

Figure 8.27 Measured 800 MHz Multipath Signal Variations.
Source: Cox, et al., *IEEE Proc.*, 1987.

Frequency diversity involves the transmission of the same circuit on two different frequencies sufficiently separated such that again their fading characteristics are uncorrelated. For the frequencies under discussion here, the separation between the two frequencies would have to be on the order of 1–2 MHz [30]. Frequency diversity is employed on wideband microwave point-to-point systems. The use of multiple frequencies to transmit the same message to a mobile user, however, would be intolerably wasteful of spectrum in a mobile-radio application. The concept of frequency diversity, however, is relevant in discussion of wideband spread-spectrum systems, some of which employ an ingenious form of frequency diversity. (See Chapter 12.)

8.4.3 Supplementary Base Stations

To fill in large holes caused by shadowing from hills or buildings, the conventional answer has been to deploy a satellite base station to radiate the same signal from a different angle. (See Figure 8.28.) The principle of what is in effect base-station diversity has been built into today's cellular architecture through the use of corner-sited antennas with directional transmission to provide alternative paths for reaching subscribers in shadowed areas within the cells [31]. (See Figure 8.29.) The advantage of this architecture is that the number of cell-sites is not greatly increased, which holds down costs. According to calculations presented by Jakes *et al.*, the use of multiple base stations, with the ability to select and hand off the mobile within the cell to the best one, should improve average signal-to-noise ratio by something on the order of 10 dB. It should also eliminate the truly deep shadows. A penalty associated with this approach, however, may be an increased likelihood of hand-off which overburdens the central system processor. Given the fuzziness of cell boundaries for an FM system in a fading environment, we now face the prospect of a fuzzy zone at the center of every cell, where the signals from competing corner-sited base stations may be very close in average strength, and stationary vehicles finding themselves in a temporary but deep fade may undergo an undesirable hand-off [32].

8.5 DIGITAL COUNTERMEASURES

In a digital system, the inherently robust format and the opportunity to apply intelligent signal processing open up new avenues for counteracting the effects of the mobile environment. The most important of these are:

1. Robust voice coding;
2. Robust modulation;
3. Adaptive equalization;
4. Error correction.

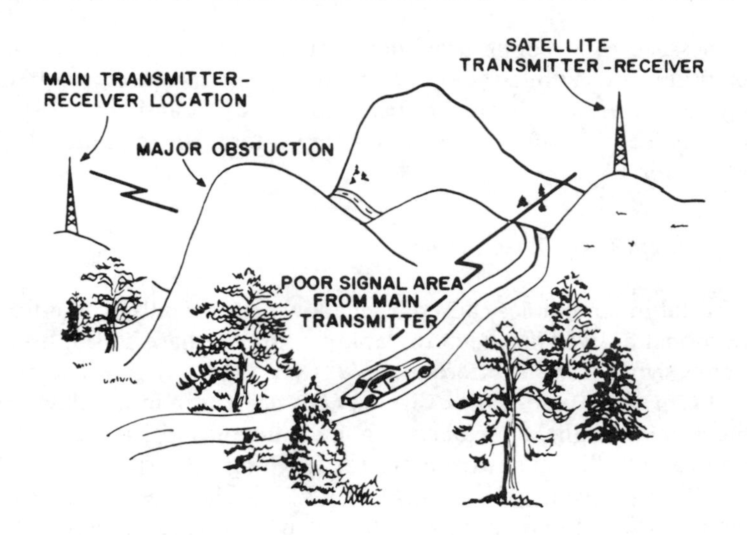

Figure 8.28 Use of Supplemental Base Stations to Fill-In Blocked Coverage.
Source: William C. Jakes, *Microwave Mobile Communications,* p.378.

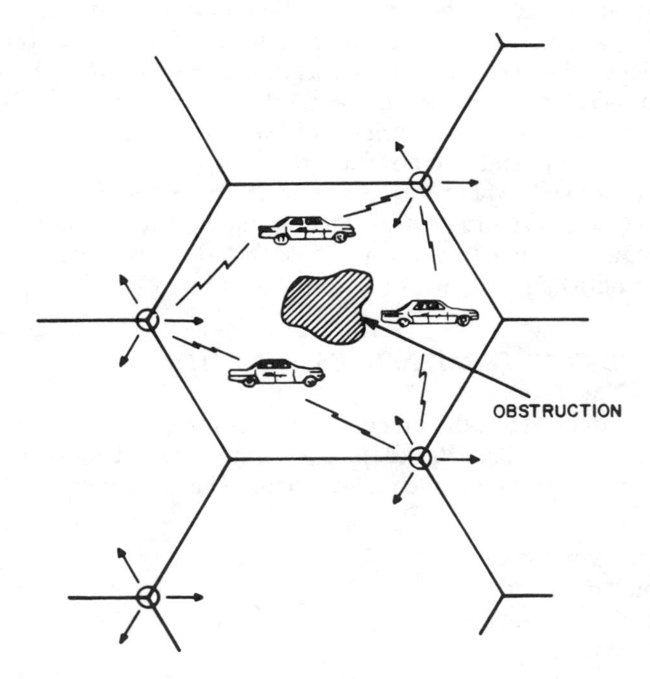

Figure 8.29 Corner-Siting of Cell-Site Transmitters.
Source: William C. Jakes, *Microwave Mobile Communications,* p.379.

The use of robust techniques is based on the principle of making the transmission more error-resistant by limiting the damage caused by errors. As noted in Chapter 6, adaptive equalization is a generic signal processing technique designed to help recover the physical level of the code. Error correction is another generic technique which focuses on recovering the logical level of the code.

8.5.1 Robust Voice Coding

Conventional PCM was designed for wireline digital systems where the system engineer can employ regenerative repeaters to assure virtually error-free transmission. PCM is actually rather vulnerable to transmission errors. Some bits are relatively more important than others. This means that some errors can have a much greater effect than others.

In the dirty mobile-radio channel, regenerative repeaters are not feasible and error-free transmission is not a practical design goal. In fact, the mobile channel often operates at error rates that would overwhelm an ordinary PCM coder. There are other coding strategies, however, which are inherently more robust. For example, delta modulation, which is not a form of modulation as we have used the term here but a voice-coding technique (see Chapter 12), uses only 1-bit words, and each bit is equal in weight to all other bits. In an error-filled environment, delta modulation should outperform standard PCM.

The digital mobile-radio designer, however, is constrained by another opposing objective: the search for lower bit rates to reduce the signal bandwidth. Today's most advanced coding strategies use complex algorithms to reduce the bit rate dramatically. These methods, however, actually increase the relative importance of certain bits and thus could heighten the vulnerability to errors. The consideration of coder robustness is intimately bound up with the question of error correction.

The search for robust voice coders is aided by the characteristics of the human speech processor itself, i.e., the ear and the brain. We are all able to communicate in very noisy acoustic environments — for example, an individual listening to a conversation at a crowded cocktail party. As researchers have developed an understanding of the mechanisms underlying speech production and interpretation, many clues have emerged for the design of coders that will simultaneously satisfy both objectives of the digital mobile-radio designer — imperviousness to errors and low bit rates. We will delve more into this area in Section 12.2.2.

8.5.2 Robust Modulation

The same general considerations apply to the second coding stage, or modulation. The vagaries of the mobile channel impact some modulation schemes much more than others. We have already noted that the violent oscillations in signal amplitude due to Rayleigh fading render amplitude modulation schemes almost inoperative. The relative size of multipath-induced frequency variations is considerably less than amplitude or phase alterations: hence the superiority of FM to AM in a mobile environment.

Digital modulation schemes — which are quite numerous, as we shall see in Section 12.2.1 — face the same winnowing. Some techniques which work well and are highly spectrum efficient in fixed radio applications are simply unable to cope with the mobile channel conditions. Again, the digital mobile-radio designer must satisfy two conflicting criteria, spectrum efficiency *versus* robustness in an error-dominated channel. For example, as noted earlier, 16-level phase modulation is capable of transmitting twice as much information in a given bandwidth as 4-level phase modulation. The 16-level demodulator, however, is required to detect phase differences of only 22.5°, while the 4-level demodulator works with 90° differences. Since the phase alterations produced by the mobile environment are the same for either case, the relative error rate for 4-level modulation is much lower than for 16-level at a given signal-to-noise ratio.

8.5.3 Adaptive Equalization

The digital demodulator is designed to detect the incoming pulses, which in radio systems are usually discrete variations in some characteristic of the carrier, and to assign them their proper value to allow subsequent processing to proceed. The detection process is fundamentally quite simple. At regular intervals the demodulator samples the received waveform and quantizes it, assigning it to one of a small number of possible values. For example, in a 4-level phase modulation scheme, each sample must be assigned to one of four values. If the receiver is properly synchronized, it should see the samples clustering at these four levels. In between samples, the value of the incoming waveform varies over the entire range. The universe of possible waveforms, such as will be seen if a random data stream is transmitted and a large number of pulses superimposed, is usually drawn in a characteristic diagram known as an *eye pattern.* Figure 8.30 shows an eye pattern for 4-level modulation. The sample is taken and the pulse measured at the middle of the eye pattern, where, in principle, the decision will always be fairly easy.

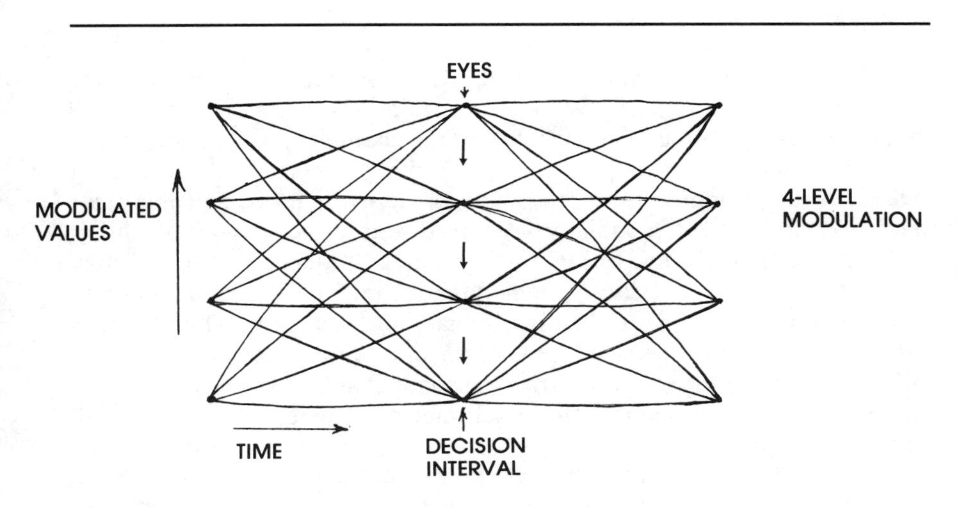

Figure 8.30 Four-Level Eye Pattern.
Source: William C.Y. Lee, *Mobile Communications Engineering,* p.381.

As we have seen, the hazards of the mobile environment often produce an increase in intersymbol interference, smearing the pulses together. (See Figure 8.31.) This tends to close the eye pattern and to make the demodulator's decision much less accurate. In the 1960s, R. W. Lucky of Bell Labs was the first to devise a signal-processing technique — known as *equalization* — capable of counteracting this problem [33]. In the following decades, the study of equalization techniques has grown into a vigorous field with a rich literature [34].

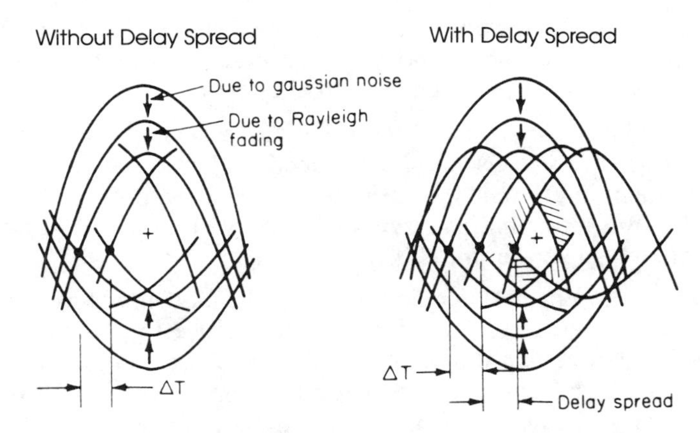

Figure 8.31 Degradation of the Demodulator's Eye Pattern Due to Factors in the Mobile Environment.
Source: William C.Y. Lee, *Mobile Communications Engineering,* p.381.

The fundamental principle of adaptive equalization is to obtain, prior to the commencement of transmission, certain information about the distortion and attenuation characteristics of the channel, and then to employ this information to correct the received signal, to reshape the incoming pulses, so as to improve the resolution of the eye pattern and the error performance of the demodulator. A typical way of accomplishing this is to begin the transmission with a brief "training" sequence: a known data sequence is transmitted, and the demodulator compares the actual received signal with what it knows to be the original input sequence. The observed error is therefore channel-induced. From these data, the system derives sufficient information to characterize the channel and uses this information to set its processing algorithms to subtract the expected error from all subsequent samples.

If the channel characteristics are expected to be fairly stable over the duration of the transmission, as they may well be for most wireline channels, the initial training may suffice. If, however, the channel characteristics are expected to continue to change during the course of the transmission — as in mobile radio — then the initial training must be supplemented by an ongoing adaptive equalization algorithm that is capable of measuring the channel characteristics and continuously tuning the equalizer parameters to maintain good pulse shaping and low error rates. The basic idea of *decision feedback equalization* (DFE) is also quite simple. The equalizer is assumed to start out with the right setting — a correct analysis of the expected error — immediately after training is completed. The first sample will be analyzed and extracted from the incoming signal. Once the first pulse has been decided and its value is known, the error between that value and what was actually received can be recomputed. This actual error is now compared with what had been the expected error. *This* difference — between expected error and observed error — can now be used to update the settings of the equalizer circuitry.

> The basic idea is that if the value of the symbols already detected are known (past decisions are assumed to be correct), then the ISI [intersymbol interference] contributed by these symbols can be canceled exactly, by subtracting past symbol values with appropriate weighting from the equalizer output [35].

The concept of using past decisions to aid in the making of future decisions is the central concept for all adaptive signal processing and can be generalized beyond equalization *per se* [36]. It is one of the overarching advantages of digital transmission in dirty environments. As the official AT&T history of engineering in the Bell System observes:

Feedback control information was obtained by comparing decided bits against the receiver's incoming, distorted analog pulse train in order to ascertain an error component. This technique of using prior decisions to determine an ideal upon which to base adaptation became known as decision direction; it subsequently became the basis not only for equalization but for the recovery of timing and other necessary receiver parameters [37].

The effects of equalization properly applied can be dramatic. Figure 8.32 is derived from a study by Bell Labs of the effects of using equalization on an 8-level signal. The chief challenge in designing equalization algorithms for mobile-radio applications appears to be the requirement for continuous adjustment to very rapid and large magnitude changes in the channel characteristics. The trend today is toward fully digital software implementations of equalization algorithms on new, fast digital signal-processing chips.

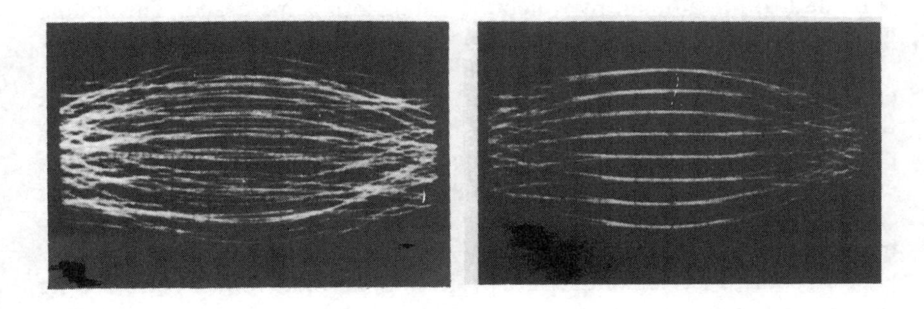

Figure 8.32 Eight-Level Eye Pattern: Before and After Equalization.
Source: AT&T Bell Labs, *A History of Engineering and Science in the Bell System: Communications Sciences,* 1925–1975, p.423.

8.5.4 Error Correction

Once the incoming pulses have been sampled and decided to the best of the demodulator's and the equalizer's capabilities, there is still the possibility that errors in detection may have occurred. In the mobile channel, it is likely that even post-equalization bit-error rates of 10^{-2} to 10^{-3} may be experienced in some link conditions. At this stage, it is possible to cleanse the received bit stream further by means of *forward error correction* (FEC) techniques.

FEC is a *very* complex field, and it is neither possible nor necessary here to present even a conceptual survey of different approaches [38]. All FEC techniques are based on the principle that by encoding the bit sequences prior to transmission in such a way as to introduce certain redundant information, the presence of errors can be detected by the receiver, which knows the coding principles, and in many cases actually corrected at the receiver by making use of this redundant information. In this sense, the simplest form of FEC is gross multiple transmission of the same information more than once. This is done in some of today's cellular signaling, for example. The enormous ingenuity of the field of error coding, however, has been expressed in the ideal of finding more efficient techniques that will allow the maximum error correction with the minimum additional information.

The desire to minimize the additional information is very appropriate in the bandwidth-limited radio environment. All FEC techniques increase the bandwidth of the transmission. Indeed, the hypothetical coding gain is related to the amount by which the bandwidth is increased. The coding gain is the amount by which the signal-to-noise ratio may be *reduced,* or the amount of noise increased, while still maintaining a desired BER criterion, assuming FEC is employed. For example, Bhargava calculates that a coding gain of 11.2 dB in SNR is "theoretically possible" for binary phase modulation (BPSK) [39]. While such a gain cannot be realized with current algorithms, Bhargava nevertheless feels that "it is safe to say that coding systems (delivering 2–6 dB) will be used routinely in digital communication links [40]."

In short, FEC techniques are promising, but, like robust voice-coding and modulation techniques, their use in mobile-radio systems is constrained by (1) the computational complexity of implementing advanced algorithms, which is a function of the contemporary hardware, and (2) the bandwidth limitations we have referred to throughout.

REFERENCES

[1] John C. Bellamy, *Digital Telephony,* New York: Wiley, 1982, p. 71.

[2] Bell Laboratories, *Engineering and Operations in the Bell System,* AT&T Bell Laboratories, 1977, pp. 604 ff.

[3] William C. Jakes, ed., *Microwave Mobile Communications,* New York: Wiley, 1974, pp. 11, 189.

[4] *Ibid.,* p. 161.

[5] *Ibid.,* p. 201.

[6] *Ibid.*

[7] *Ibid.*, pp. 112 ff.; also David R. Smith, *Digital Transmission Systems*, New York: Van Nostrand Reinhold, 1985, pp. 361 ff.
[8] Jakes, *op. cit.,* p. 65.
[9] Jules LeBel, "Mobile Radio Signal Statistics in Non-Urban Environments," *Proceedings of the 37th IEEE Vehicular Technology Conference,* Tampa, June 1–3, 1987, p. 135.
[10] Jakes, *op. cit.,* Chapter 2.
[11] *Ibid.*, pp. 105–106.
[12] *Ibid.*, p. 107.
[13] *Ibid.*, p. 110.
[14] *Ibid.*, p. 298.
[15] *Ibid.*, p. 110.
[16] Donald C. Cox, Hamilton W. Arnold, and Philip T. Porter, "Universal Digital Portable Communications: A System Perspective," *IEEE Journal on Selected Areas in Communications,* Vol. SAC-5, No. 5, June 1987, pp. 764–773.
[17] *Ibid.*
[18] Jakes, *op. cit.,* p. 46.
[19] *Ibid.*, p. 162.
[20] *Ibid.*, p. 383.
[21] *Ibid.*, pp. 215–217.
[22] Bellamy, *op. cit.,* pp. 193, 489–493, 295.
[23] William C. Y. Lee, *Mobile Communications Engineering,* New York: McGraw-Hill, 1982, p. 17.
[24] See Smith, *op. cit.,* pp. 381 ff.
[25] Stan Roelefs, "Fade Margin Requirements for Microwave Systems," *Mobile Radio Technology,* December 1986, pp. 52–61.
[26] Lee, *Mobile Communications Engineering, op. cit.,* p. 276.
[27] Jakes, *op. cit.,* p. 379.
[28] *Ibid.*, p. 362.
[29] Cox *et al., op. cit.*
[30] Jakes, *op. cit.,* p. 312.
[31] *Ibid.*, pp. 377–386.
[32] Richard C. Bernhardt, "Macroscopic Diversity in Frequency Reuse Radio Systems," *IEEE Journal on Selected Areas in Communications,* Vol. SAC-5, No. 5, June 1987, pp. 862–870.
[33] S. Millman, ed., *A History of Engineering and Science in the Bell System: Communications Sciences (1925–1980),* AT&T Bell Laboratories, 1984, pp. 422–424.
[34] Shahid Qureshi, "Adaptive Equalization," *IEEE Communications Magazine,* March 1982, pp. 9–16; Thomas J. Aprille, "Filtering and

Equalization for Digital Transmission," *IEEE Communications Magazine,* March 1983, pp. 17–24; T. A. C. M. Claasen and W. F. G. Mecklenbrauker, "Adaptive Techniques for Signal Processing in Communications," *IEEE Communications Magazine,* November 1985, pp. 8–19; Curtis A. Siller, "Multipath Propagation," *IEEE Communications Magazine,* February 1984, pp. 6–15.

[35] Qureshi, *op. cit.,* p. 13.

[36] Claasen and Mecklenbrauker, *op. cit.*

[37] Millman, *op. cit.,* p. 424.

[38] Vijay K. Bhargava, "Forward Error Correction Schemes for Digital Communications," *IEEE Communications Magazine,* January 1983, pp. 11–19.

[39] *Ibid.,* p. 12.

[40] *Ibid.,* p. 13.

Chapter 9
DESIGNING FOR FREQUENCY REUSE

Frequency reuse is the second most important design challenge to the mobile-telephone systems architect. As described in Chapter 3, the real payoff envisioned from the cellular architecture was the escape from spectrum limits through frequency reuse. Reuse was the engine of the cellular "perpetual spectrum machine"; through reuse and cell-splitting, the cellular architecture promised to create virtually unlimited capacity from relatively modest spectrum allocations.

In systems with reuse, however, the radio engineer must confront the problem of interference. Until cellular telephony, no radio system had ever attempted to design for frequency reuse from a systems perspective. In the old IMTS world, interference was simply undesirable, like noise. The goal was to minimize it, or to maximize the signal-to-interference ratio. The basic remedy was geographical separation. Operators using the same frequencies were kept far enough apart to stay out of each other's way. For example, the standard separation for IMTS transmitters was 70 miles — well beyond the curvature of the earth in most terrains — although the normal practical communication distance for such systems was probably no more than half of that distance.

With the advent of cellular architecture, systems engineers had to reorient their thinking about interference. The goal, as we shall explore below, was no longer simply to maximize the signal-to-interference ratio; in fact, people soon began to realize that, in a sense, the real goal of the cellular-system engineer was to *minimize* the signal-to-interference ratio. Put another way, the cellular system should be capable of operating in an extremely high interference environment.

It turns out that designing a system capable of reusing the same radio frequency many times within a small geographical area, such as a city, poses a set of problems that early cellular architects did not entirely foresee. Cell-splitting has not worked well; there appear to be hard limits, imposed by interference, on just how small the cells can be engineered. Difficult interference effects, compounded by the vagaries of the mobile environment dicussed in Chapter 8, have capped cellular capacity well below initial targets.

One of the most attractive aspects of digital techniques is the likelihood that digital systems may function better in high-interference environments, may be able to tolerate a higher level of interference than today's analog systems. Indeed, some observers believe that this is the *premier* advantage of digital systems. Better interference performance could allow the next generation of cellular systems to surpass current capacity limits significantly, even without, or in addition to, any improvements in bandwidth efficiency. (See Section 15.2.)

9.1 THE ENGINEERING IMPLICATIONS OF FREQUENCY REUSE

In an IMTS system operating at microwave frequencies, the system boundary is ultimately provided by the curvature of the earth. (See Figure 9.1.) Within the radio horizon, the IMTS operator attempts to achieve as high a signal-to-noise or signal-to-interference ratio as possible. The IMTS-type system, like a broadcast radio station, is designed to fill the entire coverage area, every nook and null, with abundant signal. From a system-engineering standpoint, the optimization is simple: the high-power transmitter provides a large *fade margin,* to ensure that the required signal levels will be achieved at the receiver, even in the deepest of Rayleigh fades.

On paper, the concept of frequency reuse seems also quite simple: instead of drawing large circles, we contract the compass and draw lots of smaller circles. On the flat surface of the map, the logic seems impeccable. Instead of wasting spectrum by occupying huge coverage areas with high-power transmitters, lower-power transmitters are used to bring down the coverage area to much smaller cells. As we discussed in Chapter 3, the pure mathematics of frequency reuse are staggering: potentially hundreds of times the capacity of single-cell IMTS coverage, without additional spectrum.

It took longer to recognize that frequency reuse — defined here as the transmission of multiple signals on the same frequency *within* the radio horizon — drastically altered the radio-engineering problem. Now, instead of a simple optimization — the transmitter against nature — the radio engineer had to *balance* many transmitters against each other. Obviously, the old solution of cranking up the power no longer made much sense. The objective was to maximize the total number of radio-telephone circuits that could be utiiized.

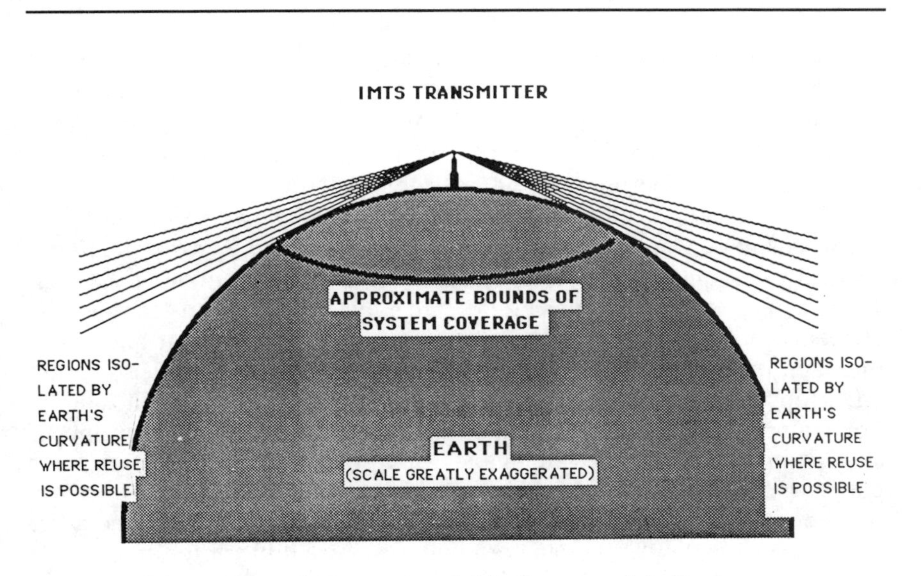

Figure 9.1 IMTS-Type Mobile System Bounded by Curvature of the Earth.

The IMTS transmitter is like an orator speaking in a huge theater: he has to boom to be heard at the farthest limits of the hall. In this simile, the walls of the theater are like the limits imposed by the curvature of the earth. Next door, in another theater, another orator booms out; both are well shielded from one another almost no matter how loudly either declaims. With cellular, on the other hand, the simile must be modified: the orator now finds himself at a crowded cocktail party, surrounded by many others trying to shout at the same time. Lung power *per se* becomes at best irrelevant [1] and at worst counterproductive from the system point of view. In fact, the cocktail party analogy is not a bad one: in such a situation, how well we hear someone who is talking to us depends not on how loud they are speaking but on (1) how close they are and on (2) how many other people are talking nearby. It shifts from a transmission to a reception problem. In a quiet corner we can follow a conversation quite easily; in the middle of the room conversation may become impossible even with shouting. In fact, imagine trying to engineer the communication at such a cocktail party mathematically so that we could pack the most people into the smallest space and they could still communicate. Also, some speakers have very faint voices, corresponding to the portable low-power cellular units, while others have loud voices, like the mobile units. That offers some idea of the difficulties facing cellular-system engineers.

9.2 INTERFERENCE-LIMITED SYSTEMS

In the IMTS architecture, the system is said to be *noise-limited.* That is, like the orator in the theater, the main problem to be overcome is the natural or man-made noise of the environment.

By contrast, a cellular system is *interference-limited* [2]. Each transmitter has to cope not only with the characteristics of the environment but also with the signals being simultaneously produced by a number of transmitters. The effects of interference are typically much larger than the effects of noise [3]. Moreover, interference generated by transmitters of the same sort may, depending upon the modulation, exhibit a capture effect on the receiver that exacerbates the impact on the circuit quality. For example, in many analog modulation systems, interference may appear as a competing, partially intelligible voice signal, or, in television, as a partly visible "ghost" of another transmission. Intelligible interference can be likened to cross talk on the wireline network. Or, to lean once more on the previous simile, we can note that it is more difficult to listen to one conversation if a second person is talking simultaneously than if the same amount of acoustical energy is present simply as random noise. The brain itself has a "capture effect" for speech over noise, and in certain situations this works against efficient communication.

Actually, there are several kinds of interference with which the cellular system must cope. *Adjacent channel interference* occurs when signal energy from one channel spills over into an adjacent channel or when the filter on the receiver is too "loose" and captures energy from a broader band than it really needs to [4]. (See Figure 9.2.) In principle, it is possible to control adjacent channel interference completely through filtering of both the transmitter and the receiver to ensure, in the first case, that no energy is transmitted outside the desired band and, in the second case, that the receiver filters out any unwanted out-of-band energy. (See Figure 9.3.)

Adjacent channel interference becomes a system-design issue because of its economic impact. Generally speaking, the tighter the filter requirements, the more expensive the implementation. The FCC generally enforces relatively tight transmission characteristics, known as the *emissions mask.* (See Figure 9.4.) Receiver filtering, however, is often regarded as an economic penalty to be avoided through spectrum-management techniques that introduce certain inefficiencies in order to keep costs down. For example, broadcast television never uses adjacent channels in the same city. Enormous amounts of TV spectrum are destined to sit idle because a general policy decision was made not to impose the cost burden of tight filtering upon every television receiver. In today's cellular systems the same

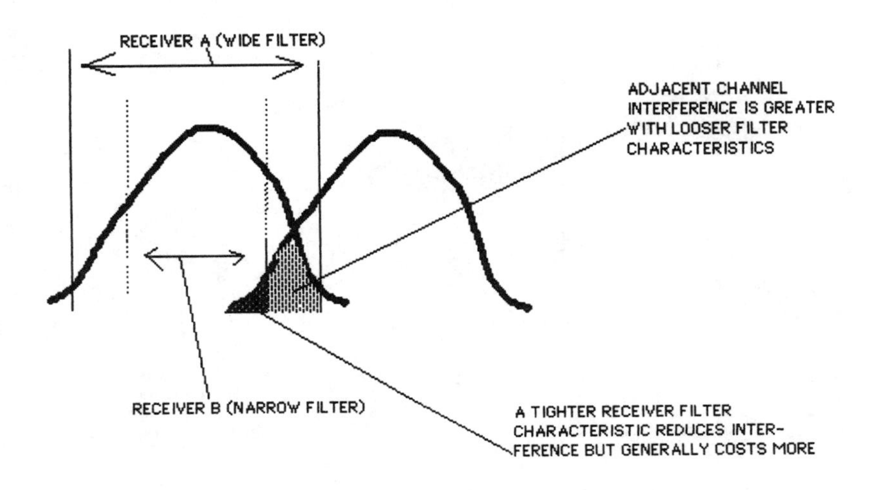

Figure 9.2 Adjacent Channel Interference and Receiver Filter Characteristics.

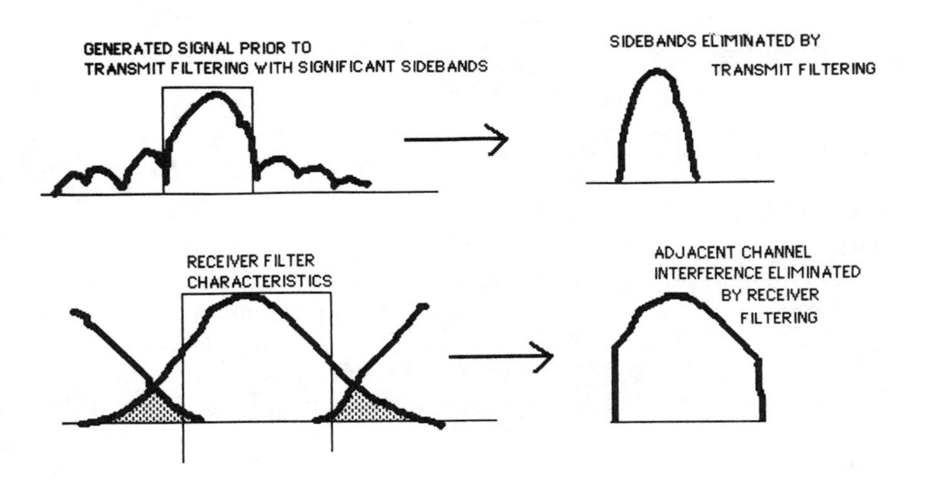

Figure 9.3 Examples of Transmit and Receiver Filtering.

is true: adjacent channels cannot be used within a given cell because of very loose receiver filtering. This loose filtering was permitted under cellular standards, as opposed to IMTS, to reduce mobile-unit costs. In the next generation of cellular systems, improved filtering techniques, including digital filtering, on both transmitters and receivers may well allow for

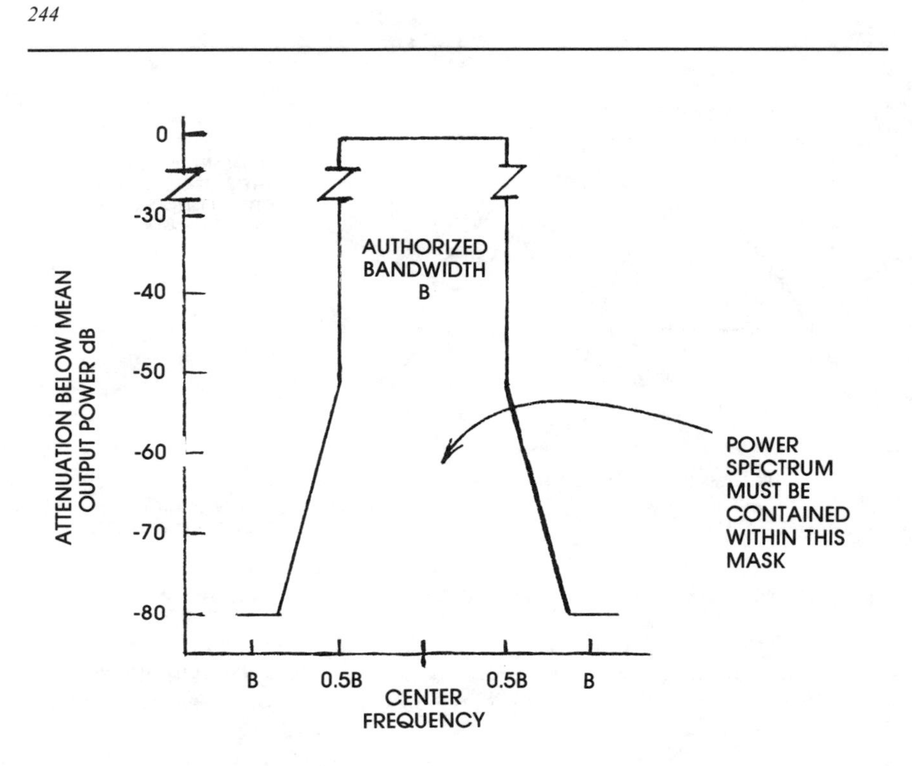

Figure 9.4 Example of an Emissions Mask.

better control of adjacent channel interference without such a severe cost penalty.

The dominant form of interference, however, is *cochannel interference,* which, since it occurs in the same frequency band as the desired signal, cannot be removed by filters [5]. Cochannel interference becomes, in fact, the single most important constraint upon system performance in the reuse situation [6]. The carrier-to-cochannel interference ratio, often simply abbreviated as the *C/I ratio,* is the fundamental parameter in calculations of reuse factors. If a C/I standard of, say, 24 dB is established, based on voice-quality objectives, this will determine the degree of reuse and of spectrum efficiency that can be achieved in a cellular architecture (see Section 9.5).

9.3 COCHANNEL INTERFERENCE AND THE MOBILE ENVIRONMENT

When the complexity of the mobile environment is added to the complexity of the cochannel-interference problem, the prediction of radio performance and coverage characteristics becomes so difficult that only by extreme simplification can we even begin to discuss the matter. Many studies of cellular reuse architecture simply ignore the mobile environment altogether. For example, the following is taken from the conclusion of a recent paper on alternative cell configurations for digital cellular:

> The analysis above is based on . . . very simple and idealized channel and interference models . . . The analysis above was carried out under the idealized assumptions of flat fading. Uniform transmission conditions were assumed for all cells. No delay spread was considered. Perfect timing and synchronization was assumed with coherent detection and ideal maximal-ratio combining. It was furthermore assumed that perfect synchronization for the time-division retransmission scheme was established. The analysis was confined to local mean values of signal and interference at isolated points. Consequently, the effects of shadow fading were not taken into account, and no results were obtained for overall signal-to-interference statistics throughout the entire cellular areas. The effect of channel occupancy (the fraction of time that a channel is in use) on interference was not considered. All of the above problems and others have to be taken into account in a refined system analysis [7].

Reading this caveat, one must ask: Can a "refined system analysis" ever be accomplished, taking all of these factors into account? How many supercomputers would it take? The next, even more ominous, question: Are the simplified analyses really worth anything? Given that cellular reuse equations are being upset in the field by "small factors" like improper antenna installations, how much can we expect from *any* model that is simple enough to be constructable? The author of the article cited above is distinguished by his unusual candor in admitting the limitations of the analysis. Most cellular reuse studies do not so carefully enumerate the "simplifying assumptions" that have made their analyses possible.

Recall from Chapter 8 how the mobile environment affects the concept of cell boundaries and coverage areas. Instead of a stable, clean "signal contour," we find that the mobile environment is a Swiss-cheese world of countless fades, within which signal levels may be diminished enormously — fades of factors up to 10,000 or greater are quite frequent. For any given criterion signal level, the actual coverage area defined by that criterion would look like a ragged blotch surrounded by, in fact dissolving into, a galaxy of large and small islands which grow less frequent only gradually as distance increases (see Figure 9.5). Moreover, the precise pattern of this galaxy is constantly shimmering and changing. Finally, the patterns of fades from two different transmitters, e.g., two cell-sites, are uncorrelated.

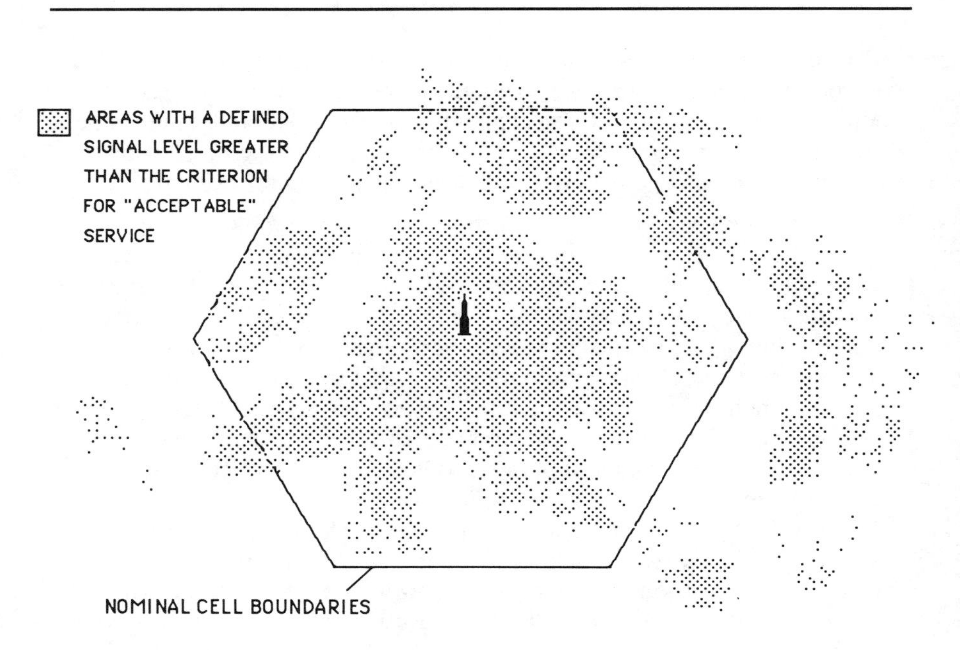

Figure 9.5 The Fuzzy Cell.

Thus if we imagine two cells side by side, transmitting on (reusing) the same frequency, the boundary would look at any instant in time something like Figure 9.6 — two "galaxies" interpenetrating. A mobile unit traveling near the boundary of the two cells would find itself in an impossible situation — the short-term fluctuations of each signal would be far larger than the difference in average signal levels (see Figure 9.7). The receiver would lock onto first one transmission, then the other, hopping back and forth at a rapid rate. Useful communication would be overwhelmed by the capture effect of cochannel interference.

This is why it is not possible to reuse frequencies in adjacent cells: the boundary conditions would be characterized by nearly balanced signal levels from competing transmitters and no communication would be possible. (See Figure 9.8.) This interference zone is very wide because of the statistical and constantly changing nature of the mobile environment.

The cell boundary must be defined by the system such that it does not fall in the balanced signal zone (see Figure 9.9). This is usually conceptualized as geographical separation — removing the competing transmitter to a greater distance. It may also be viewed as the shrinking of the effective coverage area. In either case, the goal of the system engineer is to ensure that any mobile unit operating within the bounds of its cell will

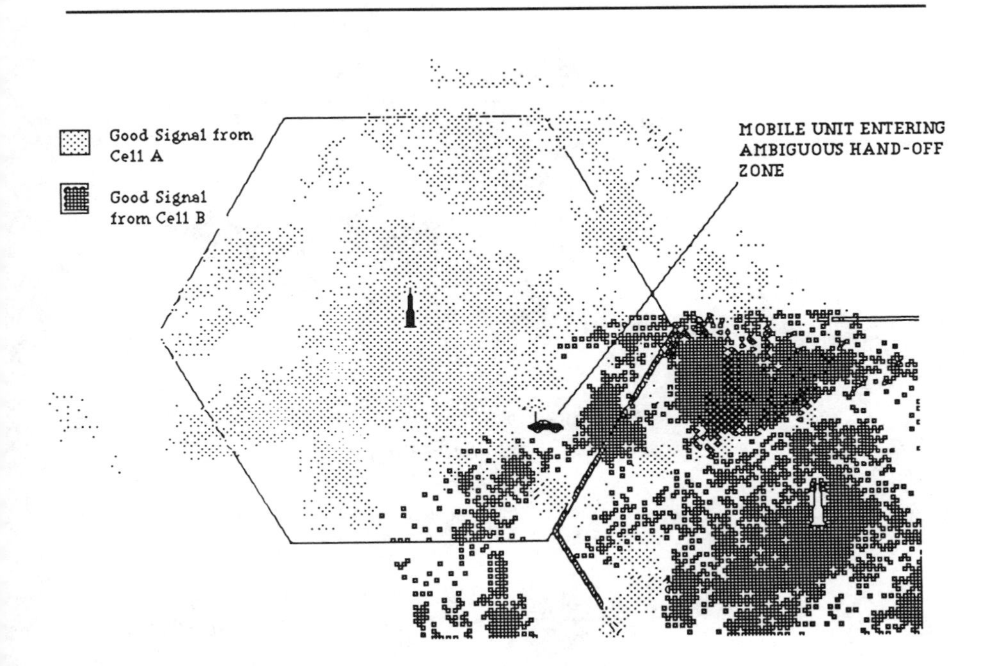

Figure 9.6 Two Adjacent Fuzzy Cells.

"always" receive the desired signal at a level higher than any potential cochannel interfering signal by a specified amount. ("Always" is defined statistically, of course.) *In effect, it is the isometric contour of cochannel interference, not of signal level, that defines the cell boundary.*

The amount of separation between cells using the same frequency is thus driven by:

1. The C/I ratio that is required to achieve the desired transmission quality, plus
2. The fade margin that is necessary to take care of statistical fluctuations in desired signal level induced by the mobile environment.

Both components of the reuse equation are important, and both may be influenced to some degree by the system architect's choice of technologies. The C/I ratio is set, in part, by the type of modulation and transmission techniques employed (see Section 9.5). The fade margin is determined by the maximum fades likely to be experienced, given the particular set of countermeasures being employed by the receiver. For example, if no countermeasures are employed, e.g., no antenna diversity, the maximum frequent fade may be 40 dB — which sets the fade margin

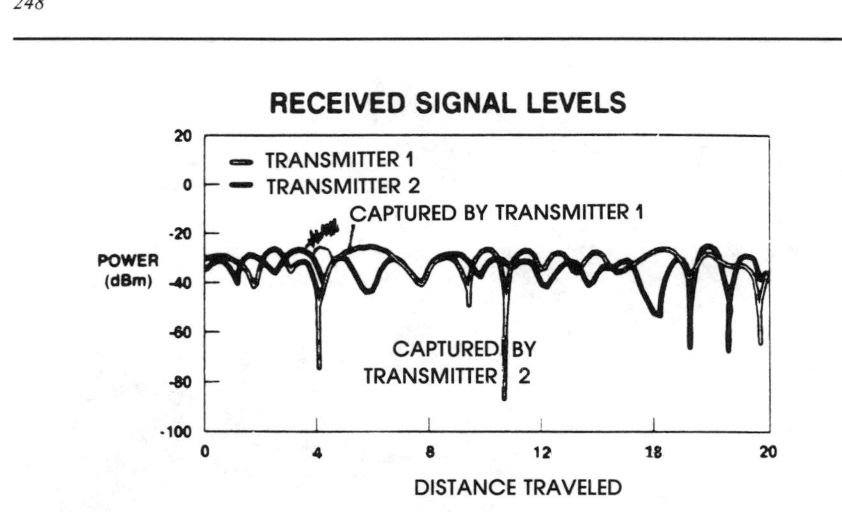

Figure 9.7 Signals Levels Near the Crossover Point between Two Co-Channel Transmitters.

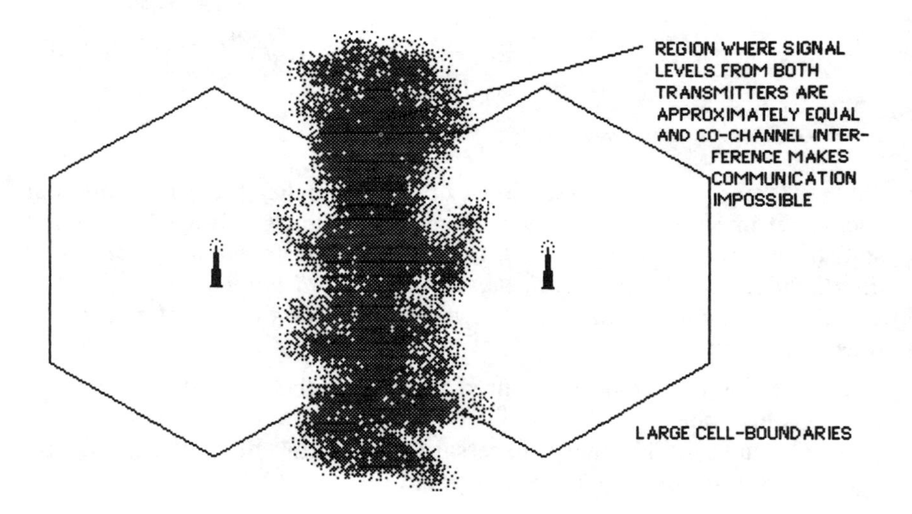

Figure 9.8 Why Frequency Reuse Is Not Possible in Adjacent Cells.

at no less than 40 dB, since the desired signal may be in a deep fade while the undesired signal is not. On the other hand, as noted in Section 8.4.2, certain antenna diversity configurations have been shown, in principle, to be capable of reducing the maximum Rayleigh fade to as little as 3–4 dB. At this level, shadowing, or terrain-induced slow fading, becomes the dominant effect of the mobile environment [8]. The variation due to slow fading plus residual Rayleigh fading under such circumstances appears to

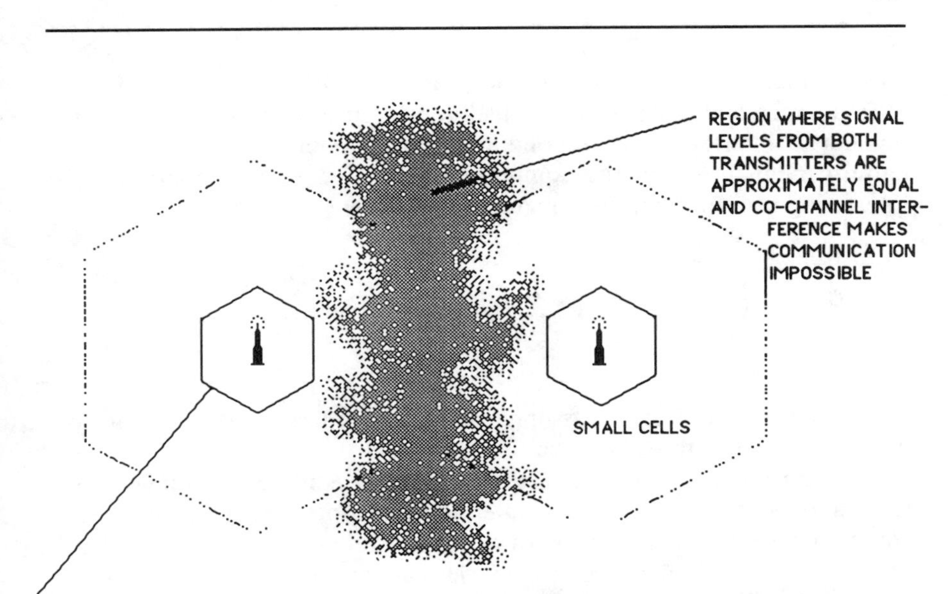

Figure 9.9 Redefinition of Serving Areas to Achieve Nonadjacent Cells.

range up to a factor of 100 (20 dB) or so — a substantial improvement over the factor of 10,000 in uncompensated Rayleigh fading [9]. Based on these limited experimental observations, it seems likely that widespread implementation of advanced diversity techniques could substantially improve the cochannel-interference engineering problem by removing the most highly variable component of the mobile environment. Cellular operators, however, have been reluctant to enforce expensive diversity solutions, "antenna farms," upon individual mobile users. This is one of the sources of interest in digital modulation techniques that appear to be more robust in the presence of cochannel interference and would not require intrusive and costly measures at the mobile receiver.

9.4 THE RELATIONSHIP BETWEEN GEOGRAPHICAL SEPARATION AND THE REUSE FACTOR

As should be intuitively apparent, it is not absolute separation between cell-sites using the same frequencies which is important, but the separation relative to the cell radius defined by the system engineer. This is normally defined as a ratio between the *distance* between cochannel transmitters and the cell *radius* (which circumscribes the "usable" coverage area of the cell). Lee calls this the D/R ratio [10]. According to Lee, D/R is dependent upon the cochannel interference criterion (C/I) and the number of interferers, corresponding to the number of cells in the reuse pattern minus one. For the nominal hexagonal case where there are six interfering transmitters, the equation for D/R is:

$$D = R\sqrt[4]{6\left(\frac{S}{I}\right)}$$

where S = signal and I = interference.

The fourth root reflects an assumption of a path-loss factor equal to the inverse of the fourth power (see Section 8.2.1).

Using a C/I criterion of 18 dB — which is equal to a desired signal level approximately 63 times greater than the interfering signal — this equation results in a D/R ratio of approximately 4.4. In other words, if the cell radius is 8 miles, the frequencies can be reused by another transmitter located approximately 35 miles away. (See Figure 9.10.) The intervening cells must use other frequencies. The ratio holds true, according to Lee, for any defined cell radius, given the C/I criterion. If the usable cell radius is reduced to 2 miles, the reuse distance is also reduced to about 9 miles. At a 0.5-mile radius, the frequencies may be reused by a transmitter only 2.2 miles away.

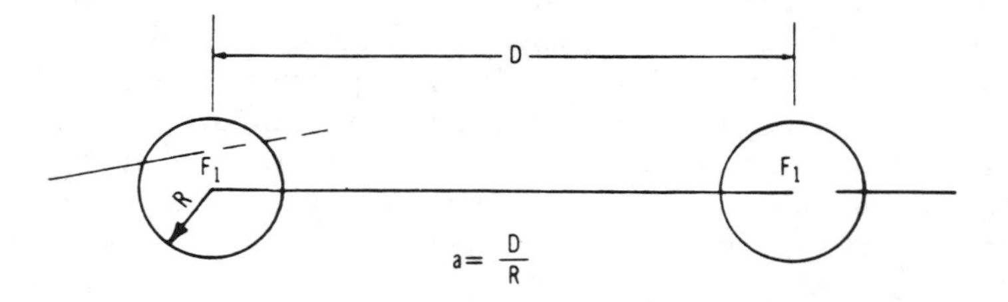

Figure 9.10 Illustration of D/R Ratio for Reuse Calculations.

Source: William C.Y. Lee, *Mobile Communications Design Fundamentals,* p.185.

Recent anecdotal information from some cellular operators, however, seems to indicate that the ratio may not hold for very small cells [11]. As very small cell radii are created, the interference effects appear to increase significantly. We may speculate that since Lee's equation effectively incorporates a statistical idealization in the C/I factor (e.g., it is a threshold which will be exceeded 99% of the time) which actually reflects environmental variations that tend to remain constant regardless of cell size, at smaller cell radii and closer transmitter separations the environmental variations described in the previous chapter will appear relatively more significant. It should also be noted that the precise interference calculations are considerably more complex than this simple equation. The interested reader is referred to Lee [12] for fuller treatment.

The reuse separation distance is important because it determines the general reuse pattern within the cellular network. For example, if the C/I criterion is set at 24 dB, Lee's equation will generate a reuse separation distance of 6.23 cell radii. Under such a criterion, the frequencies in an 8-mile radius cell would not be available for reuse at less than about 50 miles. Thus the C/I criterion determines how densely the frequencies may be packed, or how much a given spectrum band may be exploited within a geographical area. The effects can be profound. For example, the effect of raising the C/I criterion from 18 dB to 24 dB in the examples above is to reduce the number of circuits per megahertz per square mile by about 50%.

9.5 THE EFFECT OF REDUCED C/I REQUIREMENTS FOR DIGITAL SYSTEMS

The uncertain promise of Lee's equation, which, although it may not be completely general, does reflect the thinking of system engineers on the vexing issue of cochannel inteference, is that if a modulation technique can be employed which requires a lower cochannel interference criterion (C/I) than today's FM systems, cochannel transmitters can be packed more tightly in a cellular network, and the reuse factor can be improved.

This is widely viewed as a major potential benefit of shifting to digital transmission. Because of the threshold effect of digital communication — the fact that the digital demodulator is required to detect only the presence or absence of discrete pulses which may take one of only a small number of possible values — digital transmission tends to be more robust in the presence of interference than analog transmission, or at least this is the general belief among radio engineers. (The caveat reflects the fact that experience with cochannel interference on a large scale in digital radio systems is still quite limited; it may be that, as with analog cellular, certain systemic interference effects — for good or ill — will arise in dense system implementations that cannot be fully foreseen today.) At the very least, however, the perceived degradation resulting from interference in a digital system should be less severe for a given C/I ratio, because the interference will never be perceived as intelligible. In an analog system, a competing (interfering) signal will often intrude as an intelligible or partially intelligible voice signal. In a digital system, interference, like noise from the environment, is simply ground down into bit errors. The BER may be higher or lower, as a result of one or another effect of man or the environment; the result is the same. The analogy can be drawn with cross talk in the analog wireline network. Analog cross talk is often intelligible and is very disturbing, even at low energy levels; in speech pauses, for example. In a digital system, by comparison, "even if the crosstalk is of sufficient amplitude to cause detection errors, the effects appear as random noise and, as such, are unintelligible [13]."

The potential improvement in cochannel-interference tolerance may be substantial. In analog FM systems, the minimum C/I ratio is usually stated as 17–18 dB, and sometimes considerably higher [14]. In digital systems, the ability to withstand higher levels of interference results in published C/I objectives of as low as 9–13 dB [15]. Applying Lee's equation, a 13-dB C/I would result in a reuse separation ratio (D/R) of 3.3 — an almost 80% improvement in telephone circuits per square mile. A C/I of 9 dB would give a D/R of 2.6 — nearly three times as many circuits per square mile compared to analog FM. Such low C/I ratios — or, put another way, the ability of digital systems to tolerate such high levels of interference — would mean that frequencies could quite likely be reused in *every other cell.* A three-cell reuse pattern could probably be used with the same circuit quality as today's seven-cell patterns.

There are a great many published papers discussing the quantitative relationship between C/I criteria and reuse factors. The careful reader is advised to beware, however, of overly complex and theoretical geometry; the field experience with reuse engineering is dismal. Cochannel interference is not well understood, theoretically or empirically, especially in large systems with many interferers and exotic modulation techniques. As one observer comments, "a well-founded radio network planning methodology does not yet exist [16]."

Nevertheless, the ability of digital systems to tolerate higher levels of cochannel interference could go a long way toward solving one of the major problems with today's cellular architectures. The ability of any proposed transmission standard to handle cochannel interference thus becomes second in importance only to the issue of the mobile environment itself, and in many ways, as we have seen, the two issues are vitally interrelated.

REFERENCES

[1] William C. Y. Lee, *Mobile Communications Design Fundamentals,* Indianapolis: Howard W. Sams, 1986, p. 143.

[2] *Ibid.,* p. 141.

[3] *Ibid.,* p. 142.

[4] *Ibid.,* pp. 143 ff.

[5] *Ibid.,* p. 184; see also William C. Y. Lee, "Comparing FM and SSB in Cellular Systems," *Communications,* November 1985, p. 99.

[6] William C. Y. Lee, "How to Evaluate Digital Cellular Systems," paper presented at the Federal Communications Commission, Washington, D.C., September 2, 1987.

[7] C. E. Sundberg, "Alternative Cell Configurations for Digital Mobile Radio Systems," *Bell System Technical Journal,* Vol. 62, No. 7, September 1983, pp. 2057–2058.

[8] William C. Jakes, ed., *Microwave Mobile Communications,* New York: Wiley, 1974, p. 379.

[9] *Loc. cit.*

[10] Lee, *Mobile Communications Design Fundamentals, op. cit.,* p. 184.

[11] Remarks of William C. Y. Lee at EIA Industry Meeting on Digital Cellular Standards, Washington, D.C., December 3, 1987.

[12] Lee, *Mobile Communications Design Fundamentals, op. cit.*

[13] John C. Bellamy, *Digital Telephony,* New York: Wiley, 1982, p. 72.

[14] Lee, *Mobile Communications Design Fundamentals, op. cit.,* p. 254.

[15] P. Porzio Giusto, "Digital Modulation Schemes for Narrowband Mobile Radio," in *Proceedings of the International Conference on Digital Land Mobile Radio Communications,* Venice (June 30–July 3, 1987) pp. 28–36; also Gerald Labedz, Ken Felix, Valy Lev, and Dennis Schaeffer, "Hand-off Control Issues in Very High Capacity Cellular Systems Using Small Cells," *ibid.,* pp. 360–366.

[16] A. Gamst, "Remarks on Radio Network Planning," *Proceedings of the 37th IEEE Vehicular Technology Conference,* Tampa, June 1–3, 1987, p. 164.

Chapter 10
OTHER DESIGN CONSIDERATIONS

Conquering, or at least coming to terms with, the mobile environment is essential for any *mobile* communication system. Overcoming the design challenges of the interference-limited environment to achieve at least some measure of frequency reuse is essential for any *cellular* communication system. Once these two essentials have been satisfied, the system architect begins to enjoy a greater degree of design freedom. What additional design considerations we accept depends upon what kind of cellular mobile system we intend to create. What is cellular telephony intended to do and for whom?

The design of any large-scale mobile telephone system will be called upon to satisfy many additional criteria which may be viewed as constraints or as objectives and accorded greater or lesser weight depending upon the overall perceived function of mobile telephony within the telecommunication network. Conceived as a specialized niche — which seems to be the implicit point of view of many in the industry today, recognizing that even several million users still constitute only a niche — cellular radio perhaps may give short shrift to some of the concerns listed in this chapter. For example, if we accept a definition of mobile telephony as an urban car phone, mainly for commuting businessmen, optimized for short voice messages, then the issue of compatibility with digital network services may seem rather remote. On the other hand, for those who tend toward the view that mobile telephony is the vanguard of a much broader wireless-access concept, the issue of adapting to emerging ISDN standards cannot be so lightly dismissed.

Reflecting upon the shortcomings of the current analog cellular architecture, we can identify four areas in particular that the next technological generation must improve upon:

1. *Modularity and Geographical Flexibility*. Today's systems have probably been overoptimized for very large cities; greater modularity in

system design, construction, financing, and greater flexibility in serving different population patterns is regarded as an important correction in the next generation of architectures.

2. *Cost.* High system costs have hobbled the current generation of cellular systems; the emerging architectures must address this problem.
3. *Compatibility with Future Network Services.* Lack of communication privacy and difficulties in handling data are only the superficial symptoms of the general incompatibility of today's cellular systems with long-term trends in the evolution of the public switched-telephone network.
4. *Competition.* Today's architecture was designed during the era of telecommunication monopoly; the next generation of systems must assume, at least in the United States, a continuing high level of competitiveness and deregulation in the telephone world in general and in mobile telephony in particular.

10.1 DESIGNING FOR MODULARITY AND GEOGRAPHICAL FLEXIBILITY

Mobile telephony has always had an overwhelmingly urban market. This is the result of many factors including raw population density, average income levels *versus* mobile service pricing, locations of prime business customers, locations of financial centers, and so forth. The basic fact is that a few large metropolitan agglomerations hold the vast majority of potential mobile telephone users in a very small geographical area. There are about as many people, 38 million, in only the *three* largest metropolises — New York, Los Angeles, and Chicago — as there are in the *entire United States west of the Mississippi River, excluding California and Texas, 19 states with a population of 39 million people* [1].

In the precellular era, this led to a terrific imbalance in spectrum requirements. In New York City there was appalling congestion and a five-year waiting list. In York, Pennsylvania, there were unused frequencies. Like ordinary real-estate prices, the price and even absolute availability of spectral real estate was extremely imbalanced between center and periphery. The paradox of spectrum feast and famine in markets virtually side by side with one another was a continuing consternation to policymakers and spectrum managers.

In fact, today's cellular systems were designed principally to alleviate the spectrum congestion in a few large cities. The idea that mobile telephony might become a much broader, mainstream communication service, reaching across the nation much like television or the telephone itself, has only emerged after the fact, so to speak. *Cellular was never originally intended for the smaller markets, and the architecture reflects that fact.*

We are now witnessing Part Two of this awful paradox. Even with 50 MHz of spectrum and frequency reuse, the major cities — New York, Los Angeles, Chicago, and perhaps half a dozen others — are already mired in a new capacity crisis. Congestion is returning to IMTS levels, and service is degrading to the point that some operators have become alarmed that the cellular industry could find itself poisoned by its own success in its most promising markets, much as CB radio went through its boom-and-bust period. On the other hand, cellular operations in many smaller cities and rural areas — where there is *still* a considerable amount of unused and freely available IMTS spectrum — seem to be running on wishful thinking. The cost obstacle that has hampered IMTS in smaller markets has not been overcome by cellular; if anything, cellular's costs are higher than IMTS when the system overhead is factored in.

Is the cellular impasse a problem of the major urban systems only? Much of the current discussion of the need for a quick fix on capacity limitations seems to reflect a continuation of the same thinking: the next generation of architectures may be driven primarily by the problems of New York, Los Angeles, and a few other large systems that have rapidly reached saturation. I believe this would be a mistake. By hyperoptimizing the systems design for very large systems, cellular has painted itself into about five corners at once. It has created the top-heavy system overhead, a nuclear power plant–like utility cost structure, and a difficult growth path, the sprint to the break-even point, which are detailed in Chapter 4. It may be difficult to apply Band-Aids to urban systems without driving an economic and technological wedge between the major markets and the smaller markets — and yet the smaller markets may find it simply too expensive to follow another round of technological upgrades practically before they have completed construction of the current systems. If mobile telephony is strictly viewed as a metropolitan service, then such a course of action may be acceptable. If it is intended to become a nationwide mainstream communication service and to benefit from a degree of interoperability — in short, if it is to be viewed as a *network* — then the next generation of architecture must reflect greater attention to modularity of design and flexibility of implementation.

10.1.1 Example: Adjacent Channel Utilization

The "big city" logic is today pervasive in cellular architecture. For example, in a large metropolitan area, it may be readily assumed that a dense multicellular system will develop rather quickly. Frequency reuse will be applied almost immediately. Only a limited subset of the total available channels will be used in any one cell. Such was the quite natural reasoning of the first generation of cellular architects. The next logical step, however, was a misstep. Since only a subset of channels would be in use, the use of *adjacent channels* in the same cell could always be avoided through proper spectrum planning. In other words, in a four-cell reuse pattern, for example, Cell A would be assigned Channels 1, 5, 9, 13 . . . and so forth, Cell B would be assigned Channels 2, 6, 10 . . . and so forth. Suddenly, a new possibility emerged: since adjacent channels would "never" be used in the same cell, why not relax the front-end receiver filtering requirements on the mobile units? This would reduce the costs of the mobile unit considerably. In fact, the emissions masks in cellular frequencies were made generally considerably more liberal than the IMTS emissions masks. So in the end, because the systems designers assumed that adjacent channels *would* never be used, the technology was designed so that adjacent channels *could* never be used.

I believe that most cellular engineers now recognize the shortsightedness of this decision. For example, in a single-cell configuration that would be appropriate for many small markets and rural areas, cellular radio is terribly inefficient in its use of the spectrum. It is less spectrum-efficient than IMTS by a factor of three or four. In fact, this statement applies in some measure to any system — even a system with a few cells — in a market too small for frequency reuse.

A more serious problem is that loose receiver filtering inhibits the ability fully to implement *dynamic channel allocation* or *channel borrowing* schemes. Channel borrowing is the idea, simple to state, difficult to implement, that as traffic flows to and from different cells at different hours of the business day, channels could be reassigned dynamically to the cells with the greatest need. Instead of a fixed, equal spectrum allocation to all cells — which means that cells covering the downtown or commuter arteries will clog quickly, while capacity goes awasting in suburban cells — a dynamic allocation scheme would allow the system to "breathe" with the flow of traffic, shifting channels back and forth between cell sites as necessary, putting the spectrum resource where the need is greatest. Promising claims have been made for the potential capacity relief to be gained from dynamic allocation [2]; one of the chief obstacles to the full application of such concepts, however, is the relaxed receiver filtering. If today's mobile units are designed to operate on the assumption that there will never be a same-cell transmission less than three channels away, the dynamic allocation will not be able to pack users on adjacent channels in the "hot" cells.

The principle of allowing for adjacent channel utilization will greatly further flexibility and spectrum efficiency, as well as carrying capacity, in the next generation of cellular systems.

10.1.2 Example: Modularity and Growth

Another important problem for today's operators is the large critical mass of customers who must be rapidly recruited to cover the heavy and integral system overhead. As described in Section 4.1.2, this makes cellular a very different type of start-up operation from, say, an IMTS system or a paging system. These other mobile services can generally follow an easier growth path, where revenues and costs are kept in closer balance. Growth and the investment required to support it is more incremental. This makes it easier for an operator to get started and reduces the capital required by cutting down the period of negative cash flow that must be supported during start-up.

I believe that the top-heavy structure of today's cellular again reflects its legacy as a large-urban solution. In New York or Los Angeles, the demand is so great that the top-heavy economics probably ultimately make little difference to the operator's fortunes. In smaller markets, this may not be true. Less expensive network architectures and greater modularity in system growth, which could permit fewer cell-sites to be built initially, could ease the burden of high start-up costs.

10.2 DESIGNING FOR LOW COST

It should be clear from the discussion in Chapter 4 that cost continues to be a major obstacle to the growth of the cellular business today. It may seem superfluous to emphasize that the next generation must address the cost problem — and yet many discussions of next-generation alternatives today are largely ignoring cost as an issue, focusing exclusive attention on the capacity problem.

As we have stressed in Chapter 4, cost and capacity are two sides of the same coin. Lower costs will produce more offered traffic from more subscribers using more minutes of airtime, which will increase the strain on capacity. Of the two, *cost is the dominant issue.* In fact, there is an easy solution to today's capacity problems which requires no new standards, no new technology, no new expenditures: Raise the prices! Higher usage prices will cause subscribers to curtail their airtime, improving circuit availability and blocking probabilities without reducing operator revenues.

Of course such a proposal would be facetious. The cost of the minimal economically justified usage level is already too high. Higher rates will cause subscribers to drop off the network — in fact, as the numbers from Chapter 4 would seem to indicate, this is probably already happening.

In short, more capacity at the same cost will not suffice. The economics of cellular must be dramatically altered if the industry is to flourish.

Another obvious point, which also turns out to be not so obvious to many participants, is that cost must be defined as *total user costs*. The cost of the mobile unit is only part of the equation. If anything, system costs and usage charges appear to be much more serious obstacles to growth.

In the first-generation architecture, decisions were generally made favoring the reduction of the mobile-unit cost. FM transmission was attractive, in part, because it promised cheap mobile units. It was a mature technology, well down the learning curve of manufacturing costs, and there were few opportunities for proprietary technical advantages that could stem the "commoditization" of the subscriber unit segment of the industry (see Section 4.1.1). In general, this promise has been met: mobile unit costs are very reasonable today compared to only a few years ago. The choice of single channel per carrier FM architecture, however, has resulted in built-in system costs on a per-subscriber basis that are today several times the cost of the mobile unit, certainly if telephone traffic patterns and grades of service are accepted as the goals for cellular. It is quite clear today that alternative architectures exist which might result in a somewhat higher mobile-unit cost, at least initially as a new manufacturing learning curve commences, but *much lower* system costs — and a lower overall user cost. Herschel Shosteck, among others, has repeatedly stressed that in computing cellular economics and price elasticity it is necessary to look at the *total monthly charge* to the user: mobile-unit amortization *plus* service or usage charges. The danger here is that the next-generation choices will be *suboptimized* once again: a technological decision will be made that will favor lower subscriber costs but higher system costs. This danger is present in part because the industry is divided between mobile-unit manufacturers and system manufacturers; with a few exceptions, strong players in one segment are not strong in the other. The tendency toward suboptimization is built in. Yet from the operators' and the consumers' point of view, it is clearly and unequivocally *total cost* that must be targeted.

10.3 DESIGNING FOR COMPATIBILITY WITH FUTURE NETWORK SERVICES

To reiterate two earlier points:

1. Digital networks are beginning to emerge;
2. Today's cellular radio is probably the last new analog telecommunication system that will ever be fielded.

Taken together, these two points suggest another important set of design criteria for the next generation of cellular radio.

Some may object that plans for digital network services are still hazy, and some interface standards are probably years away from agreement. The telephone loop is still overwhelmingly analog — and will probably remain so for decades, on a percentage basis. Whether as ISDN (integrated services digital networks) or even post-ISDN architectures, there is a latent suspicion that much of the digital network planning is still technology-driven rather than market-driven.

Nevertheless, telephone planners today are almost universally committed to tracking, if not pioneering, the evolution of digital networks. It is almost inconceivable that the next generation of cellular technology can evolve without reference to these plans. Today's analog systems are shut out of that evolution. While it is not *in itself* a fatal flaw, once the decision has been reached to take the step toward a new architecture, the requirement for compatibility with digital network services *is* virtually absolute.

But there is a problem. The emerging digital-network concepts like ISDN are inherently based on the assumption of very large bandwidths in copper or fiber facilities. For example, the residential ISDN transmission standard, known as 2B + D, for two bearer channels plus one packet-data channel, is based on a data rate of 144 kb/s divided into two 64-kb/s circuits and one 16-kb/s data circuit.

Radio is bandwidth-limited. It is difficult to see how to build a 144-kb/s capability into a digital cellular subscriber unit. At the least, it would require a much wider channel than today's cellular allocations provide. ISDN still bears the marks of its wireline parentage, and the grafting of these standards onto a wireless access system like digital cellular is not straightforward.

One approach which is being advocated by some involves a redefinition of ISDN standards to accommodate much lower bit rates. For example, AT&T has proposed a concept called cellular access digital networks (CADN), in which 2I + C replaces 2B + D [3]. Not surprisingly, the I, or information channel, is a cut-rate version of the B channel, and the C, or control channel, corresponds, at lower bit rates, to the D channel in conventional ISDN terminology. The proposed nominal rate for the I channel ranges from 10 to 16 kb/s and for the C channel, from 1 to 5 kb/s.

Will this be adequate? An alternative approach is based on ensuring that the next generation of cellular systems will be capable of supporting the current 2B + D standard, defined at the user's option in either of the following ways:

1. Functional 2B + D, two telephone-grade voice circuits plus one 16-kb/s packet data-circuit; or
2. A 64-kb/s clear channel, equivalent to one complete ISDN B channel, the basic ISDN building block.

This approach is implementable today within the current cellular channel structure [4]. It is likely that digital network compatibility will prove to be a particularly thorny requirement; the tendency will be to give it short shrift, pleading the "obvious limitations of radio" [5]. The full emergence of mobile communication on a par with today's wireline networks, however, will require a carefully thought-through approach to digital integration.

10.4 DESIGNING FOR A COMPETITIVE MARKETPLACE

The concept that the next generation of cellular radio will operate in a more competitive environment actually embraces at least two interpretations of competition that system architects must be prepared for:

1. Future proofing; and
2. Operating in a competitive marketplace.

10.4.1 Future Proofing

The analog standard for cellular radio took 10–15 years to develop — and was arguably obsolete the day it went into use. Within one year, the operators were petitioning for new spectrum. Within three years, the need to abandon the analog standard was widely discussed.

The danger of the Band-Aid approach to the next generation is that exactly the same thing will happen: a new technical standard will evolve over at least two to three years and will itself be rapidly overtaken by (1) market growth resulting from its very success and (2) new digital techniques that will render existing capabilities patently obsolete. It is extremely unlikely, in my opinion, that *any* digital standard could be defined today that would be able to remain the standard for more than from three to five years. Digital technology is moving too fast, and is too fertile, for serious anticipation of locking in another multidecade standard like FM radio. Moreover, the apparent underlying interest in *affordable* mobile communication, as indicated by, for example, the CB radio phenomenon, is large enough that it probably will overwhelm any small capacity gains in a relatively short time, assuming that prices continue to fall.

This area merges into the more general question of standards and the transition from today's systems to the next generation — which is the main topic of Part VI of this book. The general challenge is: How can the design of the next generation of systems be kept open to new technology? Does this invalidate the concept of a single standard? Are some alternatives more adaptable to future developments than others? We shall return to these questions in Section 15.7.

10.4.2 Competition Among Operators

As we discussed in Chapters 3 and 4, a good case can be made that analog cellular architecture inherited the fatal assumption that it would be operated as a monopoly service along the lines of the old AT&T wireline network. The top-heavy economics, the profligate use of spectrum, the trunking efficiencies — all are optimized for a single operator who can (1) maximize operating economies of scale and (2) avoid destructive rate competition, especially in the early years. Although it has been a long time since anyone openly advanced the natural-monopoly argument applied to cellular radio, in fact the economics of today's cellular systems are tainted with natural-monopoly reasoning.

The arguments by AT&T and others during the 1970s and early 1980s that unrestricted entry would be unfeasible turned out to be correct — although many would argue that they were self-fulfilling. Nevertheless, it seems unlikely that we shall backtrack toward the monopoly era. Competition in the operating arena is here to stay. In fact, the more interesting question is whether the next generation of cellular technology can *enhance* the competitive trend, allowing potentially *more than two* operators per market, especially in the larger markets. The latent possibility of more than two operators is indicated by the flourishing reseller segments in the larger markets. Most mobile services other than cellular function very well under conditions of much less restricted entry. These same services have proven to be much more open to new technologies.

Another key criterion, therefore, for the evaluation of the next generation of architectures must be: How well adapted is any given proposal to operating in a less restricted, more competitive, operating environment?

REFERENCES

[1] U.S. Department of Commerce, Bureau of the Census, *Statistical Abstract of the United States,* 1987, Washington, D.C.: U.S. Government Printing Office, 1987, pp. 28–29.

[2] James F. Whitehead, "Cellular System Design: An Emerging Engineering Discipline," *IEEE Communications Magazine,* February 1986, pp. 8–15.

[3] E. S. K. Chien, D. J. Goodman, and J. E. Russell, "Cellular Access Digital Networks," *Proceedings of the International Conference on Digital Land Mobile Radio Communications,* Venice (June 30–July 3, 1987), pp. 84–93.

[4] George Calhoun, "USDN: A Concept of ISDN for Wireless Access Applications," to be published.

[5] B. J. T. Mallinder and E. Haase, "'Future-Proofing' the Pan European Cellular System," *Proceedings of the International Conference on Digital Land Mobile Radio Communications,* Venice (June 30–July 3, 1987), pp. 114–123.

Part V
TECHNOLOGICAL ALTERNATIVES FOR THE NEXT GENERATION

At certain moments in the history of any industry, one particular technological solution may dominate all others, producing a uniform *de facto* standard which tends to entrench itself over long periods. The gasoline engine and its near cousin, the diesel, became such a standard in the early years of this century for most forms of land and air transportation. Similarly the analog variable-resistance telephone became the standard for wireline communication and FM became the standard for mobile radio. Such conditions favor the growth and consolidation of oligopolistic or monopolistic industry structures.

At other moments, scientific or economic breakthroughs unleash an expanding set of technological options. The industry becomes dynamic and heterogeneous, marked by turmoil and opportunity. This is happening today in the telecommmunication field, with the revolution in digital techniques propelling the process. The set of technological options is growing, rather than shrinking, with each passing year. In particular, the future of mobile telephony is more indeterminate, more open, than ever before.

The next generation of cellular radio, therefore, is emerging not as a new, monolithic solution, but as a range of alternative futures which may well coexist and will certainly continue to evolve. There are a few very general architectural alternatives — wideband *versus* narrowband, frequency division *versus* time division, *et cetera* — which are in some sense mutually exclusive and thus tend to define the various camps within the industry. There are also a great many specific alternatives at various levels of analysis:

1. *Alternatives for the radio link* between the individual mobile subscriber and the cell-site base station;

2. *Alternatives for the cell-level system design,* to accommodate a large population of users accessing multiple telephone circuits within a single cell;
3. *Alternatives for the network-level system design,* encompassing a large number of cells over a wide geographical region.

The comparison or evaluation of these alternatives is multidimensional, and there is lively disagreement over the relative importance of different dimensions. Is system cost more important than spectrum efficiency? What exactly is spectrum efficiency? What price are we willing to pay for ISDN compatibility, or for future technological flexibility?

The industry is not ready for a final analysis. Neither the alternatives nor the criteria have evolved to the point where comprehensive, definitive recommendations can be fashioned. Indeed, coping with this disturbing indeterminateness may be the greatest challenge to all parties in the next decade.

Chapter 11
THE BROAD TECHNOLOGY ALTERNATIVES

In this chapter we are concerned with the relatively small number of choices among truly mutually exclusive alternatives for the next generation of architecture. The more striking and perhaps the more important implication of digital technology, however, is how few of these hard choices there really are.

Most of the specific techniques we shall examine in Chapter 12 do not pose irrevocable design choices. Because of the way digital systems are constructed, through block processes and defined interfaces, the system designer can to some degree regard specific voice-coding techniques, for example, as interchangeable parts within an overall architecture.

Perhaps more than the individual techniques, it is this heightened flexibility of digital systems that will represent the greatest change for the designers of the next generation of systems. Choice and flexibility are a two-edged sword for the system designer. On the one hand, the more choices there are, the greater our design freedom, and, in principle, the greater our ability to optimize. On the other hand, more choices — especially choices that are not mutually exclusive — mean more uncertainty about finding the optimal strategy. Indeed, place all of this freedom in the context of continuing, rapid technological evolution and it becomes likely that the optimal strategy will be redefined over time. This is indeed a different world for those used to the long stasis of FM in the world of mobile radio. Not only are there more technical options, but also these options do not get winnowed down as tentative choices are made. We start to be concerned not only about Technique A *versus* Technique B, but also about Technique A combined and integrated with Technique B. As processors become faster and cheaper by leaps and bounds, more often the question is not which of two algorithms (for voice coding, say) is better, but how soon will we be able to integrate the best features of both?

This digital flexibility is apparent on two levels. First, we see it in the way digital systems — hardware and especially software — are put together. Second, we see it in the way digital information (such as a digitized voice signal) is organized. A brief discussion of both aspects of digital flexibility is important as an introduction to our consideration of the technological alternatives.

11.1 THE FLEXIBILITY OF DIGITAL DESIGN

Digital systems are often conceptualized and presented in block-diagram form. Each block represents a relatively self-contained process, like voice coding, or line coding, or equalization. The arrows connecting these block processes represent simplified and often standardized inputs and outputs, which can be viewed as *interfaces* between separate processes and much more complex processes. The larger, or higher-level, blocks can usually be broken down into lower-level blocks, and internal interfaces can be defined at each level. At a very low level, the blocks may represent functional components of hardware and software, such as random access memories (RAMs), or the microprocessor (and its sublevels, the registers, the accumulator). The same type of block structure is usually applied to well-designed software; subroutines function as the block processes which plug in to the main program through simplified input-output specifications.

The purpose of construing digital systems in this fashion is to allow a high degree of compartmentalization in the design, to permit the division of labor necessary for complex systems. For example, in software development, if a particular subroutine has been blocked off in this way, it becomes possible to tinker with that subroutine, troubleshoot it, fix it, rewrite it, substitute an entirely new algorithm — as long as the input-output specifications, the interface, remain the same, the subroutine can usually be worked on independently of the main program. The task can be partitioned; different subroutines can be assigned to different programmers or teams of programmers. At higher levels, different companies may divide their efforts in the same way; one company may specialize in voice-coding algorithms, while another focuses on modem design. People with different backgrounds, from entirely different disciplines, may be involved in the development of different blocks — and yet confident that their designs will integrate because the interfaces are known and specified. Obviously the system-integration function becomes crucial; but the block-process and interface approach is designed to smooth the integration path as much as possible.

The digital systems architect is thus able to approach his problems with a set of highly interchangeable parts. He has a large number of different parts with different characteristics, but all can be plugged into each other, like a giant Tinkertoy set, because of the simplified and standardized interface specifications. For example, the digital mobile-systems designer can view the wide range of voice-coding algorithms, or the equally impressive array of equalization techniques, as so many potentially substitutable parts; his choice of a particular coding technique will not constrain his selection of a modulation block, or an encryption block, and so forth. (Like all idealizations, this may be something of an overstatement; it is, however, essentially true. Today the evaluations of next-generation coders are taking place largely unconstrained by the parallel question of which modulation technique should be applied. See Section 12.2.3.)

Moreover, any new part that is invented — a new voice coder, for example — can be plugged into the existing design as long as the interface characteristics are satisfied. Digital systems are inherently much more open to new technology than analog systems, where the designs tend to be more integral and introducing a new technique can be as tricky as an organ transplant.

11.2 THE FLEXIBILITY OF DIGITAL FORMATS

The flexibility of digital systems can also be appreciated by looking at the way digital information itself is organized before, during, and after transmission. We can distinguish between the *content* of a message, the *medium* of the message, and the *organization* of that message. For example, the content, e.g., the text on this page, may be stored as variations in the reflectivity of a paper surface, i.e., as printed matter, or as magnetic marks on a computer tape or disk — different media for the same message. Each medium encodes not only the message content, but also certain organizational information to allow the message to be stored, accessed, retrieved. The same information content, in the same medium, can be organized in different ways. Different ways of organizing this information are called *formats.*

For example, the printed page contains certain obvious format information: usually a page number and perhaps a chapter heading at the top of the page. It also contains other, more implicit, format information: the spaces between words and lines, the indentations for paragraphs, the arrangement of the message in lines to be read from left to right and top to bottom, are but a few examples.

Formats are of vital importance for computer-stored information. The information may be identical, the medium identical, and yet a Macintosh disk file differs from a Tandy disk file in its format and the one machine cannot always read the other's files.

In analog systems, the content and the organization of the message are often integrated such that it is hard to separate them, even conceptually, for communication purposes. Semantics and syntax interact heavily. On the other hand, one of the sources of tremendous flexibility in digital systems is the ease with which content and organization can be separated and dealt with distinctly. For our purposes here, the specific binary values of digitally transmitted messages — the 1s and 0s — are the content. The way they are grouped together, blocked for processing, framed with various signaling and control bits, timing bits, and so forth — that is the organization or format of the message. One of the nice features of a digital architecture is that it can allow for a great variety of formats — different structures for organizing the same digital information — to operate within an overall system design. An analogy can be drawn between formatting for information and *packaging* for conventional goods. Within broad limits, the same information can be repackaged in different ways, depending upon the particular hazards or requirements of the block process, e.g., transmission in a TDMA radio channel, which it is about to undergo.

The characteristics of digital formats, then, are:

1. Formats are *transparent* to the end user — generally the user does not even know what format the information is in at any given moment or what reformattings it may have undergone during its passage through the communication system;
2. Formats do not *modify* the information content; casting the message into a particular format does not constitute signal processing in any substantive sense;
3. Formats can be introduced or modified by the system designer with considerable freedom, at least within block processes;
4. Formats are easily *translatable* — two incompatible digital systems can generally talk to one another by means of a small translation program, sometimes incorporated in a specially designed device, usually without imposing any internal modifications on either system.

The last point is most important for digital flexibility. Translation from one format to another is generally easy and inexpensive. For example, the T-carrier format used to organize multiplexed PCM transmissions in North America is different from the format used in Europe. (See Figure 11.1.) Yet once the information is in digital form, effecting a translation from the North American format to the CCITT format is not difficult. (Strictly speaking, the "bit-robbing" in some T-1 frames is a modification of message content; it is minor, but a reminder that even in digital systems the line between content and format is not completely sharp.)

The ease of format translation is one of the chief attractions of digital communication systems. It allows for continuing technical evolution and

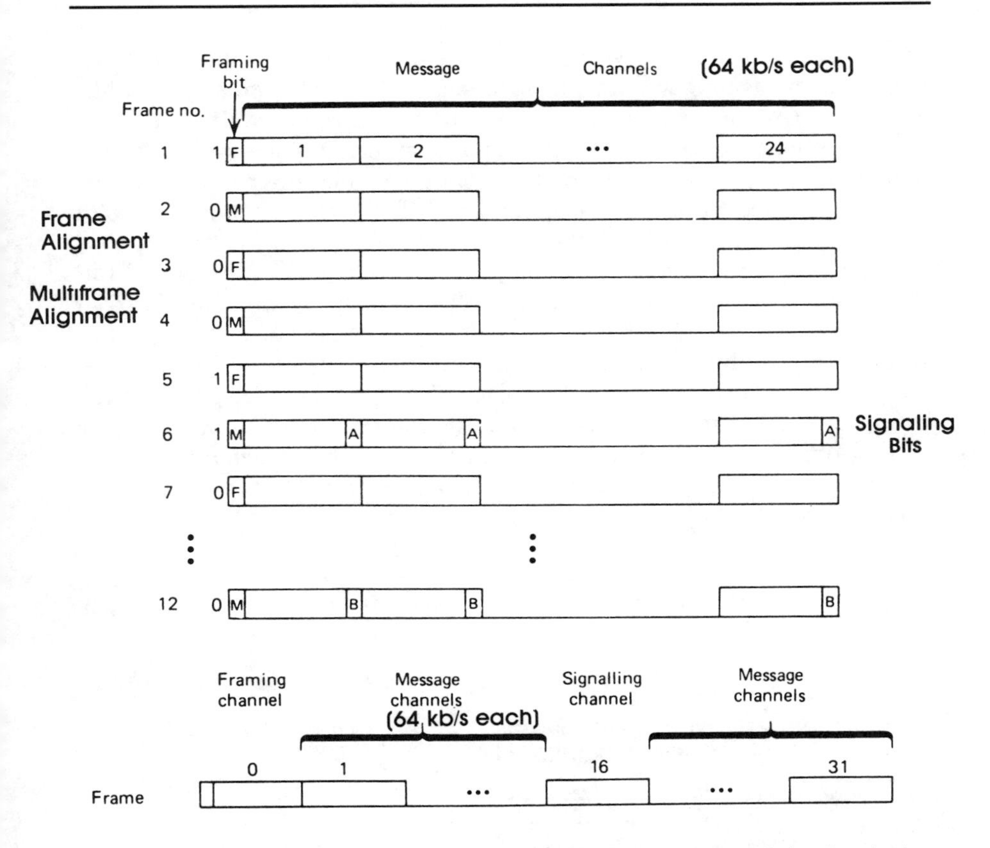

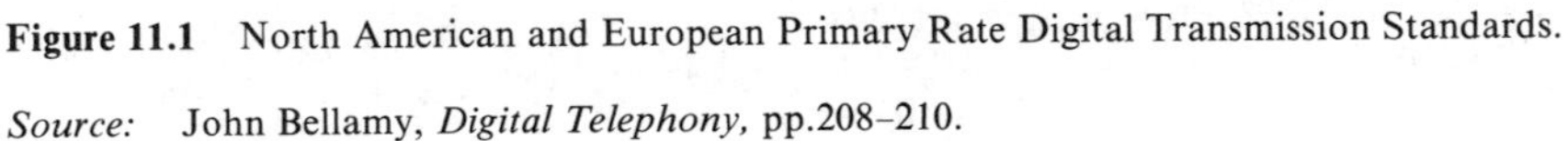

Figure 11.1 North American and European Primary Rate Digital Transmission Standards.

Source: John Bellamy, *Digital Telephony,* pp.208–210.

experimentation within system elements and subsystems, by defining standard *interfaces* that allow system designers to make sure different elements are interoperable. For example, the computer industry has always tolerated a very high degree of variety among different hardware and software systems, because it is very easy to convert any proprietary format to a recognized standard like RS-232C and then transmit data from one system to another. Once this conversion is made, virtually any two computers in the world can talk to each other — indeed, two computer users can communicate without either one knowing anything about the hardware or operating systems in use at the other end.

Similar interface standards have evolved within the telephone industry. This has been liberating for the systems designer. The developers of new central-office switches, or new subscriber-carrier systems, or new microwave systems, or new PBXs, are free to experiment with new digital technologies, as long as they translate back to a recognized standard like T-carrier PCM, or even analog VF, at the interface point. In fact, there is already much greater variety in digital communication systems than is generally recognized. For example, in a digital subscriber-carrier system designed for the North American market, like the Bell system SLC-96 or Northern Telecom's DMS-1, there is a multiplex link between two points connected by conventional wire pairs. (See Figure 11.2.) The digital formats used over the internal link are all nominally within the T-carrier family, but actually differ in the details from manufacturer to manufacturer, and are often proprietary. What this means is that a Northern Telecom remote terminal will only work with a Northern Telecom central-office terminal. The two are a system. The interface from the COT to the central-office switch, however, uses a standard format, which may, in fact, be an analog format, to ensure that such subscriber-carrier systems can interface with any type of central-office equipment.

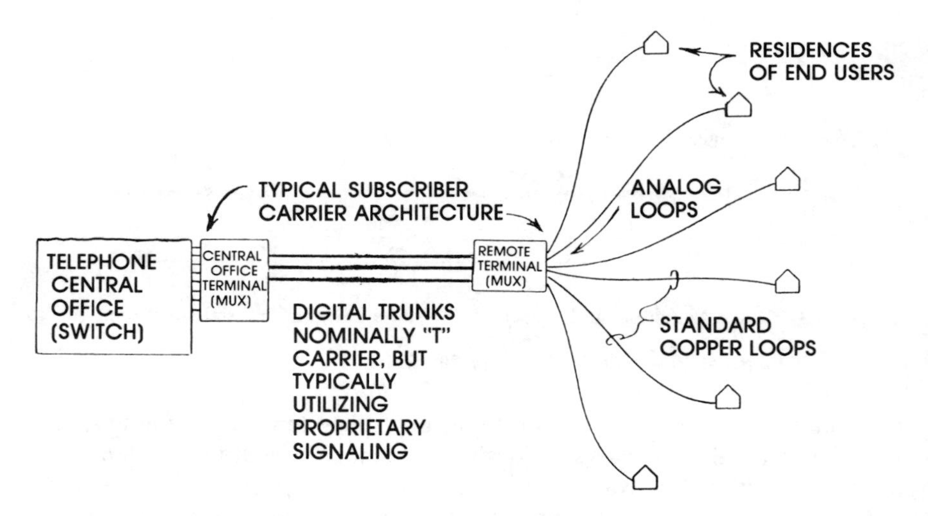

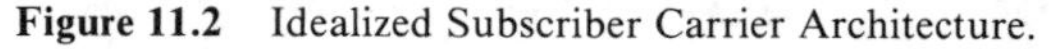
Figure 11.2 Idealized Subscriber Carrier Architecture.

A point for future reference: This vision of a single central-office switch supporting different types of subscriber-carrier systems, each operating with its own proprietary internal transmission formats, suggests one model of how the future of digital cellular radio may evolve, preserving heterogeneity *and* interoperability. We shall return to this issue in Part VI.

Thus, many of the specific technological alternatives for digital mobile telephony that we shall discuss in Chapter 12 are not really mutually exclusive, despite the sometimes heated debates which often seem to assume that they are. If suitable interface standards emerge, different voice coders, even different modulation schemes, may be viewed as to some degree interchangeable options within the overall design of the system.

Mobile-radio system designers, however, are perhaps unaccustomed to this degree of freedom. In the past, the development of analog radio systems, including broadcast radio and television as well as the first generation of cellular radio, has been characterized by protracted and acrimonious struggles to define a single industry standard prior to large scale production. This has fostered a mindset among regulators and industry planners that tends to assume that such single-standard architectures are the inevitable and necessary precondition for a successful deployment of such systems. In Part VI, we shall explore and critique this mindset in light of the possibilities of modern digital technology. The point to be emphasized here is that the discussion in the following chapters of some of the alternative digital techniques need not be conducted under the severe aegis of either-or thinking.

In fact, there are only a handful of truly mutually exclusive, irreversible choices that the designers of digital mobile-telephone systems must make at the outset. These decisions define a small set of very general architectures which are probably not compatible with one another in any easy way. Within any one of these architectures — for example, a digital, narrowband TDMA format — there are a wide variety of alternative techniques that may be interchanged and may coexist to a considerable degree.

These few, very broad technology decisions tend to divide the telecommunication industry into several camps, and it is around these broad issues that the greatest amount of regulatory and political heat will certainly be generated. It is around these issues that the greatest international frictions will emerge. It is around these issues, for example, that the debate over the Pan-European digital cellular standard has largely focused.

In my view there are, at most, four such nubs of real controversy:

1. *Analog versus digital technology;*
2. *Wideband versus narrowband channelization of the radio spectrum;*
3. *Choice of multiplexing technique: Frequency division versus time division versus code division;*

4. *Nature of the interface to the existing telephone network: centralized control versus distributed control.*

These are the grand issues where emotions may run rampant. The ease of format translation and the promise of rapidly evolving technology tend to dampen the cosmic overtones in arguments over coding algorithms. If, however, we opt for a wideband system, for example, for the next generation of digital mobile telephony, it will likely preclude further development of narrowband alternatives. If we opt for digital transmission, we are unlikely to return to analog. These are the decisions that weigh the heaviest on the architects of the next generation.

11.3 ANALOG *versus* DIGITAL

It should be clear by now that the choice of analog or digital technology is a fundamental, irreversible decision that will define the next generation of cellular systems. As we have argued at length in Chapter 4, it was the choice of analog technology by early cellular designers in the 1960s and early 1970s which, more than any other single factor, put mobile telephony in the box it is in today.

Is there any reason to consider analog technology for the next generation of mobile telephony? Admittedly, I believe the answer to this question is fairly obvious. The next generation will be digital, for all the reasons presented here at considerable length. In the past year or so, the industry consensus on the advisability of going digital has solidified.

The analog mobile radio, however, still has its partisans. There is talk of further channel-splitting, to multiply the number of cellular channels by reducing the bandwidth from 30 kHz to 15 kHz or to 12.5 kHz, still utilizing FM transmission. There has also been considerable interest in single-sideband analog techniques, which theoretically could allow for channel bandwidths on the order of 5 kHz or so. Both alternatives are discussed further in Chapter 12.

Nevertheless, the vast majority of mobile-telephone engineers today would vote unhesitatingly for digital technology. Strangely, this body of opinion reflects not so much the current capabilities of digital technology — which are still more or less on a par with analog techniques in terms of economics and spectrum efficiency — as the *faith* in what has been called the *digital (r)evolution* [1]. It is also fueled by a pessimism regarding the potential for squeezing further improvements from existing analog techniques. These techniques, especially FM radio, are the mature result of decades of development and refinement. They are far down the learning curve of manufacturing cost and component design. They are close to the theoretical limits of analog-communication efficiency: we may anticipate an improvement in spectrum efficiency by another factor of two or four, at best. Such small improvements will not suffice to bring mobile telephony to large numbers of users under likely spectrum-allocation scenarios. Similarly, an analysis of the *structure* of the costs of analog systems also discloses that further cost reductions are not likely to be dramatic, especially on the systems side, where the single-channel-per-carrier architecture limits the savings that can be realized. In fact, the costs of an analog system are likely to begin to climb as the rest of the telephone network goes digital. Solutions to the problems of privacy and medium to high-speed data transmission, not to speak of ISDN compatibility, will *add* costs to analog systems.

The bias toward digital is based on the belief that digital techniques *will* undergo rapid and dramatic improvements in performance and decreases in cost in the near future. Costs *are* declining so rapidly in almost all areas of digital electronics that such optimism seems well-founded. For example, a standard generic measure of digital cost is the price for memory. Another generic indicator is the falling price of processing power. In terms of *performance (versus* cost), the advances have been even more rapid. The microprocessor has challenged conventional assumptions about the limits on hardware.

Many telephone and radio engineers, however, are still adjusting to the digital (r)evolution. In one sense, digital communication has been around for quite a while, since the early PCM systems during World War II. Certainly it has been common in the telephone network since the mid-1960s. This is, however, somewhat misleading. As recently as the late 1970s, the Bell system was still bringing out *new* analog technology, and most telecommunication engineers still regarded digital technology as a largely foreign province. The *pervasive* influence of digital design principles in telephony is still a very recent phenomenon. In telephony circles, there is often still a generational stratification: the older, more tradition-bound, management levels are inclined to view analog *versus* digital as a conventional choice between known, specific alternatives, while the younger generation, which predominates at engineering-staff levels, believes that digital techniques hold a vastly greater *potential,* and is inclined toward digital technology on that basis alone.

In short, there is a general, if ill-defined, understanding that going digital will open up a new set of possibilities for dramatic cost and performance improvements, and it is this, in addition to specific near-term solutions for the nagging problems of cellular radio, which determines the selection of digital technology over analog for mobile telephony.

11.4 WIDEBAND *versus* NARROWBAND SYSTEMS

The second technological Rubicon involves the way in which the total spectrum allocation assigned to mobile telephony is parceled up into individual mobile-telephone circuits, or channels. There are two broad alternatives, referred to as *narrowband* and *wideband* systems, representing very different ways of defining the telephone circuits and managing the spectrum resource.

It is assumed here that any mobile-telephone system conforms to the following service objectives:

1. It utilizes full-duplex, telephone-type circuits, rather than the simplex, push-to-talk circuits commonly used in precellular mobile-radio systems;
2. It provides the equivalent of single-party service, i.e., circuits are not simultaneously shared among many users, CB-radio style;
3. It is a fully trunked, *multiple-access* system; in other words, instead of preassigning specific users to specific channels or circuits, such systems must allow every user to access any circuit; circuits are assigned to individual users at the time of call set-up (see Section 11.5).

Still, there are two very different ways of satisfying these requirements within a large allocation of spectrum, such as the 50 MHz currently set aside for cellular radio in the United States.

Narrowband Systems

Conventional architectures, both analog and digital, are *channelized.* The total spectrum allocation is carved up into a large number of relatively narrow radio *channels,* defined by carrier frequency. Each channel is actually a *pair* of frequencies: one frequency, used for transmission from the base to the mobile unit, is usually called the *forward channel* and another frequency, used for transmission from the mobile unit to the base, is usually called the *reverse channel.* (See Figure 11.3.) A user is assigned both frequencies for the duration of the call. The forward channel and the reverse channel are actually widely separated, to help the radio keep the transmit and receive functions separated. In some bands this transmit-receive separation is as little as 5 MHz. In the current cellular allocations in North America the separation is 45 MHz. (See Figure 11.4.) The width of these narrow channels is today normally either 25 kHz or 30 kHz in each direction. In the North American cellular system today, the width of the channels is 30 kHz. Each circuit thus requires 60 kHz, 30 kHz for the forward channel and 30 kHz for the reverse channel. The 40 MHz (40,000 kHz) originally allocated to the cellular service could therefore be carved

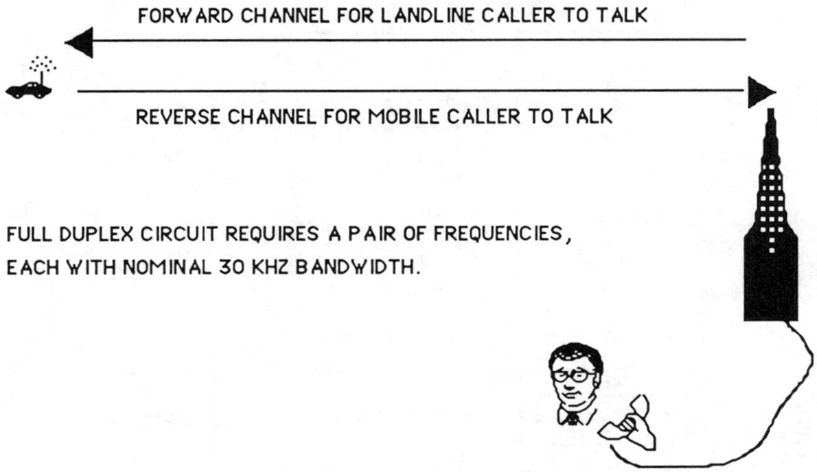

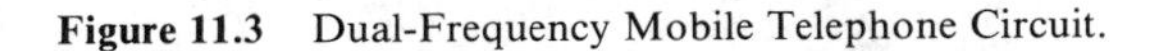

Figure 11.3 Dual-Frequency Mobile Telephone Circuit.

up into 666 frequency pairs (40,000 divided by 60), each serving one full-duplex telephone circuit. (Note that for convenience these channels are often referred to as 30-kHz channels for purposes of bandwidth analysis — even though there are actually 60 kHz utilized by each circuit.)

Narrowband channelized systems have a number of common characteristics, regardless of the precise channel bandwidth. First, they imply the necessity of sharply defined emissions limitations on individual transmitters. In a narrowband system, the entire transmission by a given mobile must be confined within the narrow, specified bandwidth; otherwise adjacent channel interference will be created. The Federal Communications Commission defines precise emissions *masks* to indicate how far the signal must be attenuated as a function of the deviation from the center frequency. (See Figure 11.5.) The tightness of these bandwidth limitations becomes a dominant factor in the evaluation and selection of radio-modulation techniques. (See Chapter 15.) It also influences the design of the transmitter and receiver elements, especially the filters, which can greatly affect the cost of the mobile unit. As a rule, the narrower the bandwidth, the tighter the filter requirements, the more expensive the radio. It was partly to lessen this cost burden that the FCC permitted a relaxed receiver mask for cellular radio, on the assumption that adjacent channels would never be utilized in a large, multicelled system. (See Figure 11.6.)

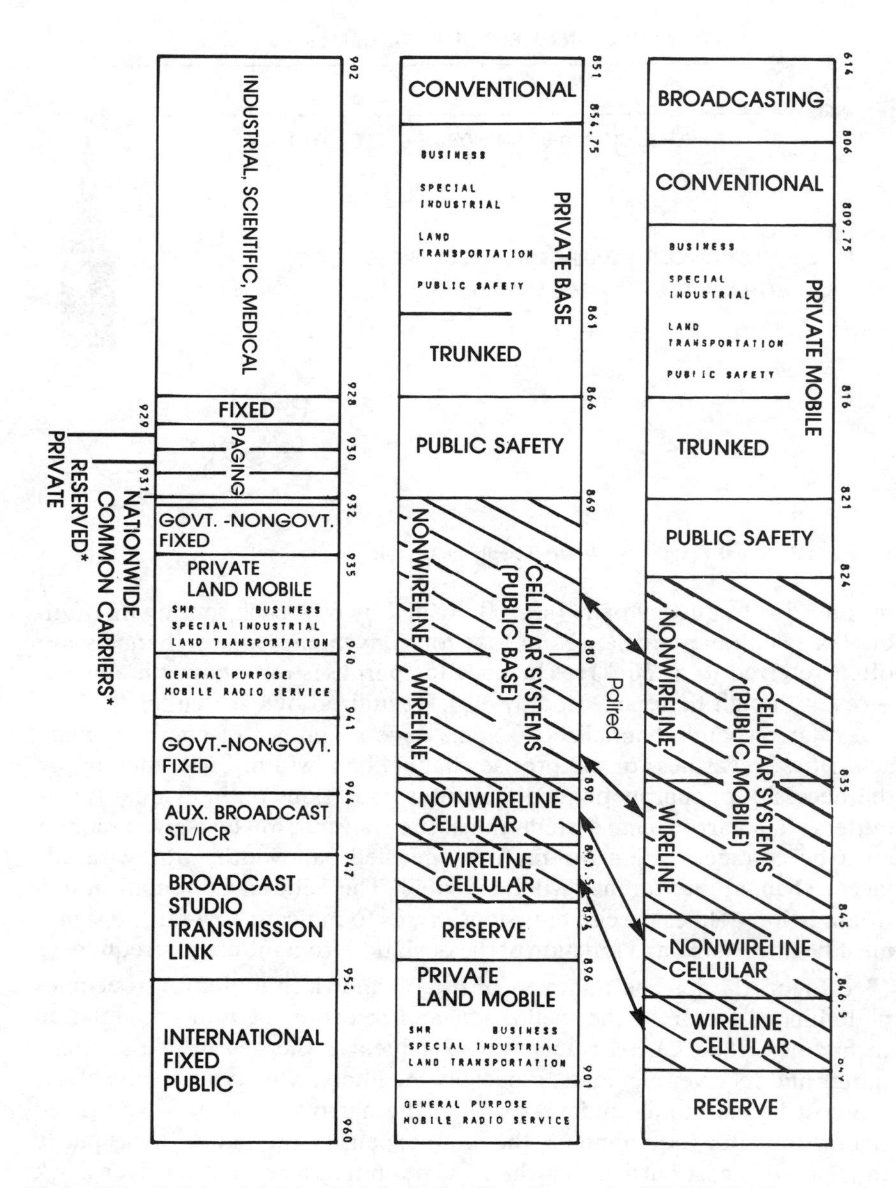

Figure 11.4 Cellular and Other Mobile Radio Allocations in the 800 MHz Region.

Source: RCR Publications, Land Mobile Frequency Chart.

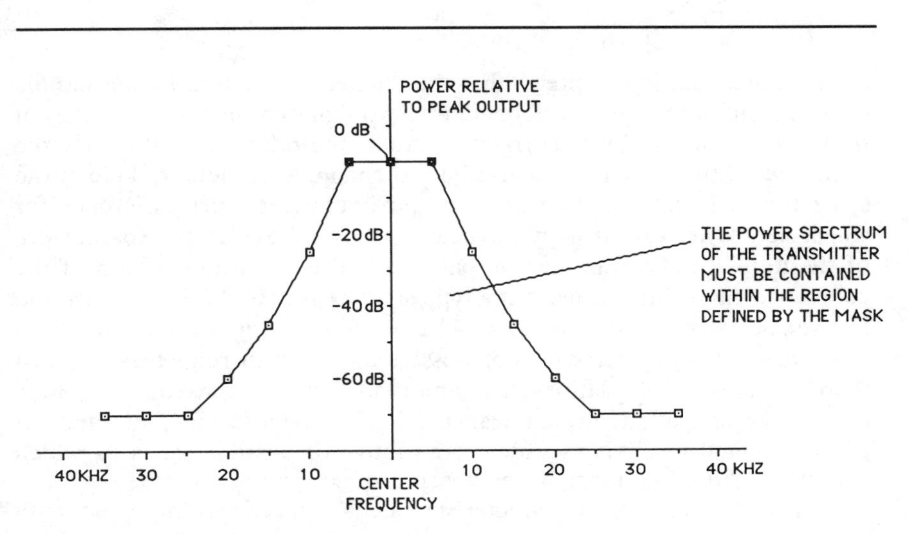

Figure 11.5 Emissions Mask for Narrowband Channels and Adjacent Channel Interference.

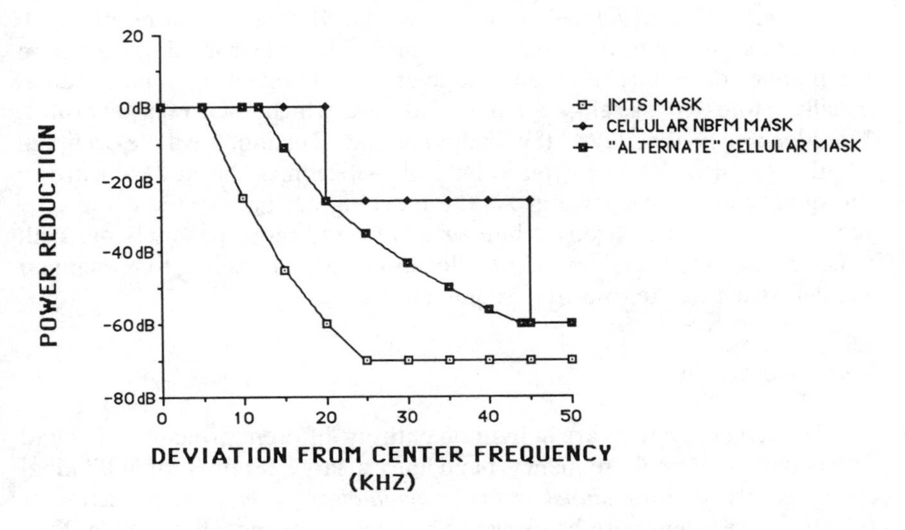

Figure 11.6 Conventional IMTS Mask Overlaid with Cellular Receive Mask.

A second important characteristic of narrowband transmission is that the entire transmission occurs within what is called the *coherence bandwidth,* also called the *correlation bandwidth* [2]. In fact, some authorities argue that this is the defining feature of narrowband systems — "narrow" means "within the coherence bandwidth [3]." What does this mean? We have pointed out in Chapter 8 that the fluctuations caused by the mobile environment are frequency-dependent. Two signals at sufficiently different frequencies will exhibit uncorrelated fading and other properties. By the same token, two signals on near-adjacent channels will tend to fade at the same time. As Jakes defines it, "the maximum frequency difference for which signals are still strongly correlated is called the coherence bandwidth of the mobile radio transmission path [4]." The precise definition of the coherence bandwidth varies, but a typical figure for 800-MHz mobile-radio frequencies is given by Jakes as 640 kHz [5]. The significance of this is that narrowband systems do not possess any built-in robustness against mobile-channel degradations, as some wideband systems do. If a fade occurs, the entire narrowband transmission will be affected. The range of possible signal levels is as wide as the range of possible fades — which may be 40–50 dB or more in the absence of countermeasures.

The third important characteristic of channelized systems is that they are by their nature *blocking* systems. In a given configuration, a given base transmitter operating with a given number of channels can only handle a number of simultaneous calls equal to the number of channels. If there are 50 channels and 50 calls in progress, the 51st caller will be blocked. A statistical model of the blocking system can be constructed, based upon the number of mobile users and the average amount of telephone traffic, usually expressed in call-seconds or Erlangs, which each one generates. The *blocking probability* — the likelihood that all channels will be occupied when a given user tries to place his call — becomes the key measure of the quality of mobile service. As the blocking probability rises, the only recourse in a channelized architecture is to add more channels per cell, which means additional frequency allocations, or, if possible, to reengineer the cell structures to improve frequency reuse.

Wideband Systems

Wideband systems are built upon entirely different principles. Instead of dividing the total frequency band into a large number of individual channels, *the entire channel is made available to every user.* Each user transmits over the entire band, which may be many megahertz wide. Furthermore, a large number of individual users can all use the same wideband channel *at the same time.* Such systems are known generically as *spread-spectrum* systems and they involve some truly ingenious techniques. A more detailed discussion of spread-spectrum communication is given in Section 12.3.

One of the chief attractions of wideband transmission is the fact that the transmission bandwidth exceeds the coherence bandwidth, which means that a multipath-induced fade does not affect the entire signal. Overall, the impact of Rayleigh fading could be, potentially, diminished considerably. This inherent robustness in the face of mobile-channel degradations is one of the reasons that wideband systems have attracted interest. Another potential benefit of wideband systems is possibly greater resistance to interference effects in a frequency reuse situation [6], although there is disagreement on this point.

Indeed, there is disagreement on many points concerning wideband applications to mobile telephony. Wideband systems are derived from decades of military communication-system work, much of it still classified, and, as such, the wideband approach is enveloped in a certain mystique. Spread-spectrum systems exhibit many characteristics that seem exotic from the standpoint of conventional channelized systems. For example, unlike narrowband systems, there may be no hard limit on the number of mobile users who can simultaneously gain access to a spread-spectrum base station. There may be no blocking *per se.* Instead, as more users pile on top of each other, the noise level for each user gradually increases — the circuit quality deteriorates "gracefully," until, it is argued, a kind of feedback principle comes into effect: individual users who are involved in nonessential calls begin to shorten or postpone such calls until off-peak hours when the circuit quality available to them will be better. Priority calls in which the users can easily tolerate a "gracefully degraded" circuit — calling the fire station, for example — are never blocked. It is a bit like the self-regulating aspects of the urban rush hour. As peak-hour travel conditions become really awful, some drivers are encouraged naturally to stagger their arrival and departure times. Whether or not one agrees with the application of this principle to a mobile-telephone system, it is certainly a very different way of looking at the capacity issue from conventional, channelized, narrowband architectures.

Another strange possibility opened up by wideband architecture is the prospect of actually overlaying a wideband system upon existing, fully loaded, channelized, narrowband systems. Apparently, under certain assumptions, it may be possible to create a wideband network right on top of today's cellular systems, using precisely the same spectrum.

This may suggest that the wideband *versus* narrowband dichotomy is not a hard either-or choice. I believe that this would be a misleading inference, however. It is uncertain whether, as a practical matter, spread-spectrum transmissions could be superimposed upon existing FM transmissions without objectionable mutual interference. Indeed "uncertain" is probably the key word with respect to wideband technologies. To my knowledge, no commercial wideband system has ever been built; working wideband-system experience has been with military systems where performance objectives and criteria are totally different from those of commercial telephony. The application of such techniques to the design of mobile-telephone systems is still somewhat speculative. There are significant disagreements in the literature over theoretical calculations for fundamental performance parameters of proposed wideband mobile systems. Since no one has ever built a spread-spectrum radio for the commercial market, however, the economics of such systems are also highly uncertain.

For this reason, it appears likely that the immediate next generation of mobile telephony will be based on narrowband architecture. There are, however, still a number of strong proponents of wideband techniques for the next round. Also, it must be said that, as the potential of spread-spectrum techniques is explored and the current uncertainties clarified, there is the possibility that wideband systems will gain greater favor.

Addendum: Ping-Pong Architecture

For the sake of completeness, it should be noted that there is a third possible channel structure — really a variation of narrowband channelization — in which the base and mobile-unit transmissions would both take place on a single frequency in separate time intervals. Such systems, which have been used in wireline digital-loop transmission, are known as *Ping-Pong* systems, *burst-mode* systems, or *time-compression multiplexing* systems [7]. Conceptually applied to a radio channel, the Ping-Pong architecture would use a time-division slot structure (see Section 11.5) with alternate slots assigned to the mobile unit for transmission, interspersed with slots for base-station transmission. (See Figure 11.7.)The burst-transmission bit rate would have to be at least 2 times faster, and probably 2.25 to 2.5 times faster, than the coder rate.

A chief potential advantage of time-compression multiplexing is the elimination of the requirement for a channel pair. In principle, this could allow duplex mobile telephony to penetrate areas of the radio spectrum that, for historical reasons, have not been organized with paired channel allocations.

While Ping-Pong systems can be imagined that might function well enough for individual mobile-radio circuits, however, it is harder to envision how such systems would work in large-scale systems. The problem is controlling interference among so many transmitters all operating on the same channel. At the very least, it would seem that all mobile subscribers would have to be rigorously synchronized, or slaved to a master

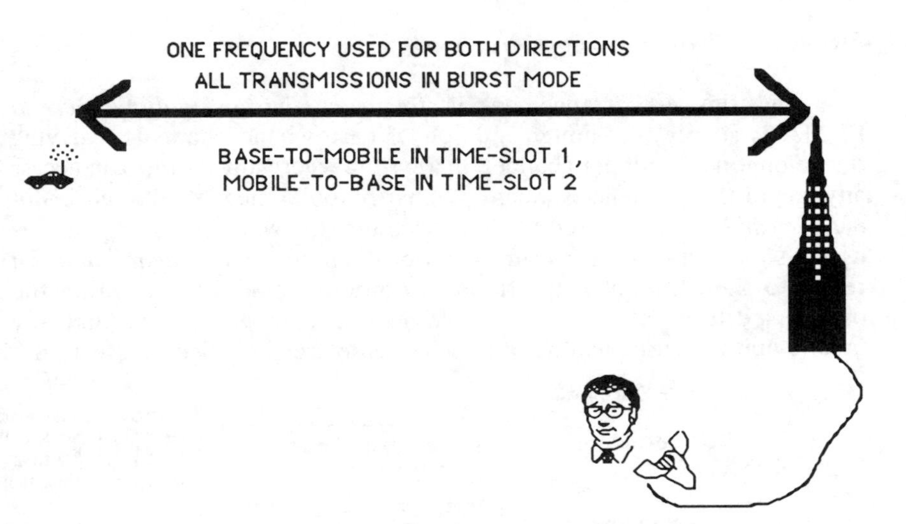

Figure 11.7 Ping-Pong Architecture for Mobile Telephony.

clock. This synchronization might well have to extend not just to all the users within a given cell, but to all the cells in the entire system. In any case, so far no one to my knowledge has proposed time-compression architecture for cellular radio.

11.5 MULTIPLEXING AND ACCESS TECHNIQUE: FDMA *versus* TDMA *versus* CDMA

Once the telephone circuits have been created, a method must be established for allocating individual circuits to individual users on demand. As specified in the preceding section, the access system should allow any user to utilize any circuit in a fully trunked system. This procedure is called *demand-assigned multiple access* (DAMA) or, simply, *multiple access*. It should be noted, however, that the choice of access technique has implications beyond merely the way in which a channel is assigned; it also affects the way in which the circuits are mapped onto radio channels and has a tremendous impact upon system economics, as we shall see in Chapter 15.

The options for implementing multiple access fall into at least three broad and, to some degree, mutually exclusive categories.

Frequency-Division Multiple Access

The simplest arrangement is *frequency-division multiple access.* FDMA is simply the implementation of narrowband channels, carrying one telephone circuit per channel, in a system where any mobile can access any one of the frequencies. Such systems are sometimes called *single channel per carrier,* abbreviated SCPC. (Note that the word "channel" here is used in the sense we are using the word "circuit.") *Frequency division* refers to the creation of the frequency-differentiated circuits within the overall spectrum allocation. Multiple access, to repeat, means that any mobile unit may use any one of the circuits so created. (See Figure 11.8.)

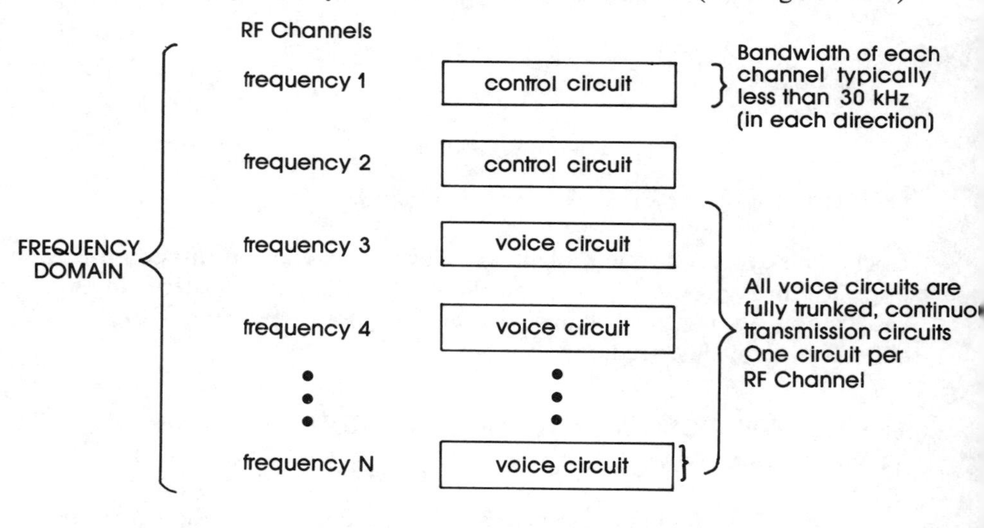

Figure 11.8 FDMA Architecture.

The mobile unit must have *frequency agility:* that is, it must be able to tune to every one of the available frequencies. The system controller must have the ability to control the assignment of each mobile to the appropriate available frequencies at the time of each call request. The mobiles must be under strict control of the base station.

Existing analog cellular systems utilize FDMA. It is also possible to have digital FDMA, where the voice is digitized and some form of digital modulation is employed, but only one circuit is established for each frequency.

Another characteristic of FDMA systems is that the transmission is continuous in both directions. This means that *duplexer* circuitry is required at both the mobile and the base-station radios, to allow the transmitter and the receiver to operate simultaneously.

Time-Division Multiple Access

Time-division multiple access is a more complex architecture. Each carrier frequency is divided up into a number of time slots, and each time slot constitutes an independent telephone circuit. The arrangement is analogous to the structure of T-carrier described in Chapter 6. The bandwidth of the carrier frequency, the bit rate in the channel, and the number of time slots created within the channel, may differ from one TDMA proposal to another.

For example, the Pan-European digital cellular design will apparently be based upon a narrowband TDMA architecture involving a carrier of 200-KHz bandwidth with a channel bit rate of approximately 200 kb/s. This bit stream is to be divided into eight time slots, each representing approximately 25 kb/s and capable of encoding one telephone voice circuit each. There would be thus eight circuits per carrier in the GSM proposal. (See Figure 11.9.) To take another example of TDMA architecture, International Mobile Machines Corporation in the United States has developed a wireless point-to-multipoint subscriber carrier system employing a 20-KHz carrier, transmitting at a 64-kb/s channel rate, with four time slots on each channel utilizing 16-kb/s coding for the voice. IMM has also conducted tests of this architecture in a mobile configuration. (See Figure 11.10.)

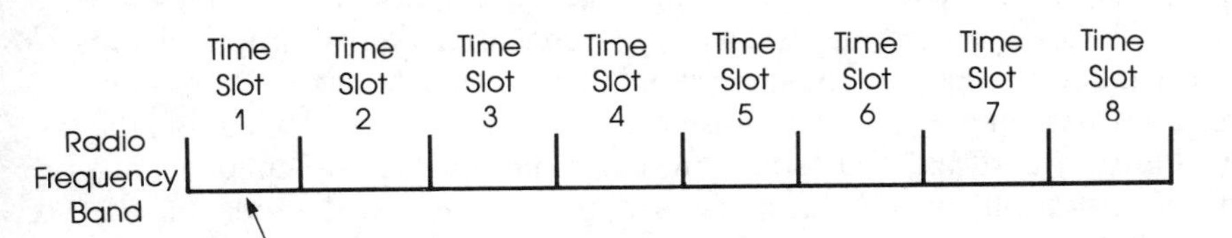

Note: As of this writing, many details of the transmission standard have yet to be finalized. The European system is expected to employ frequency hopping.

Figure 11.9 One Version of the Emerging European Standard for Digital Cellular Transmission — TDMA.

Both systems are actually combined FDMA and TDMA systems, in the sense that users have access to all frequencies and all time slots within

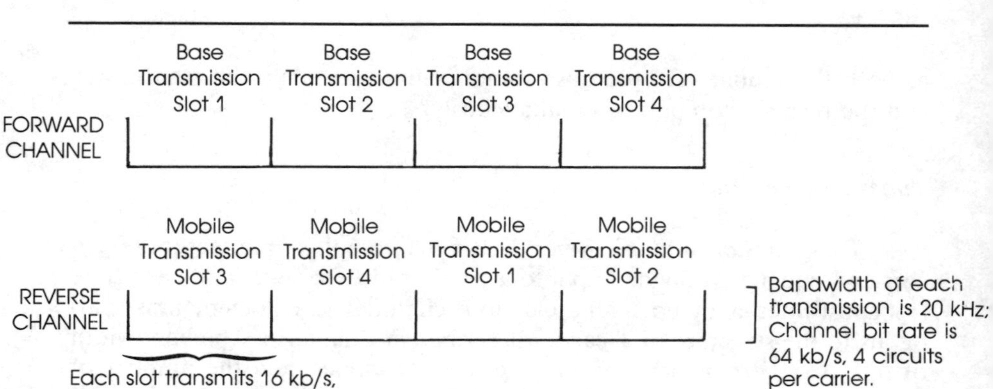

Figure 11.10 Another TDMA Channel Structure.
Note: Base and mobile transmissions for any given circuit are offset, which allows us to eliminate the duplexer circuit in a mobile unit.

frequencies. Effectively, such systems create a time-frequency matrix in which each cell represents a telephone circuit. The IMM system matrix is shown in Figure 11.11. The management of multiple users accessing circuits on each frequency channel is accomplished by tightly synchronized transmission sequences. Each user is assigned a transmit slot and a receive slot, which are offset. The bit stream which is continuously generated by the IMM coder at a 16-kb/s rate is buffered for three slots and then transmitted during the fourth slot at the 64-kb/s channel rate.

This noncontinuous, burst-transmission mode allows the simplification of the radio, particularly through the elimination of the duplexer circuitry. This can be done because the transmitter and the receiver are not operative during the same time slots. In place of a duplexer, the system uses a fast switch between the transmitter and the receiver. This offers the possibility of cost and size reduction in the mobile radio-telephone unit. The base station, which transmits in a continuous TDM mode on the forward channel, must synchronize four operating subscriber units on the reverse channel, each bursting in sequence for one time slot only. (See Figure 11.12.) These illustrations based on the IMM system are similar in overall structure to the arrangements for any digital TDMA system. (See Figure 11.13.)

Another result of noncontinuous TDMA transmission may be a lessened cochannel-interference problem (see Chapter 15).

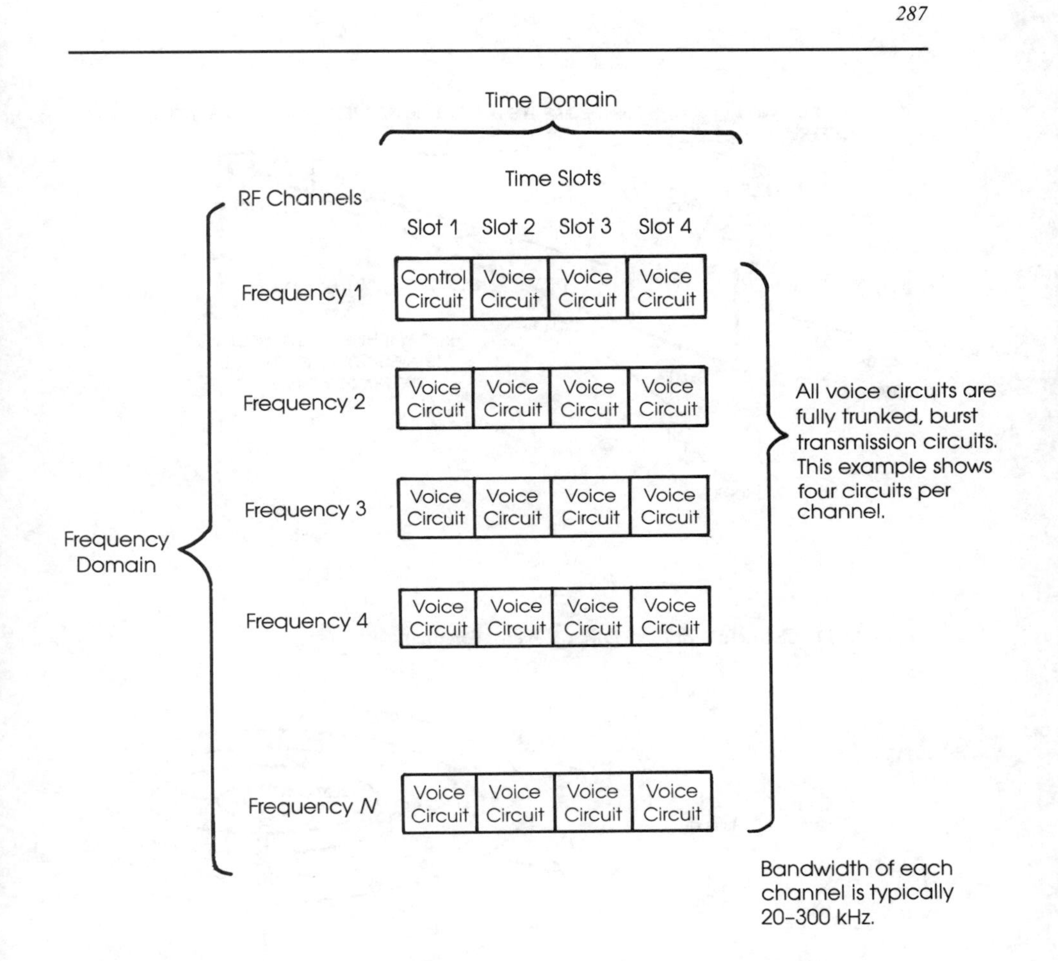

Figure 11.11 TDMA System Architecture — the Time-Frequency Matrix.

Code-Division Multiple Access

Code-division multiple access is the form of multiple access employed by spread-spectrum wideband systems. It is a very different concept, based on the principle that each user is distinguished from all others by the possession of a unique code which he imprints upon his transmission in one of several ways, and which the receiver also uses to select the proper

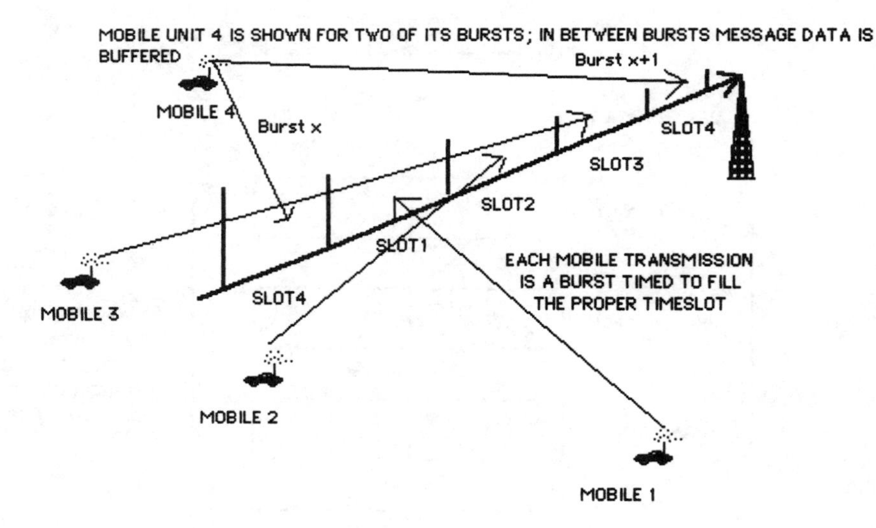

Figure 11.12 Structure of the TDMA Reverse Channel.

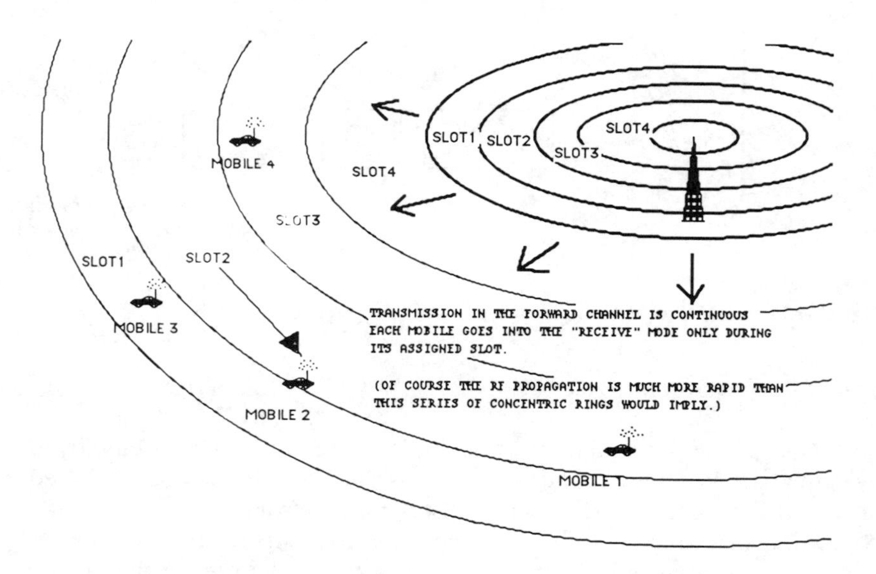

Figure 11.13 Idealized Structure of the TDMA Forward Channel.

transmission from among the many simultaneous spread-spectrum circuits in operation. The reader is referred to Section 12.3 for an exposition of code-division multiplexing and other aspects of spread-spectrum systems.

Summary of Access Techniques

Thus, there are several possibilities for the choice of multiplexing technique [8]. (See Figure 11.14.) As noted above, the Pan-European digital cellular standard appears to be headed toward a digital TDMA architecture [9]. In North America, both AT&T [10] and Motorola [11] — the leaders in the first generation of cellular-systems architecture — have given early indications of a preference for a digital FDMA architecture for the next generation. IMM has fielded a TDMA system. In the discussions preceding the tentative GSM standard, strong French and German industry groups have favored the wideband CDMA approach with mixed TDMA elements [12].

	ANALOG	DIGITAL
FDM (SCPC)	TODAY'S CELLULAR RADIO SINGLE SIDEBAND	NORTH AMERICAN PROPOSALS AT&T (1988) MOTOROLA (1988)
TDM	NONE	EUROPEAN GSM STANDARD NORTH AMERICAN PROPOSALS IMM (1988) NTI (1988) NEC (1988)
CDM	NONE	GERMAN & FRENCH PRE-GSM PROPOSALS INCORPORATED CDM ELEMENTS

Figure 11.14 Classification of Next-Generation Cellular Proposals.

Leaving wideband systems aside for the moment, the choice between FDMA and TDMA is not as clear-cut, nor the consensus as well defined, as is the case for the analog *versus* digital or the narrowband *versus* wideband decisions. The TDMA *versus* FDMA issue appears to be shaping up as the major remaining architectural dispute with respect to the next generation of cellular systems. The relative strengths and weaknesses of the two approaches are discussed below in Chapters 13 and 15.

11.6 SYSTEM CONTROL STRUCTURES: CENTRALIZED *versus* DISTRIBUTED CONTROL

The fourth broad choice facing designers of the next generation concerns the method by which mobile-telephone circuits are assigned to specific callers upon request and the procedures for achieving the proper signaling and control interface to the existing wireline network.

11.6.1 Nets

One decision relates to whether the system is designed to allow *net* operations. Nets involve the creation of multilateral communication links that join multiple users in the same circuit. Often where individual mobile units are allowed to communicate with one another directly, without the intermediation of the base station, except possibly during call setup. (See Figure 11.15.) In a standard mobile-telephone system such as today's analog cellular systems, net operations are not permitted. All circuits are bilateral only, and calls are linked through the central cell-site base station at all times. (See Figure 11.16.)

Net operations may encompass a number of variations in communication topology. For example, fleet dispatch — where a base unit can communicate with all users simultaneously — is a netlike operation that is essential for many business services, such as taxi fleets, as well as for police and other emergency communication. Another net feature is the ability of each user to break in and broadcast to the entire population of users, or to some defined subpopulation. This is also vital for many police and emergency functions.

Net operations are permitted in many types of mobile system. CB radio is one kind of net. The now-defunct proposal from General Electric for a personal communications radio service was another netlike operation. In the United States, special mobile-radio services (SMR) are authorized to provide either interconnected telephone service or fleet-dispatch service. Net capabilities are required in many areas of government and business communication, and to the extent that mobile telephony anticipates the absorption of some of these users within an integrated system, the incorporation of net capabilities is a valid design question. Moreover, I believe the very concept of a net operation will undergo a transformation, as digital

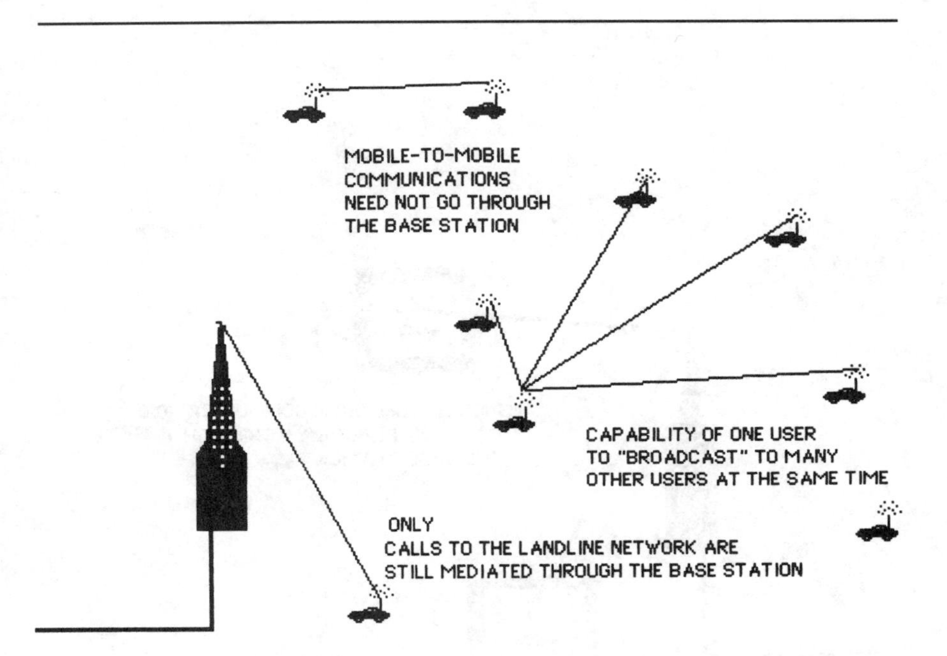

Figure 11.15 Model of Net Operations.

systems come in. Nets, as conventionally understood, are a kind of lower-grade service in the eyes of mobile-telephone engineers. In a digital system capable of transporting voice and high-speed data among many users simultaneously, however, a net begins to look surprisingly similar to a local area network in which the nodes happen to be mobile.

There are a number of advantages to net operations. For one thing, a net is inherently more spectrum efficient than a base-controlled system. For intra-system, mobile-to-mobile calls, a net utilizes only one frequency pair, while a conventional system requires two frequency pairs. A net is also inherently more robust than a base-controlled system, which makes it preferable for certain military or defense-related applications. If the central base station is incapacitated, a net will remain operational, although interface to the telephone network will most likely be impaired [13]. (See Figure 11.17.) A net architecture can potentially reduce system costs as well, by simplifying and reducing the amount of equipment necessary at the base station. Finally, in some net architectures, each mobile unit can act as a relay for communication between the network interface point and more remote subscriber units, again providing an additional path redundancy and greater robustness. (See Figure 11.18.)

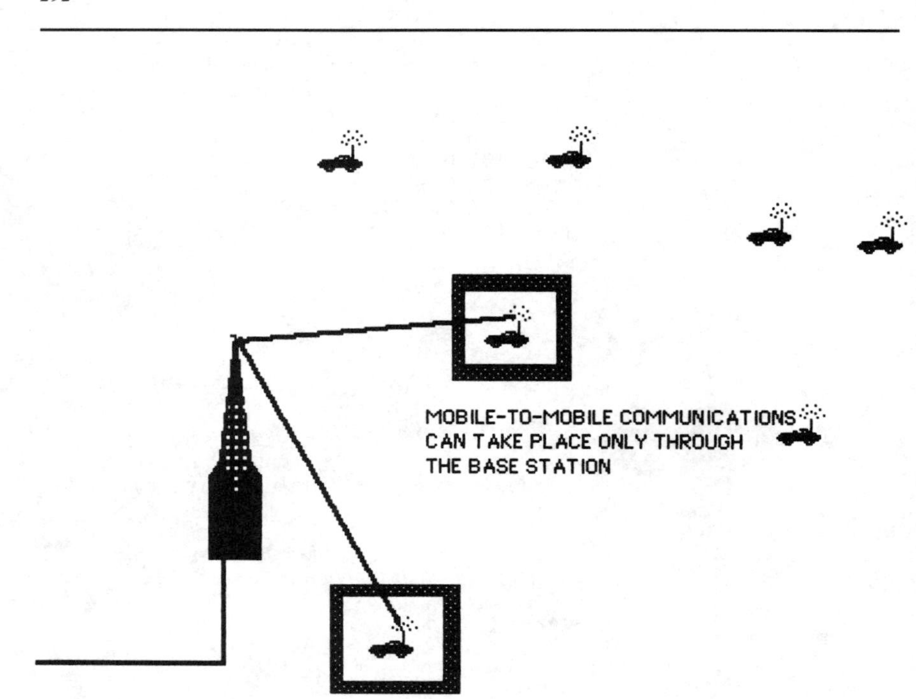

Figure 11.16 Model of Cellular-Style Base Station Control.

Nets do incur several disadvantages. For one thing, the mobiles are no longer under the strict control of the base station. In base-controlled systems, interference can be regulated by slaving the mobiles along a number of dimensions, including power, timing, and frequency assignments. Although it is a complex subject which has not been fully analyzed, it would appear that a net architecture would require a greater margin against interference due to the inability of the base station to optimize each subscriber's transmission characteristics.

It is also likely that net operations in a tightly synchronized format like TDMA or CDMA could significantly increase the complexity and the cost of the mobile units. FM nets, single channel per carrier, are fairly easy to implement and are widely used today. But a TDMA net requires much more intelligence at each mobile unit, and the few such systems that have been fielded, notably the Army's Joint Tactical Information Distribution System, are an order of magnitude more costly than current mobile-telephone units.

To my knowledge, none of the proposals for the next generation of cellular systems currently being discussed allows for net operations. It is perhaps shortsighted, but I think it likely that net operations will not be part of the next-generation architecture, and I do not intend to discuss

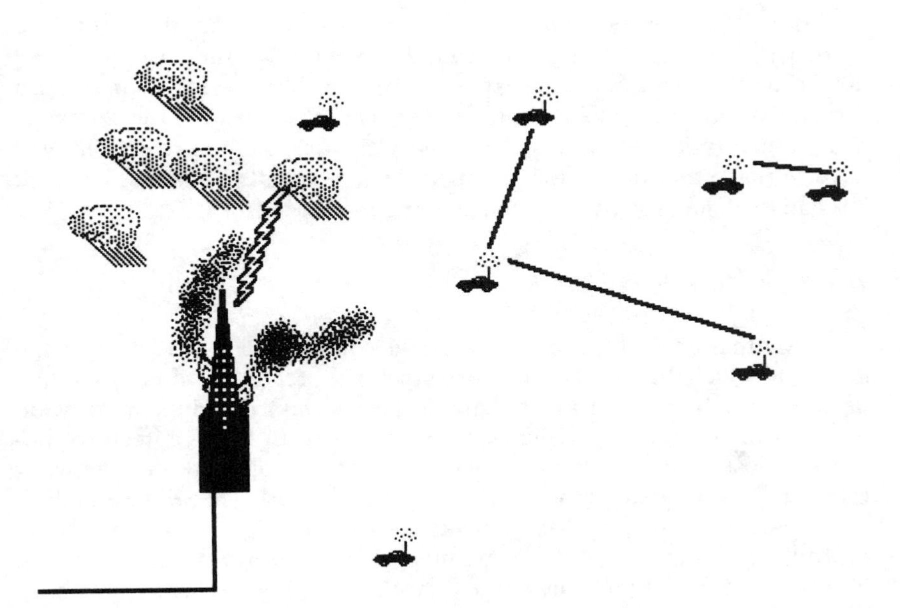

Figure 11.17 Mobile-to-Mobile Network Operations May Continue Despite the Loss of Base Station Capabilities.

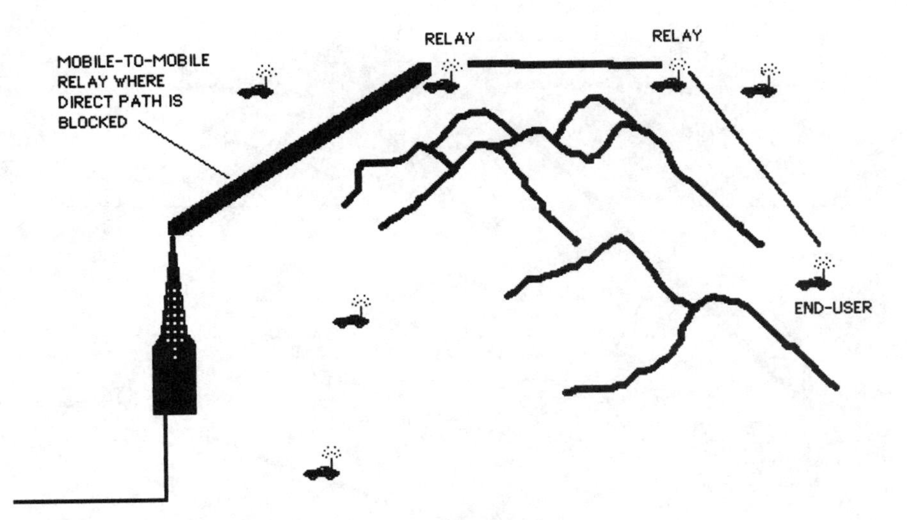

Figure 11.18 Hypothetical Mobile Relay Architecture.

them further in this work. Net operations, limited as they are mainly by assumptions about processing loads and the associated costs, are, however, susceptible to the same digital logic we have been propounding heretofore — namely, as digital technology comes rapidly down the cost curve and up the performance curve, today's assumptions about what is or is not a prohibitive processing load are likely to become less valid in the relatively near future (mid-1990s). I believe that ultimately mobile telephony will have to make allowances for net operations, just as the wireline network has had to come to grips with the emerging realities of local area networks.

11.6.2 Interface Point

Another important and very broad control decision relates to the degree to which the mobile-telephone system is integrated with the wireline network and the point at which that interface is defined and the integration is achieved. There are two diametric alternatives. In the architecture chosen by most first-generation cellular designers, a complete new network is built up, overlaid upon the existing wireline network. (See Figure 11.19.)

The alternative approach is to interface directly to the local telephone central office and to allow the existing wireline network to serve as the backbone of the mobile network as well. (See Figure 11.20.) Such an approach has been espoused by some manufacturers — most notably Novatel Corporation of Calgary.

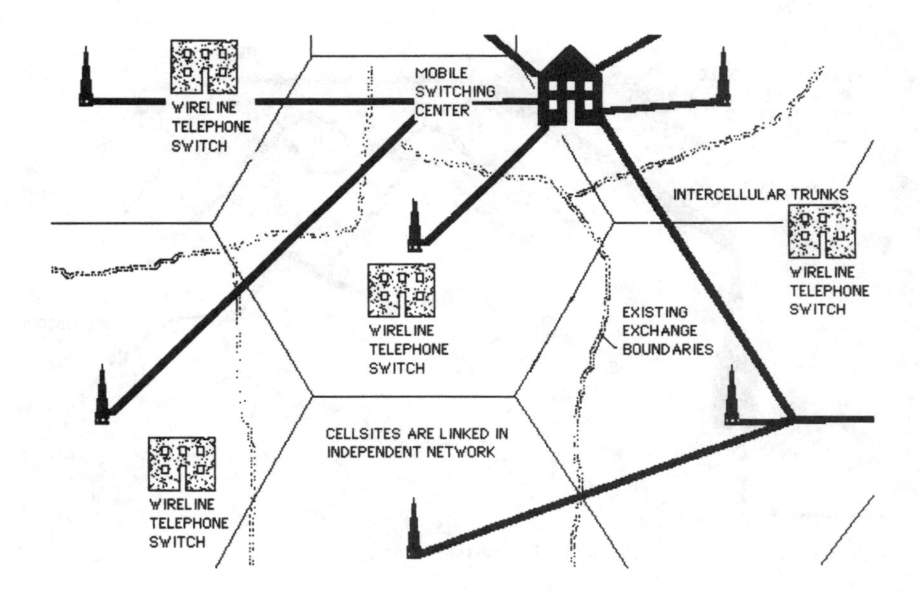

Figure 11.19 Overlay Cellular Architecture.

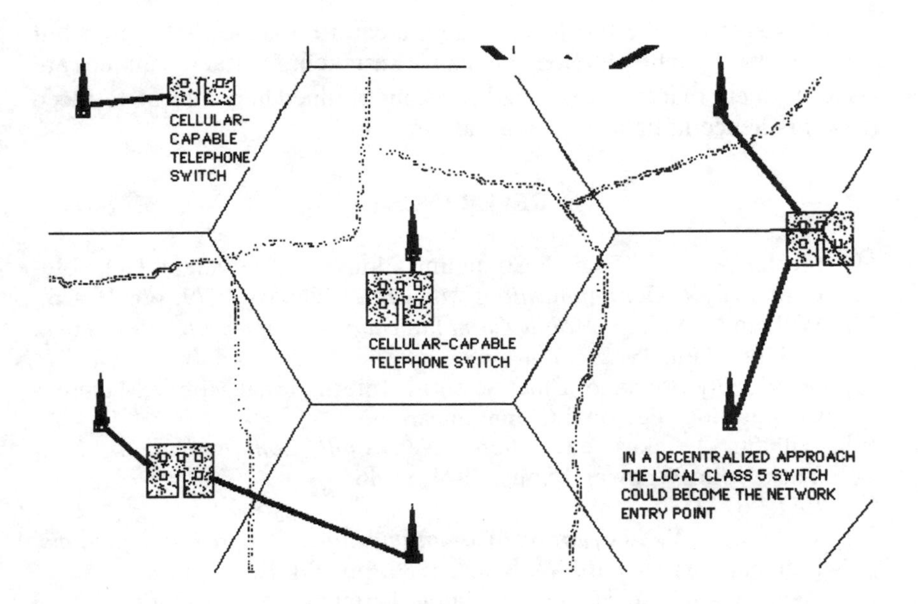

Figure 11.20 A Decentralized Cellular Network.

The use of the existing network for trunking and switching, and possibly billing, administration, and so forth, offers certain economic advantages. It also raises questions, however, about the ownership of the mobile system, when the operator is someone other than the local telephone company. These implications will be explored below in Chapter 14.

11.7 SUMMARY OF GENERAL DIRECTIONS OF THE NEXT GENERATION

All in all, if these questions were all to be resolved in the direction in which industry consensus seems to be developing today, the next generation of mobile-telephone systems would probably be based upon narrowband, digital, TDMA architecture, without net operations of any kind, and would probably use an overlay or redundant network type of architecture for control. As of this writing, however, these broad questions are still quite open, especially in North America and Japan, where next-generation planning is only now beginning. Even in Europe, where a narrowband TDMA format has been specified, some knowledgeable observers are beginning to speak of "two more generations" of mobile telephony before the end of the century [14]. This reinforces a theme to which we shall return in the final section: The notion that we are collectively taking a single step from the FM-technology plateau to a somewhat higher but equally static digital-technology plateau is mistaken. Digital techniques are in such ferment that even the broad decisions outlined here may be retraced five years hence in light of new breakthroughs.

REFERENCES

[1] M. R. Aaron, "Digital Communications — The Silent (R)evolution?" *IEEE Communications Magazine,* January 1979, pp. 16–26.

[2] William C. Y. Lee, *Mobile Communications Engineering,* New York: McGraw-Hill, 1982, pp. 44–45.

[3] D. Ridgely Bolgiano, Chief Scientist, International Mobile Machines Corporation, Personal Communication.

[4] William C. Jakes, ed., *Microwave Mobile Communications,* New York: John Wiley and Sons, 1974, p. 46.

[5] *Ibid.,* p. 51.

[6] William C. Y. Lee, *Mobile Communications Design Fundamentals,* Indianapolis: Howard W. Sams, 1986, pp. 176 ff.

[7] David G. Messerschmidt, "Digital Termination and Digital Signal Processing," in John C. McDonald, ed., *Fundamentals of Digital Switching,* New York: Plenum, 1983, p. 287.

[8] Tongze Jiang, "A Comparison between the Three Mobile Digital Communications Systems," *Proceedings of the 37th IEEE Vehicular Technology Conference,* Tampa, June 1–3, 1987, pp. 359–362.

[9] See the *Proceedings of the International Conference on Digital Land Mobile Radio Communications,* Venice, June 30–July 3, 1987, especially Session I on "Status of Studies, Experiments, and Harmonization Process in Europe."

[10] Joseph Tarallo and George I. Zysman, "A Digital Narrowband Cellular System," *Proceedings of the 37th IEEE Vehicular Technology Conference,* Tampa, June 1–3, 1987, pp. 279–280.

[11] James J. Mikulski, Remarks at the Federal Communications Commission Seminar on "The Future of Cellular Radio," September 2, 1987, Washington, D. C.

[12] K. D. Eckert, "Conception and Performance of the Cellular Digital Mobile Radio Communication System CD 900," *Proceedings of the 37th IEEE Vehicular Technology Conference,* Tampa, June 1–3, 1987, pp. 365–377.

[13] Tongze Jiang, *op. cit.*

[14] Remarks by Ericsson spokesman at Cellular Conference sponsored by Donaldson, Lufkin & Jenrette, St. Regis Hotel, New York, June 3, 1987.

Chapter 12
ALTERNATIVES FOR THE RADIO LINK

The radio link between the cell-site base station and the individual mobile-telephone user is the heart of the cellular system. The choice of radio technology will determine to a great extent the economic and performance characteristics of the cellular system.

Twenty years ago the choice was straightforward. There was one dominant radio technology: frequency-modulated analog transmission. Today there are dozens of alternatives, mostly in the digital domain, based upon an expanding portfolio of modulation methods and voice-coding techniques. There are several major families of voice-coding algorithms, for example, and within each family there are almost as many different algorithms as there are researchers in the field. Also, where once FM was the alpha and omega of modulation, today's mobile-system architect would probably consider the following techniques, among others:

Proposed Modulation Technique	*Illustrative Reference*
Single Sideband	[1]
Frequency Shift Keying (FSK)	[2]
Fast FSK	[3]
Tamed Frequency Modulation	[4]
Generalized Tamed FM	[5]
Double Phase Shift Keying (DSK)	[6]
Minimum Shift Keying (MSK)	[7]
Multiamplitude MSK (MAMSK)	[8]
Gaussian MSK	[9]
Continuous Phase Modulation	[10]
Phase Shift Keying (PSK)	[11]
Offset PSK	[12]
$\pi/4$ Shift PSK	[13]
Raised Cosine PSK	[14]
Compact Spectrum, Constant Envelope PSK (CCPSK)	[15]
Nonlinearly Filtered Quadrature Amplitude Modulation	[16]

Some of these techniques blur into one another. New variants are being spawned all the time. In fact, each technique actually comprises a subfamily of related methods. For example, phase shift keying (PSK) modulation may be based on either *coherent* detection, where there is an absolute phase reference, or on *differential* detection, where phase differences from one symbol to the next are the bearers of information, or on a mixture of the two. Differential PSK, to take it one level further, comes in several versions, depending upon the modulation level. Binary DPSK (2-level), quaternary DPSK (4-level), and 16-ary DPSK (16-level) have been proposed for mobile systems.

It would be beyond the scope of this book to sort out the overlapping alternatives completely, let alone attempt a comparative evaluation. It is important, however, to understand the general technological principles underlying the different approaches to the problem. Broadly speaking, there are three distinct super-families of radio-link technologies that are being considered for the next generation of cellular systems:

1. Advanced analog techniques;
2. Mainstream digital techniques;
3. Spread-spectrum techniques.

12.1 ANALOG TECHNIQUES

Although, in my view, we are clearly headed toward a digital architecture, it should be recognized, for the sake of completeness, that advanced analog techniques are still viewed by some as a viable link technology for the next generation of mobile telephony. Specifically, two approaches have been put forward: narrowband FM and single-sideband modulation.

12.1.1 Narrowband FM

As discussed in Chapter 2, the development of mobile telephony during the 1940s, 1950s, and 1960s was paced by several rounds of channel-splitting, which reduced the bandwidth of the standard FM mobile channel from around 100 to 50 kHz and then to 25–30 kHz. Some believe that it is quite possible now to reduce the FM bandwidth once again to 12–15 kHz, maintaining more or less the same transmission quality, although the quality question is still debated [17]. FM mobile channels of 15 kHz are being deployed in Japan [18] and in the United Kingdom [19] to help alleviate frequency congestion, and at least one major manufacturer in North America has proposed the implementation of channel-splitting to double capacity within today's cellular systems, at least as an interim move until full digital systems arrive [20]. On the other hand, split-channel architecture has recently been rejected on technical grounds for certain non-cellular mobile allocations in the United States [21]. Also, one authority on cellular technology has flatly stated that merely reducing the bandwidth in an FM system will produce *no gain in system capacity* because the narrower bandwidth channels will require higher carrier-to-interference ratios in a frequency reuse situation (see Chapter 9) in order to maintain the same quality [22]. In other words, channel-splitting may actually produce no additional circuits *in a cellular system,* unless we are willing to tolerate a significant reduction in transmission quality due to higher C/I.

In any case, channel-splitting can hardly be considered a true next-generation proposal; at best, it is a short-term Band-Aid for the capacity problem; in all likelihood, it may provide little real relief.

12.1.2 Single-Sideband Modulation

The more interesting analog approach is one that is both quite old and quite new. Single-sideband (SSB) systems have been in use for many years, first in wireline frequency multiplexed carrier systems dating back to the 1920s and 1930s [23], and later in point-to-point, long-distance radio [24]. SSB has become popular in the latter application because it is potentially far more spectrum efficient than conventional FM (by a factor of three to four). SSB microwave development began in earnest in the 1960s and 1970s, with such success, ultimately, that SSB became "the principal means for expanding [microwave] radio capacity in the 1980s," in the somewhat overstated view of the official engineering history of the Bell System [25].

The concept of single sideband is straightforward. SSB is a form of amplitude modulation. In the process of ordinary amplitude modulation, when a voice signal is impressed onto a carrier frequency, the pattern of the voice information is actually reproduced twice in frequency bands on either side of the carrier frequency. (See Figure 12.1.) Each *sideband* carries exactly the same information. It is possible to filter out one of the sidebands along with the carrier frequency without losing any information. This saves bandwidth. In principle, it should be possible to reduce the bandwidth of SSB transmission to equal that of the original baseband voice signal. In telephony, the voiceband, the actually transmitted bandwidth of the voice, is about 3 kHz. SSB can be thought of as simply a straightforward translation of the 3-kHz signal from one frequency to a higher frequency. Allowing for guardbands between channels, it should be possible to pack SSB channels together with about a 5–6 kHz spacing, five or six times more spectrally efficient than today's FM channels. (See Figure 12.2.) Indeed, claims have been made that certain forms of SSB may be capable

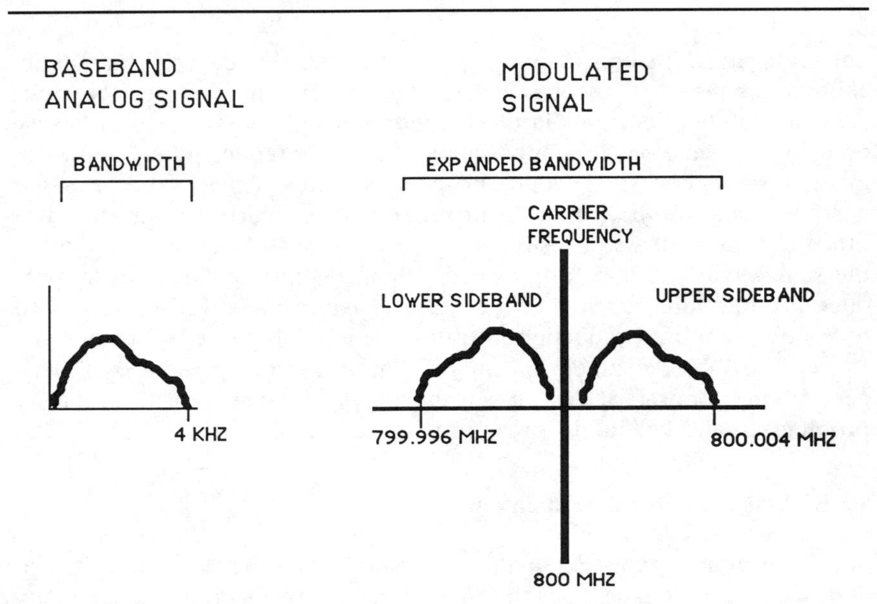

Figure 12.1 Double Sideband Modulation.

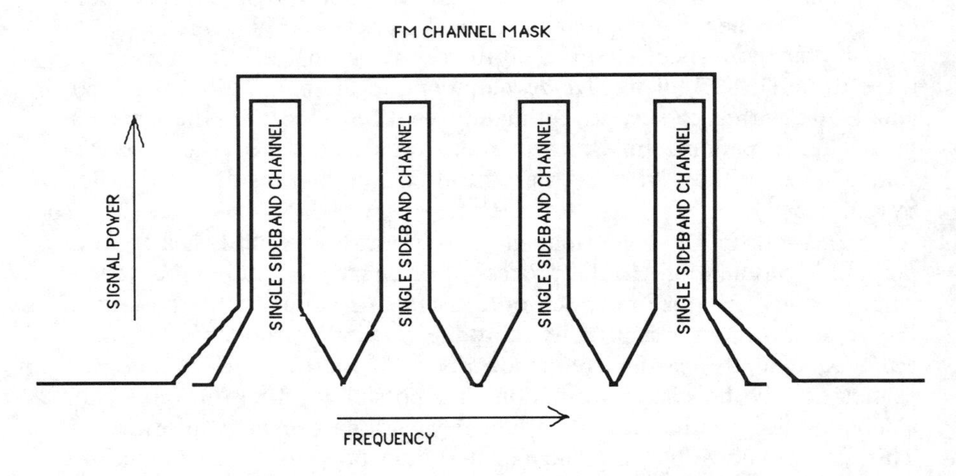

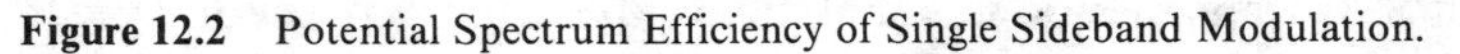
Figure 12.2 Potential Spectrum Efficiency of Single Sideband Modulation.

of a reduction in bandwidth *below* that of the original baseband voice signal, such that mobile-channel spacing of as little as from 2 to 2.5 kHz, including guardbands, would be possible [26].

Any modulation technique capable of delivering such an enormous spectrum-efficiency advantage over FM should have attracted considerable interest from the spectrum-starved mobile-radio community. The severe amplitude distortions created by multipath fading in the mobile environment, however, undermine any AM system in a mobile-radio application. The following paragraph from Jakes *et al.*, written in 1974, summarizes succinctly the then-prevailing view of the mobile industry toward this enticing chimera:

> Rapid Rayleigh fading generally has a disastrous effect on single-sideband (SSB) and AM communications systems. . . . The distortion introduced by the fading is larger than the output signal, independent of how much the transmitter power is increased. Without fading, the signal-to-noise performance of SSB or AM may not be as good as FM; however, neglecting cochannel interference, they enjoy a bandwidth advantage over FM. But with rapid fading, SSB and AM become unusable . . . [27]

In the late 1970s, concerned by the problem of congestion in the radio spectrum and perhaps also concerned that private industry was not pursuing the development of spectrum-efficient technology with sufficient vigor, the Federal Commmunications Commission itself initiated a program to evaluate a new approach to SSB, based on a process known as *amplitude companding*. The result became known as amplitude-companded single sideband (ACSB) [28]. Companding — the word is a compound of *com*pressing and ex*pand*ing — is a process whereby the amplitude of the signal is compressed prior to modulation and transmission and then expanded at the receiver to restore the original dynamic range. Figure 12.3 is adapted from a diagram of a two-stage companding process.

Companding appears to strengthen the SSB signal considerably. One of the benefits of companding the SSB signal is a greater resistance to the destructive effects of multipath-induced amplitude fades. A number of papers spawned by the FCC study effort have claimed that ACSB would prove to be sufficiently robust to function in a mobile environment. Some field studies have supported this contention, while other observers — including the major mobile-equipment manufacturers — have questioned it [29]. Initial skepticism focused on voice quality. More recently, the issue of cochannel interference in a frequency reuse situation has again raised its head. Note that in the paragraph from Jakes cited above, the bandwidth

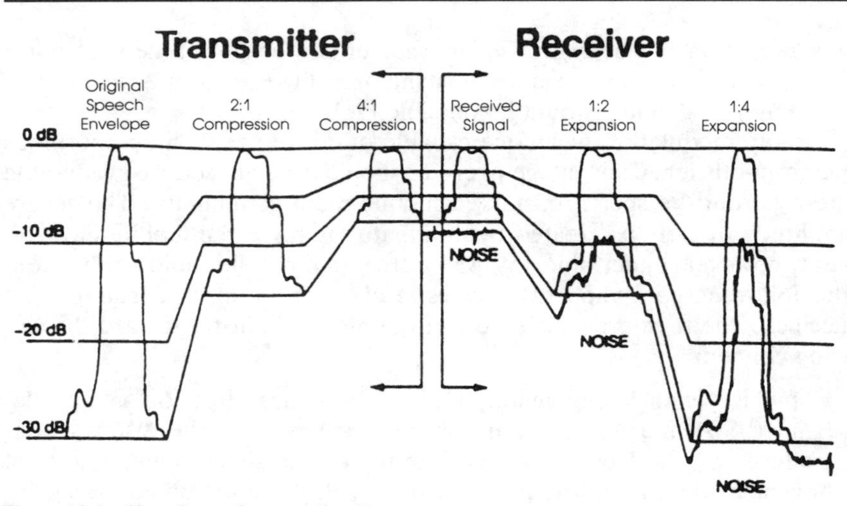

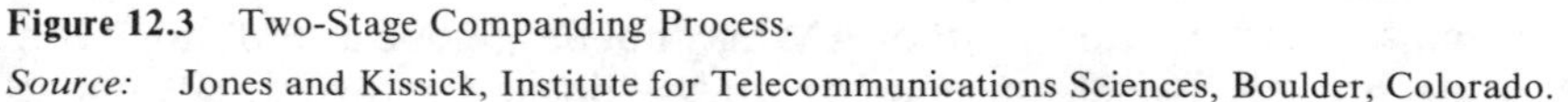

Figure 12.3 Two-Stage Companding Process.
Source: Jones and Kissick, Institute for Telecommunications Sciences, Boulder, Colorado.

efficiency was only posited "neglecting cochannel interference." In a more recent paper, William Lee argued quantitatively that in a Rayleigh-fading environment, in a cellular-reuse configuration, SSB required approximately 20 dB *higher* C/I ratio — in other words, the performance in a cellular interference-limited environment was approximately 100 times worse than standard 30-kHz FM. As a result,

> A 30 kHz FM cellular system is more spectrally efficient than a 5 kHz or 7.5 kHz single sideband (SSB) system. Also, far more SSB cells are needed to provide the same quality of service as FM cells in the same area for the same transmitted power in a cellular mobile radio system [30].

On the other hand, other experienced mobile-radio engineers have taken issue with Lee's calculations and believe that in time ACSB could surpass today's FM and alleviate the spectrum congestion [31]. Today's FM, however, is probably no longer the appropriate yardstick. Instead, ACSB must be measured against spectrum-efficient digital techniques which already achieve virtually the same spectral efficiency as ACSB, with probably superior cochannel-interference performance [32].

For a period of time during the late 1970s and early 1980s there was considerable hope that ACSB could emerge as a highly spectrum-efficient alternative to FM for mobile applications. This enthusiasm has faded as digital solutions loom. Even its ardent proponents will generally concede that ACSB is best viewed as a possible interim solution. This might change if 2-kHz channel spacing were proven feasible and cochannel interference on the narrower channels did not steal back much of the gain. As of today, that does not seem likely.

12.2 DIGITAL TECHNIQUES

As discussed in Chapter 6, a digital radio-telephone technology consists of two processes (see Figure 12.4):

1. *Voice coding:* conversion of the analog voice signal into digital form;
2. *Modulation:* impressing the digital information upon the radio signal, effected by varying some key parameter of the signal in a controlled manner.

Coding and modulation are often considered to be entirely separate arts. At the technical level this is quite true. Voice coding and modulation are logically independent processes. A voice coder produces a bit stream. The coder does not "know" whether that bit stream is destined for an 800-MHz radio transmission, a fiber-optic link, or a wireline connection. By the same token, the modulator impresses the received bit stream onto the radio carrier. It does not matter to the modulator where the bit stream came from or even what it represents. Workers in the two fields tend to concentrate on different disciplines. An expert in digital voice will usually know very little about digital radio, i.e., modulation. His world revolves around questions about how people produce speech and how they hear the speech of others. He is knowledgeable in acoustics and physiology. Discussions of digital radio, e.g., point-to-point microwave, usually ignore coding entirely, except perhaps to specify the basic bit rate that is to be assumed; the focus is entirely upon modulation techniques, detection methods, required signal-to-noise ratios, multipath effects, and so forth. The systems architect, therefore, enjoys a great deal of freedom in selecting coders and modulation techniques; in principle, any coder algorithm can be married with any digital modulation scheme.

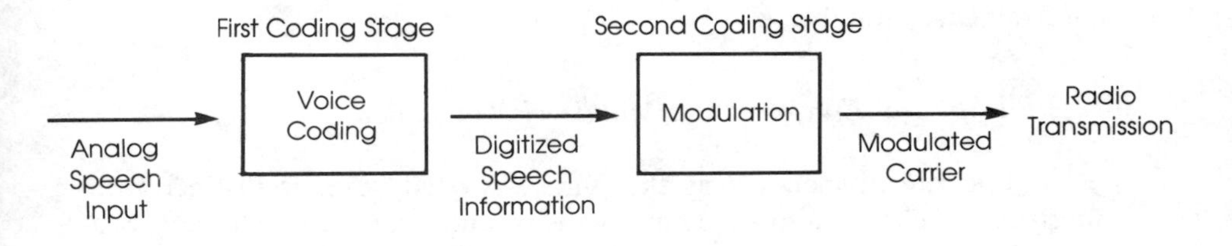

At a higher level, however, the choice of coding technique and the choice of modulation technique do interact. The forcing factor is the inherent shortage of radio spectrum, which constrains both the coder and the modulation process. The overall objective of any digital radio-telephone link within a high-capacity cellular system must be to reduce the occupied bandwidth as much as possible. This in turn sets the main goal of the coding system designer: to achieve good (acceptable) quality at the lowest possible bit rate. A more multivariate task is set for the designer of the modulator-demodulator: the modulation technique should achieve the highest possible data rate in the smallest occupied bandwidth with the best possible cochannel-interference performance.

In Sections 12.2.1 and 12.2.2, we will examine both modulation and coding in greater detail. The focus will be on the problems facing the designers of codecs and modems, given the unique problems of the mobile environment, as well as the other design challenges discussed in Part IV. In Section 12.2.3, we will consider how, from the system architect's perspective, coding and modulation decisions can be interdependent. We will begin with modulation methods, which are justly considered to be more fundamental in defining the performance of a radio-based system.

12.2.1 Digital Modulation for Mobile Radio

Ignoring economics for a moment, the choice of modulation technique for a mobile radio-telephone system is driven by several technical considerations, chief among which are:

1. The scarcity of bandwidth, leading to a need for spectral efficiency;
2. The problem of adjacent-channel interference, leading to the requirement for narrow-power spectra;
3. The problem of intersymbol interference, which imposes hard limits on the transmission rate in a mobile environment.

It should be kept in mind that all three parameters are interrelated; in the following discussion, for the purpose of clarification of the underlying issues, some conceptual liberties are taken in treating each of the three constraints separately.

12.2.1.1 Spectral Efficiency and Multilevel Modulation

"Spectral efficiency," as the phrase is used here, is distinct from *spectrum efficiency*, a more comprehensive and much more slippery concept which is treated in Chapter 15. Spectral efficiency refers to the number of bits that are transmitted in a given period of time, usually one second, over a radio channel with a defined bandwidth. Since the channel bandwidth is measured in kilohertz (kHz) or megahertz (MHz), it is possible to define spectral efficiency as the number of bits per second per hertz (Hz), sometimes loosely referred to as bits per hertz (b/Hz). (This is also called *information density* [33].) Notice that the measure says nothing about telephone circuits or traffic; it is purely a measure of modulation muscle — how many bits can be pumped through a given channel in one second. What the system architect chooses to do with those bits — how the telephone circuits are created — is another matter.

Consider the following oversimplified example. A two-level modulation scheme is used in which there are two predefined levels of signal amplitude (or frequency, or phase), representing 1 and 0 respectively. The input bit stream is encoded one bit at a time. Each pulse contains only a single bit. (See Figure 12.5.)

It turns out that under optimal conditions this modulator is capable of transmitting about 25 kb/s through a 25 kHz–wide channel, assuming a double-sideband signal [34]. The spectral efficiency is one bit per second per hertz of channel bandwidth. How can the spectral efficiency be improved?

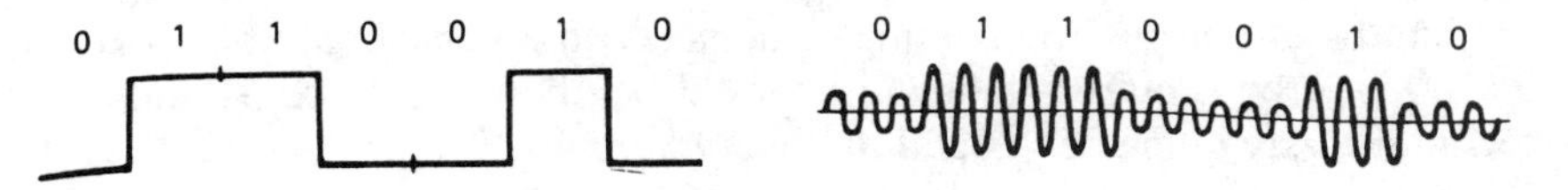

Figure 12.5 Two-Level Amplitude Shift-Keying.

Consider another ASK system in which four levels are defined instead of two. The input bit stream is encoded two bits at a time. If the two bits are 00, a pulse of level-1 amplitude is transmitted. If the two bits are 01, in that order, a pulse of level-2 amplitude is transmitted. A 10 sequence is transmitted with a level-3 amplitude pulse. An 11 sequence is transmitted with a level-4 pulse. (See Figure 12.6.) Each *symbol* now contains two bits of information. The bit rate is still (assumed to be) 25 kilobits per second, but the symbol rate is 12.5 symbols per second. Under optimal conditions, the occupied bandwidth would be reduced by 50%, to 12.5 kHz. The spectral efficiency is now 2 b/Hz.

The concept of multilevel modulation is very powerful. It is certainly possible to conceive of an 8-level system, which would transmit three bits per symbol, achieving 3 b/Hz. A 16-level modulator, encoding four bits per symbol, would achieve 4 b/Hz. Multilevel coding is a logical concept that may be applied to any of the modulation schemes discussed below.

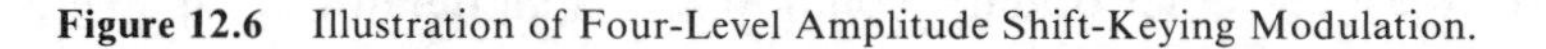

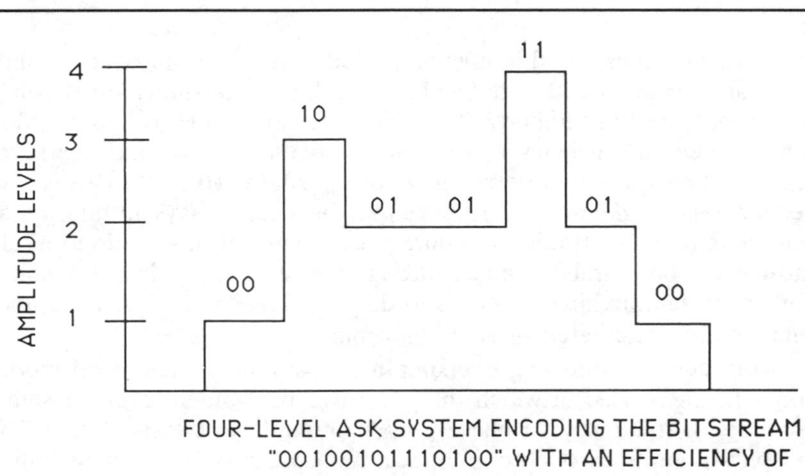

Figure 12.6 Illustration of Four-Level Amplitude Shift-Keying Modulation.

It makes it possible, in principle, to improve spectral efficiency greatly. Today, for example, 64-level modulation schemes are common enough, six bits per symbol, roughly 6 b/Hz. To transmit a 25-kb/s digital voice circuit, a 64-level system would require about 4 kHz — roughly equal to the bandwidth of the original analog signal. If the voice could be encoded at 16 kb/s, the required bandwidth would be about 2.5 kHz. Modulation techniques are being developed that would employ 256 or even 1024 levels — encoding eight and ten bits per symbol respectively [35]. Combined with anticipated lower bit-rate voice coders, it is possible to conceive of digital voice circuits, under controlled conditions, occupying less than 500 Hz of spectral bandwidth — fifty to sixty times the capacity of today's FM circuits! (But read on . . .)

It might seem that multilevel coding is the Aladdin's lamp of the radio engineer. Why not simply keep increasing the modulation level to achieve whatever spectral efficiency is desired? Alas, it is not that easy.

Consider the problem faced by the demodulator, which must detect the incoming radio signal and determine which of the prespecified levels the signal falls into at a given instant. Assume that the modulation uses phase shift keying. A two-level PSK system utilizes two opposite phase states — 0° and 180° — to symbolize 0s and 1s. The discrimination problem is relatively easy. If the detected phase is anywhere between 89° and 271°, the demodulator registers a 0; otherwise it detects a 1. (See Figure 12.7.) Such a system is robust. Distortions of the phase information which, as we have seen, can be caused by the mobile environment, can be tolerated as long as they do not distort the transmitted phase by more than 90° in either direction.

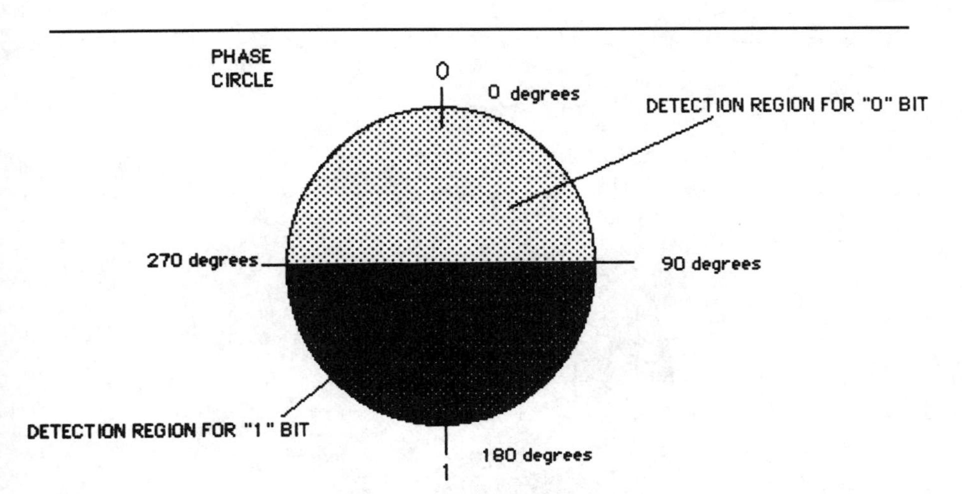

Figure 12.7 Detection Sectors for Binary PSK.

A 4-level PSK system must discriminate with twice the accuracy of a 2-level system: each pulse must be resolved into one of four 90° phase sectors. (See Figure 12.8.) To double the spectral efficiency again to 4 b/Hz, we must go up to 16 levels. Now, the demodulator must discriminate between sectors that are only 22.5° wide. (See Figure 12.9.) The next doubling, to 8 b/Hz, requires 256-level modulation, which would mean a PSK system discriminating the phase of the incoming signal into sectors only slightly larger than 1° of arc in width. (See Figure 12.10.)

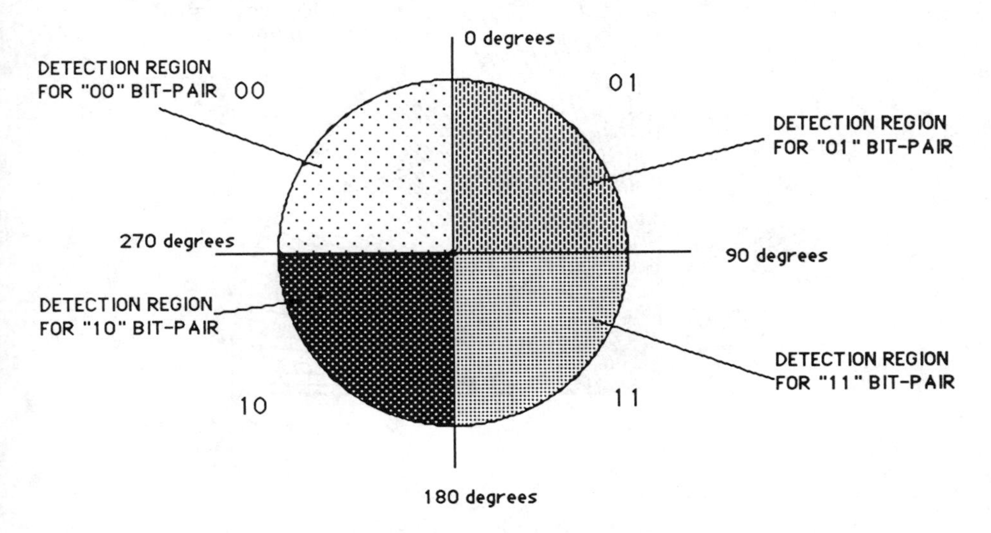

Figure 12.8 Detection Sectors for QPSK Modulation.

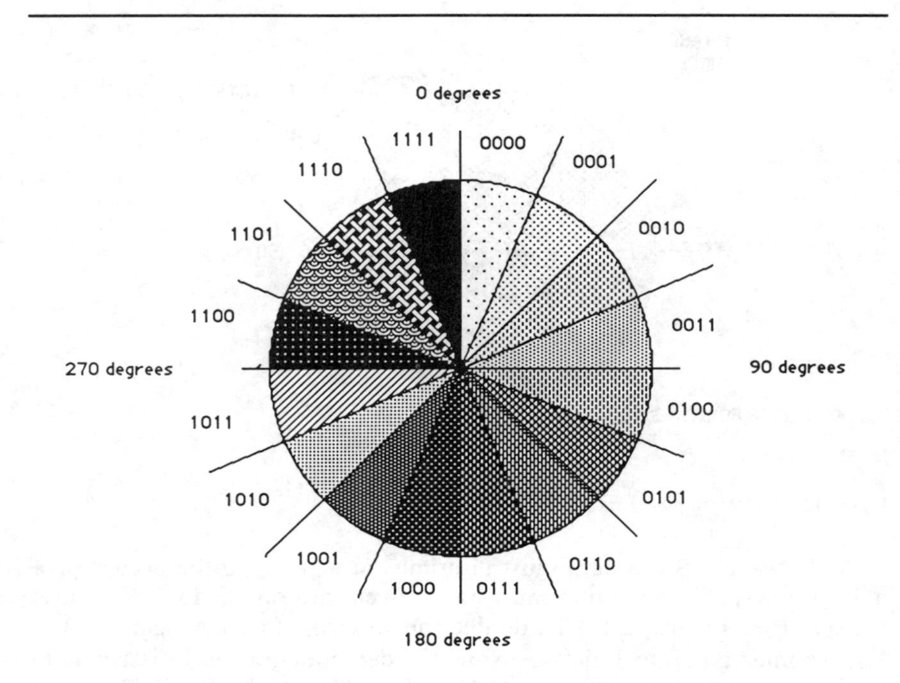

Figure 12.9 Detection Sectors for 16-ary PSK Modulation.

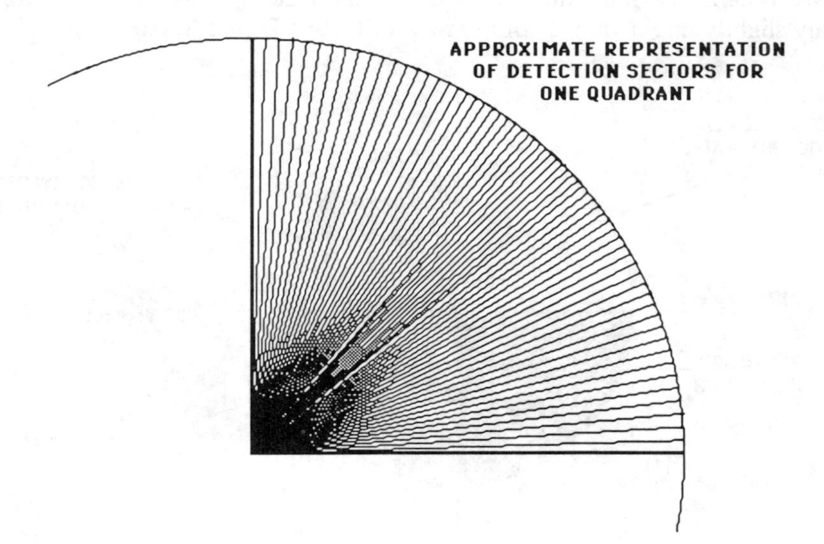

Figure 12.10 Detection Sectors for 256-ary PSK.

In short, while spectral efficiency increases arithmetically, the number of levels and the precision required at the demodulator increase exponentially. If n is the number of bits per symbol, then the number of levels equals 2^n, which also correlates with the degree of precision required in the demodulator. (See Figure 12.11.)

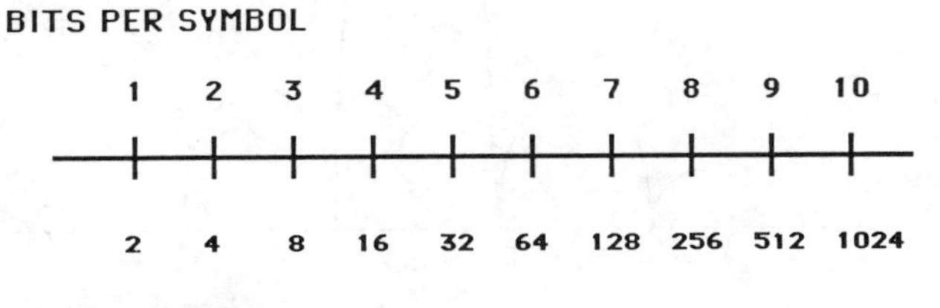

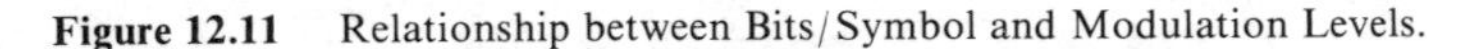

Figure 12.11 Relationship between Bits/Symbol and Modulation Levels.

Clearly, a demodulator that can tolerate less than 1° of environmentally induced phase distortion is far less robust than one that can survive a 90° shift. The chance of errors in detection is much greater when you are looking for a ½ of 1% difference than when you are looking for 50% differences. The whole issue also can be looked at as a matter of power, or signal-to-noise ratios. Finer discriminations generally dictate a higher minimum signal-to-noise ratio. For example, the difference in signal level between 4-level PSK and 16-level PSK is about 13 dB — the signal-to-noise ratio must be about 200 times better for 16-level PSK to equal the performance of 4-level PSK. This translates into higher power requirements, reduced range, and at some point into absolute limits on the ability of higher-level modulation schemes to function. The mobile environment is particularly severe, and many observers today doubt whether modulators much above 16 levels or so will ever be made to work well for mobile radio. Most field work has favored either 2-level or 4-level, for robustness.

Nevertheless, spectral efficiency or information density forms the foundation of *spectrum* efficiency in a mobile-radio system. It sets the hard limits on system capacity. There is a strong incentive to find technological solutions that will allow multilevel modulation schemes to be employed.

12.2.1.2 Narrow Power Spectrum

When a modulated radio carrier wave is transmitted, the energy it contains is distributed in a characteristic fashion about the center frequency. (See Figure 12.12.) This distribution is known as the *power spectrum.* The farther we move from the center frequency in either direction,

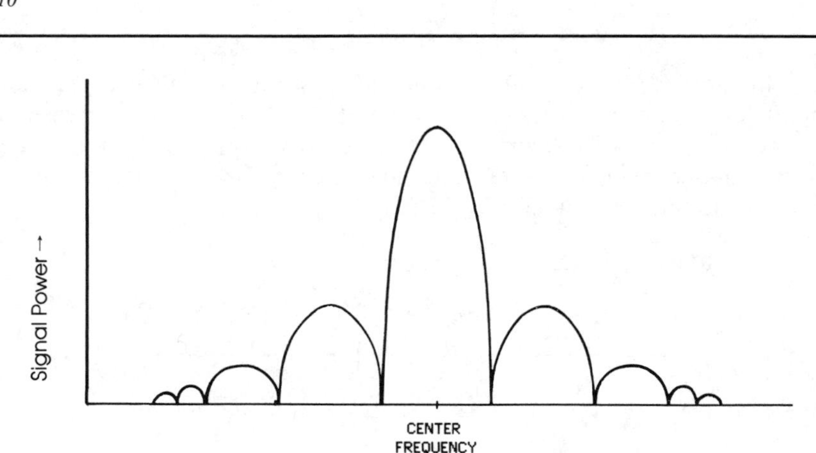

Figure 12.12 Typical Power Spectrum.

the less strong the signal. Typically, the energy is concentrated in a *main band.* Some forms of modulation, however, produce significant *sidebands.* In fact, the particular "signature" of the power spectrum, especially the size of the sidelobes, is one of the most important factors for distinguishing among different modulation proposals.

The power spectrum is a determinant of adjacent-channel interference. A modulator with a very broad power spectrum, like conventional FM, will overlap significantly with adjacent transmissions. (See Figure 12.13.) A broad power spectrum is therefore not desirable. It can be filtered to fit the mask, but such filtering can add considerable expense to the mobile unit. Indeed, first-generation cellular architects decided to allow very broad power spectra. They reasoned that since not all channels would be used in any given cell anyway, they could tolerate a much looser emissions mask by allowing each channel to splash over into the (unused) adjacent channels. The wisdom of this decision is questionable (see Chapter 4), and it is likely that future mobile architectures will strive to achieve adjacent-channel utilization through much narrower power spectra.

The width of the power spectrum is determined to a great extent by the type of modulation. Some modulation schemes naturally produce narrower spectra. For example, a class of devices known as minimum shift keying (MSK) modems produces very narrow power spectra compared to conventional FM. (See Figure 12.14.) Such narrow emissions obviate the need for expensive filters and make the control of adjacent-channel interference somewhat easier.

On the other hand, MSK modulation is not as spectrally efficient as other forms. The trade-off between narrow-emissions spectrum and spectral efficiency is one of the important choices facing mobile-systems architects.

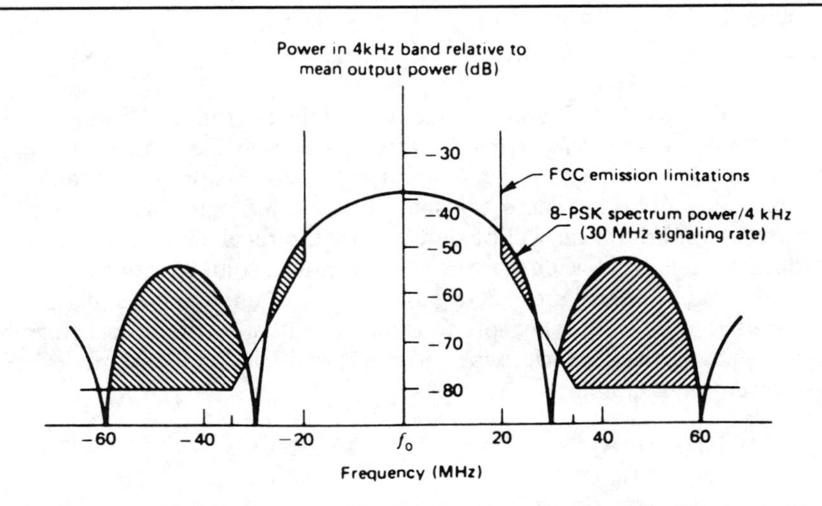

Figure 12.13 Example of a Relatively Broad Power Spectrum Requiring Filtering to Meet Mask Requirements.

Source: John Bellamy, *Digital Telephony,* p.309.

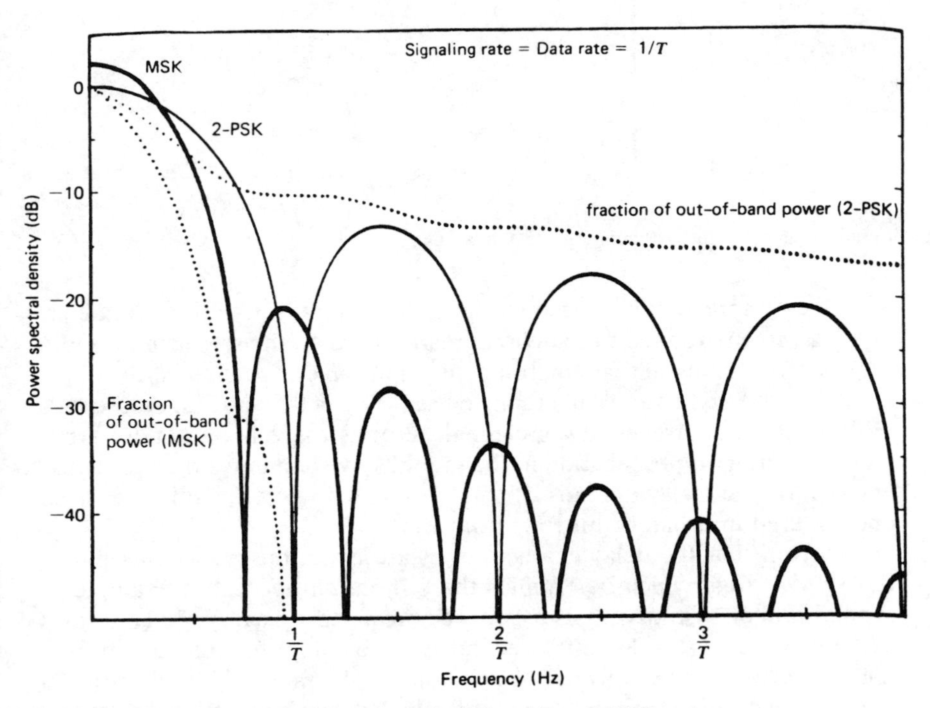

Figure 12.14 Power Spectra of Unfiltered MSK and Two-PSK Signals.

Source: John Bellamy, *Digital Telephony,* p.282.

12.2.1.3 *Intersymbol Interference*

As discussed in Chapter 8, one of the effects produced by the mobile environment is the *delay spread.* Depending upon the nature of the environmental reflectors that create multipath transmission, the speed of the mobile unit, and other factors, a sharp transmitted pulse of, say, a fifth of a microsecond duration will be detected by the receiver as a smeared and flattened bulge of considerably greater duration, sometimes up to several microseconds. (See Figure 12.15.) If it is severe enough — that is, if the transmission-induced delay spread is large relative to the average symbol time — intersymbol interference will result as the individual symbols begin to overlap one another.

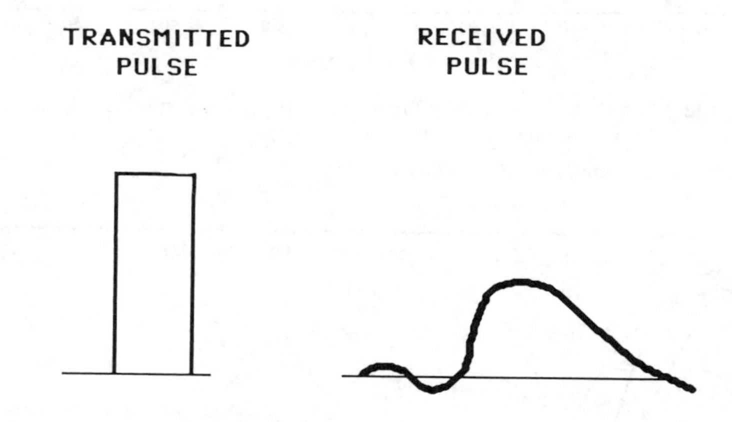

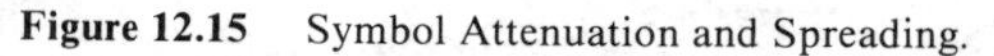

Figure 12.15 Symbol Attenuation and Spreading.

Delay spread is produced by the environment; for a given frequency and a given environment the delay spread should be the same for all radio signals propagating in that environment. In urban environments, the delay spread at 900 MHz is typically reported as between 0.5 and 5 microseconds [36]. There is, however, disagreement here, as there is on almost every matter relating to propagation in the mobile environment; other researchers report that delay spreads of "up to tens of microseconds are often encountered in urban or hilly environments [37]."

Given that the delay characteristics are fixed, the crucial variable is the *symbol rate,* which determines the *symbol duration.* For example, a symbol rate of 16 kilosymbols per second means that each symbol occupies 62.5 microseconds. A 1–2 microsecond average delay amounts to a spreading of each pulse by only 2–3%. On the other hand, a bit rate of 200 kb/s has been discussed as a standard for the next-generation European digital mobile systems. If 2-level modulation is assumed, i.e., one bit per symbol, then the symbol rate is equal to the bit rate: each symbol occupies 1/200,000 of a second, or about 5 microseconds, which means that a delay of 1–2 microseconds will create considerable intersymbol interference. (See Figure 12.16.)

To some extent this can be controlled by adaptive equalization. The impact of delay spread is also affected, obviously, by the modulation level, which helps determine the symbol duration. For example, 200 kb/s transmitted with 4-level modulation (two bits per symbol) would equate to about 10 microseconds per symbol; 16-level modulation would give about 20 microseconds per symbol. At 20 microseconds per symbol, the effect of a delay spread of a couple of microseconds is less. (See Figure 12.17.)

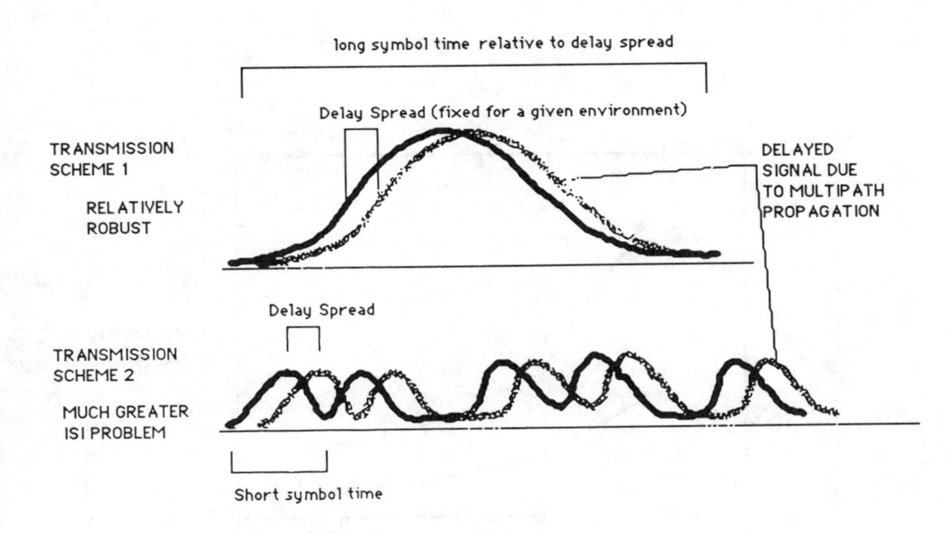

Figure 12.16 Delay Spread, Symbol Time, and ISI.

Since the delay spread tends to be fixed for a given frequency in a given environment, there is effectively a maximum symbol rate that can be transmitted. Again there is wide disagreement on the upper limit: theory and experimental results have indicated top rates of 2000 kb/s [38], 300–400 kb/s [39], "a few hundred kbits/s [40]," or 800 kb/s [41]. Figure 12.18 illustrates the relationship between transmission rate and intersymbol-induced error rates.

The consensus seems to be that rates above 200–300 kilosymbols per second will be very hard to implement and will require intensive adaptive

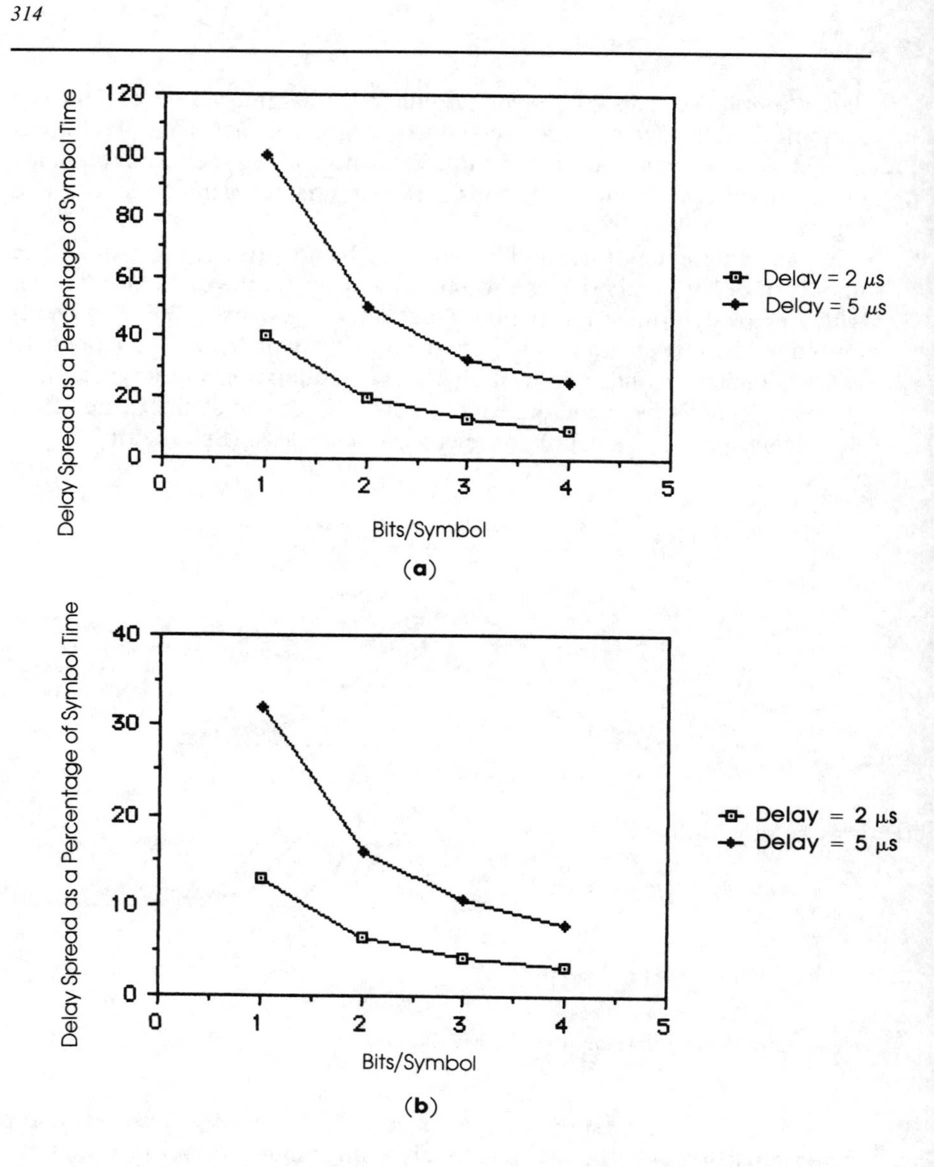

Figure 12.17 Effect of Transmission Rate on the Severity of Multipath Delay Spread. Delay Spreads and Symbol Times for Hypothetical Transmissions, with Assumed Delay Spreads of 2 μs and 5 μs; (a) 200 kb/s Transmission, and (b) 64 kb/s Transmission.

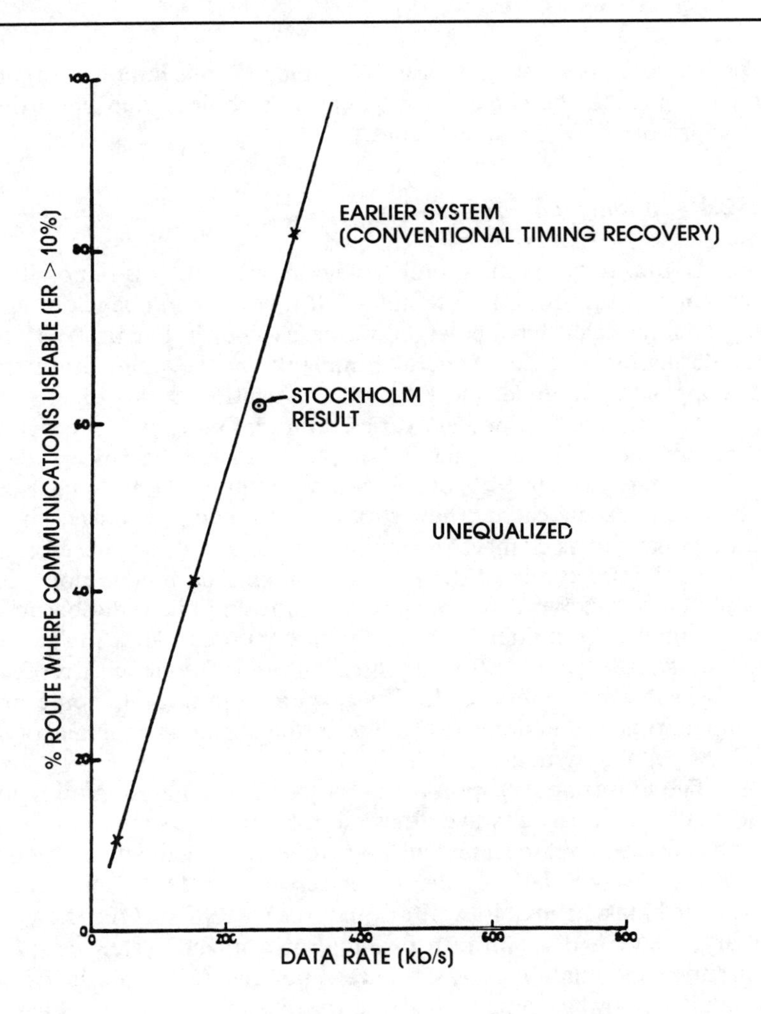

Figure 12.18 Multipath Performance for Unequalized Mobile Radio Link

equalization. As we shall discuss in Chapter 15, this limit has an important impact on both the choice of modulation technique and upon the entire structure of a TDMA architecture.

12.2.1.4 Modulation Strategies

Digital radio modulation is a field in ferment. It is impossible within the scope of this book to attempt a full review of the bewildering variety of techniques that have emerged in the past several years. We shall limit our discussion to a few general comments on the modulation strategies that appear to be under most active investigation at this time.

The modulation process varies the radio wave in some defined way to communicate the information supplied to it by the voice coder. Modulation strategies can be grouped and differentiated on the basis of which aspect of the radio carrier they work on. As noted previously, information can be encoded as changes in amplitude, frequency, or phase of the radio wave, or as some mix of these. Digital amplitude modulation, known as amplitude shift keying, has been ruled out for mobile radio because of the severe impact of multipath fading upon the amplitude of the carrier. Frequency shift keying (FSK), in which different frequencies represent the levels in the modulation code, has been incorporated in some proposals and experimental systems, as well as in the signaling channels of some of today's cellular systems.

The most popular approaches, however, are based upon modulating the phase of the carrier wave. Phase shift keying (PSK) modulation defines various discrete phase states which are used to carry the digital information from the voice coder. As already discussed, the PSK family includes 2-level, or binary, PSK, 4-level, or quaternary, PSK or QPSK, 16-level, or 16-ary, PSK, and so on. An example of a mixed system is quadrature amplitude modulation (QAM), a very popular technique in fixed point-to-point microwave systems because it is capable of yielding high spectral efficiencies; up to 64-ary QAM — around 6 b/Hz — has been implemented in practical applications. QAM utilizes both phase and amplitude information to define the modulation levels; that is, it uses both the angle of the radio wave and the strength of the signal simultaneously. The mobile environment is a severe challenge to conventional QAM; some researchers, however, have developed proposals for QAM that they believe may be feasible in the mobile channel [42].

Among the many proposals being evaluated, some are responsive primarily to the goal of spectral efficiency, while others focus on the somewhat conflicting objective of achieving a narrow power spectrum. In current discussions of the next generation of mobile-telephone architecture, there are two broad modulation strategies emerging which emphasize rather different design objectives.

On the one hand, there is the family of *linear-modulation* techniques, so called because they require a high degree of *linearity,* or low distortion, in the translation from baseband frequencies up to the carrier frequencies and amplification to full transmit power levels. Achieving sufficient linearity today usually involves a cost penalty in the design of the mobile unit, which may range between 5% and 25%. On the other hand, all other things being equal (power, bit-error rates, signal-to-noise ratio), linear methods appear to promise greater spectral efficiency than nonlinear methods. The most important linear modulation methods are based on various forms of phase shift keying, particularly differential PSK, both QPSK and higher-level PSK.

The other grouping of modulation techniques is known as *constant-envelope* or *continuous-phase modulation.* Constant-envelope modulation also avoids the linearity requirements which reduces the cost of amplifier components. Techniques drawn from the CPM family tend to have quite narrow power spectra. On the other hand, the level of spectral efficiency is somewhat lower. Among the important constant-envelope techniques being explored currently is minimum shift keying, which is a special form of two-level FSK.

It is interesting, and indicative of how rapidly things are changing in the field of digital radio, that in 1986, when industry discussions began of the need for a new generation of radio technology, the initial focus was almost exclusively upon constant-envelope techniques. MSK and related techniques were favored, due to the narrow power spectrum and the perceived cost advantage of avoiding the need for a linear amplifier [43]. By mid-1987, however, linear-modulation techniques like QPSK were in vogue. An AT&T paper gave the reason:

> Many of the digital systems investigated in the past have centered around constant envelope modulation schemes such as TFM and GMSK because they do not require the use of linear power amplifiers at cell sites. However, better spectral efficiency can be obtained by the use of linear modulation [44].

At least temporarily the driver of spectral efficiency has displaced the power-spectrum issue.

12.2.2 Voice Coding for Mobile Radio

Whereas the choice of modulation methods involves a multidimensional evaluation, the choice of voice-coding methods is driven largely by a single objective: the desire to reduce the number of bits that need to be transmitted, while maintaining "telephone voice quality." The lower the bit rate, the less bandwidth required.

A secondary goal is to select a coding technique that is *robust* over a wide variety of circuit conditions. One aspect of robustness is the ability of the coder to deal with the high input noise levels often found in moving vehicles. A second dimension involves the ability to withstand a high rate of transmission errors induced by multipath. A third is the ability to perform well with a wide range of different speakers, male and female, exhibiting a range of articulation patterns. Robustness, however, is best viewed as a constraint. The key to voice coding for digital mobile telephony is achieving a lower bit rate.

Coder development is currently one of the more exciting fields of digital design. In the past few years, breakthroughs in digital processors have allowed telephone-quality voice transmission at bit rates much lower than ever before. Only a few years ago, the question of moving from the hallowed 64-kilobit PCM standard to a 32-kilobit adaptive differential PCM technique was hotly debated. The idea of quality voice transmission at less than 20 kilobits seemed farfetched [45]. Today, all research on digital mobile systems *assumes* a 16-kilobit standard, or lower [46]. Many believe that within a few years the standard will settle around 9.6 kilobits, or even 8 kilobits. In my opinion, most current estimates of the speed of coder development are too conservative; I believe that we shall shortly see telephone-quality digital voice at 4.8 kilobits, in a form suitable for mobile systems.

12.2.2.1 Coding Strategies

How are such dramatic advances being achieved? There are several very different strategies for reducing the bit rate.

Analysis of Speech Redundancies

A balanced waveform coder, like 64-kb/s PCM, assumes nothing about the nature of the analog sound input other than that it contains no important information at frequencies above one-half the sampling rate — the Nyquist theorem. From the coder's standpoint, any input value is as probable at a given moment as any other; that is, it assumes nothing at all about the *patterns* in the input signal.

Actually, of course, human speech exhibits a high degree of patterning and redundancy, especially over short periods; up to, say, 100 milliseconds. If the sampling rate is high, a particular sample will tend to be very similar to the samples that preceded it. In fact, it is reported that the correlation coefficient between successive 8-kHz samples is generally 0.85 or higher [47]. In principle, if these redundancies can be analyzed, the coding process can be made much more efficient. Accepting the 0.85 correlation figure, we could expect to squeeze at least half of the bits out of the signal without affecting the quality — if we know which ones to eliminate.

Thus, one of the most fruitful and easiest avenues for reducing the bit rate of the coder is to analyze the waveform for inherent redundancies. *Per se,* this analysis is strictly formal; it does not require any particular foreknowledgè of the types of input patterning that may be experienced. In fact, this analysis can be extremely mechanical, as we shall see below, allowing for very economical implementations.

Analysis of Aural Performance Characteristics

The ear is a remarkable filter in its own right. The processing power inherent in the hearing mechanism can be exploited to reduce the bit rate by only transmitting the information which is *perceptually relevant.* For example, people hear low frequencies differently than high frequencies. It appears that the overall perceived quality of speech suffers more from low-frequency noise than from high-frequency noise, which often seems to blend with, or to be masked by, the high-frequency hiss of fricative consonant sounds. For example, most people can hear sounds over a range of up to 15–20 kHz. Yet the conventional analog wireline telephone is designed to cut off frequencies above about 3–4 kHz. This is acceptable because most of the speech information is contained in the lower band; our hearing for the lower frequencies is better as well. In a crude way, this represents a reduction in bandwidth that is the result of taking advantage of the hearing mechanism.

It is possible to extend this approach into the digital domain with much greater precision. The study of acoustics by the Bell System began in the 1920s and 1930s [48]. Algorithms have been developed based on extensive empirical studies of human speech and hearing which allow the coder to concentrate upon the most significant parts of the speech input signal and to ignore or downweight the less perceptually important elements. Significant reductions in bit rates are possible.

Parametric Modeling of Speech Production

In his childhood, Alexander Graham Bell once constructed an artificial speaking machine. It was actually an anatomical model of a human vocal tract, complete with teeth, throat, nasal passages, and a seven-section articulated tongue made of gutta-percha [49]. By carefully positioning the different anatomical elements, while simultaneously introducing a sound source in the "throat," Bell was able to effect a rudimentary articulation of individual English words. Others in the nineteenth century built more elaborate and apparently quite successful and intelligible speaking machines for public amusement [50].

Several generations later, armed with better techniques and materials, telephone engineers tried again. In the 1930s, Bell system researchers succeeded in constructing crude mathematical and electrical models of the vocal tract [51]. (See Figure 12.19.) They modeled the speech process as a source of sound (sound either at a given pitch or as white noise, representing voiced and unvoiced speech sounds, respectively) and a filter representing in an abstract form the passages of the mouth and nose. The characteristics of the filter were variable, corresponding to the different positions of articulation (like the settings of Bell's artifical tongue), and the whole might be represented by a series of mathematical equations. Instead of transmitting the actual input signal, that signal was measured, a mathematical model constructed, and the parameters or coefficients of the equations of that model were transmitted.

By such means, intelligible speech was achieved at very low bit rates, although quality was poor and artificial sounding. Such techniques are often called *source coders,* because they do not code the waveform; instead, they analyse the waveform to determine the characteristics of the source and then code the source characteristics, which can be transmitted much more economically.

Quantization of Block or Parametric Information

Recently a further compression of the information transmitted has been realized through quantizing the speech patterns contained in larger blocks of samples produced by waveform coders, or in the parametric information generated by source coders. These techniques actually work mainly on the primary coder output, further to compress that output. The chief method of compression is to compare the output with a previously constructed dictionary of code words and to transmit the code word in the dictionary which is closest to the output of the coder. Further details are provided in the following section in the discussion of vector quantization.

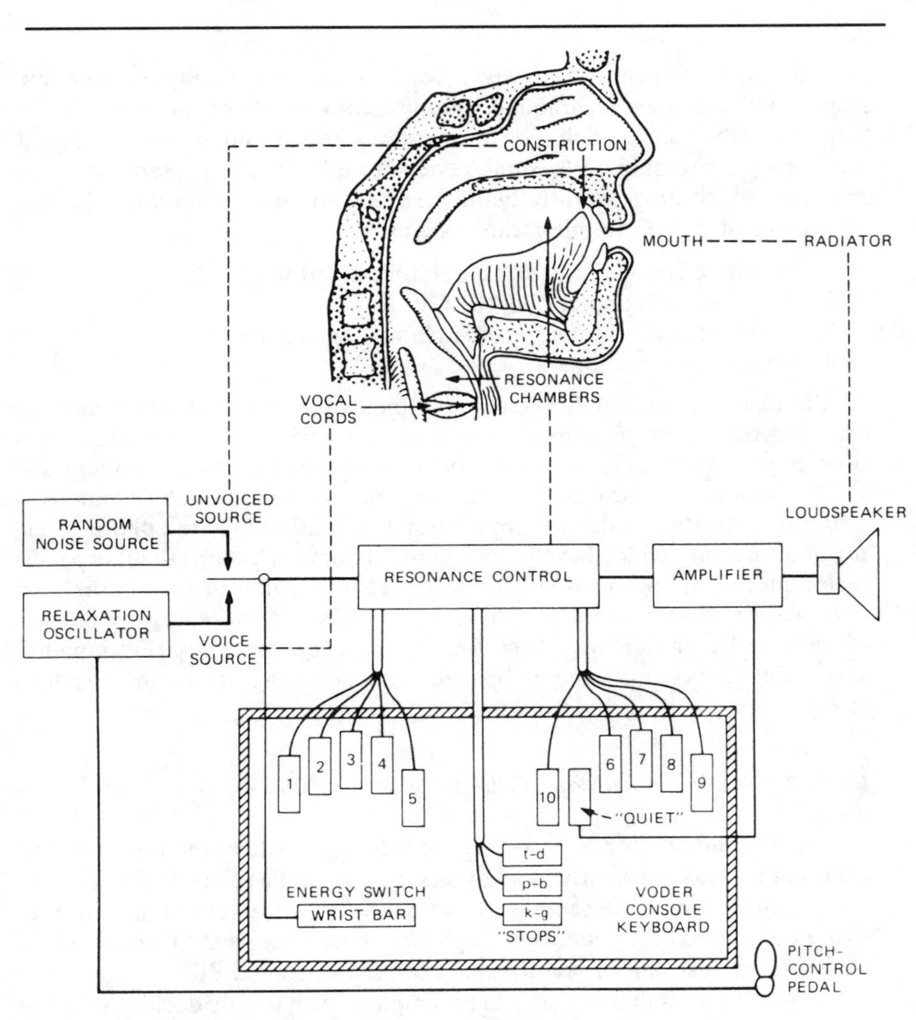

Figure 12.19 Early Bell Vocoder Speech Synthesizer

Source: AT&T Bell Labs, *A History of Engineering and Science in the Bell System: Communications Sciences,* 1925–1975 p.101.

12.2.2.2 Representative Coding Techniques

If there are a multitude of modulation techniques, there are a myriad of voice-coding algorithms. Once again, a full inventory would run well beyond the scope of this work. There are many books and a number of good summary articles available on voice-coding techniques [52].

It may be of interest, however, to delve somewhat deeper into a few specific voice-coding techniques, because they illustrate perhaps better than any other single field the tremendous range and power of digital processing techniques. We shall review briefly four representative approaches which are currently being analyzed in connection with the development of digital mobile-radio systems:

1. Continuously variable slope delta modulation (CVSD);
2. Adaptive subband coding (SBC);
3. Residual-excited linear predictive coding (RELP);
4. Vector quantization.

CVSD is a delta-modulation technique that was in favor several years ago and was one of the first modern coders to be widely implemented. It is now generally considered to have been superseded. Subband coding and RELP, and related coders, are probably the two most popular techniques under investigation today for application to digital mobile telephony. Vector quantization, which is actually a general approach more than a specific technique, is one of the areas that will likely come into its own in the next five years and will strongly influence future coder strategies. While not exhaustive by any means, these four approaches exemplify the immense processing power that is being brought to bear on digitized human speech today.

Continuously Variable Slope Delta Modulation (CVSD)

Delta modulation is based on the principle that, instead of encoding each sample independently of all others, a gain can be realized if we encode only *changes* from one sample to the next. This is a very simple way of exploiting speech redundancies. It yields a type of waveform coder which is simple, robust, and more efficient than conventional PCM.

A delta-modulation coder in its simplest form simply decides whether each sample is higher or lower in amplitude than the previous sample. If higher, it moves up one step in a preestablished quantization scale. If lower, it moves down one step. It effectively tracks the analog signal with a stepwise approximation. (See Figure 12.20.) The efficiency of delta modulation is derived from the fact that it is really transmitting only the change — the "delta" — between the current sample and the previous sample, instead of retransmitting the entire amplitude scale of each sample. In effect, a 1 bit says simply: "Add one step to whatever value you had before." A 0 bit says: "Subtract one step from whatever value you had before." The coder is assumed to have a memory of at least the previous sample.

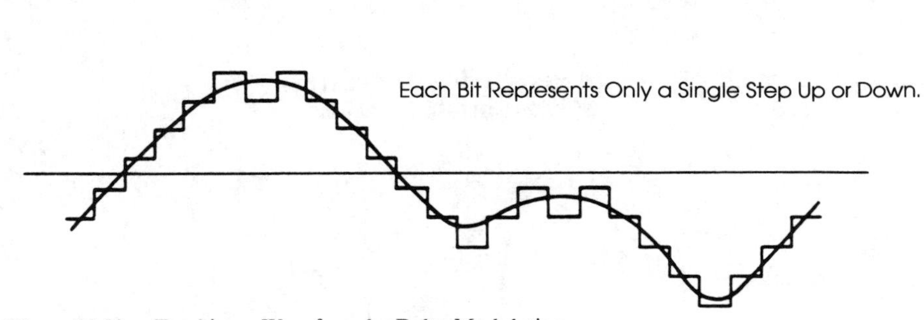

Figure 12.20 Tracking a Waveform by Delta Modulation.

Source: John Bellamy, *Digital Telephony,* p.125.

Unfortunately, however, the prospective efficiency initially proved difficult to realize, because delta-modulation coders were susceptible to two types of inherent error. On the one hand, if the signal amplitude rises or falls at a rate faster than the step size, the delta-modulation coder cannot keep up with it. It falls behind the analog signal, producing what is known as *slope overload.* On the other hand, when the signal amplitude is stable, the coder will oscillate between two states which straddle the actual value. This produces *granular noise.* (See Figure 12.21.) Unfortunately, these two forms of inherent error are complementary and are related to the step size of the delta-modulation coder. If a large step size is used, slope overload will be reduced, because the coder can keep up with a rapidly changing signal. (See Figure 12.22.) On the other hand, a large step size increases granular noise. (See Figure 12.23.) Surprisingly, granular noise is more objectionable than slope overload at equivalent power levels. This is because slope overload tends to occur during active speech, when the noise is masked by the speech itself, and may not be perceived. Granular noise occurs during periods of silence; hence it stands out. For quite a while, this Scylla and Charybdis imposed a limit on the reduction of bit rates by delta-type, difference-encoding techniques.

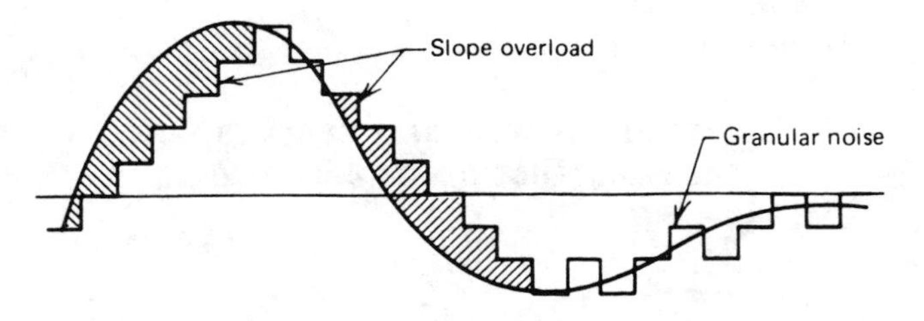

Figure 12.21 Characteristic Forms of Noise in Delta Modulation.

Source: John Bellamy, *Digital Telephony,* p.127.

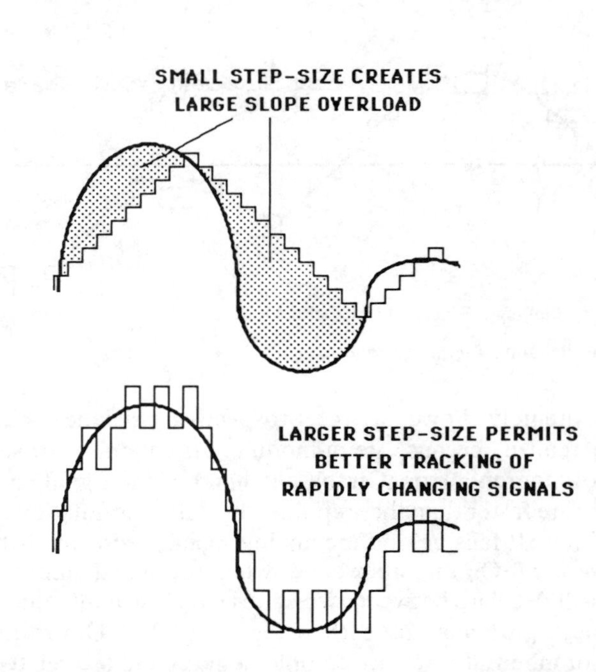

Figure 12.22 Delta Modulation: Effect of Larger Step Sizes upon Slope Overload.

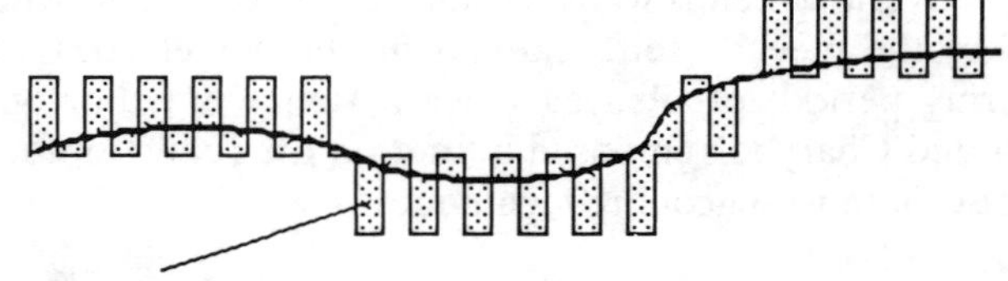

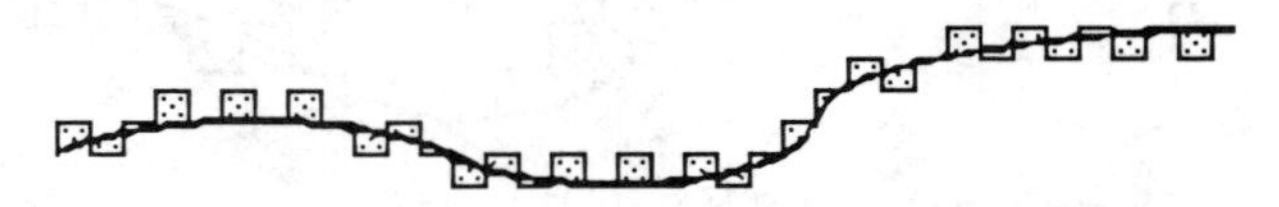

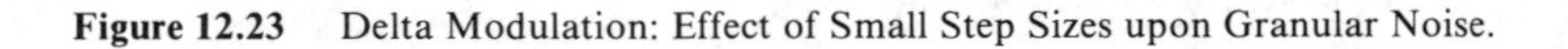

Figure 12.23 Delta Modulation: Effect of Small Step Sizes upon Granular Noise.

In the 1960s, armed with more powerful processors, researchers began to explore *adaptive* delta-modulation algorithms. The basic idea was to operate with a step size that adapted to the conditions of the signal, increasing when slope overload was detected and decreasing when the signal flattened out. For example, in a particular adaptive-delta approach that has come to be known as continuously variable slope delta modulation, the step size is increased by a predetermined amount if the coder detects four 1s or four 0s in succession, which would tend to indicate that the input signal is rising or falling faster than the coder can track it, hence slope overload is probably occurring. In between such forced increases, the step size is allowed to decrease exponentially with each sample. At some point it decreases to a very small level at which overload is detected once again, causing a new step-size increase. (See Figure 12.24.)

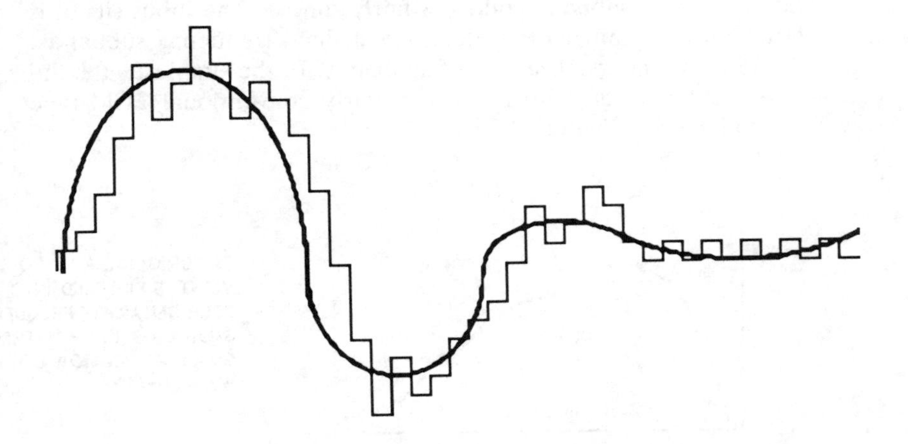

Figure 12.24 Delta Modulation with Adjustable Step Size.
Note: This simplified illustration shows four step sizes. The rule for adjustment is simple: If the delta modulator crosses the waveform, then it is tracking and the step size decreases by 1. If the delta modulator does not cross the waveform, then the step size increases by 1. This will help to minimize overload and granular noise simultaneously.

CVSD helps to overcome the inherent problems of overload and granular noise and allows the delta coder to realize the efficiencies inherent in difference-only encoding. Telephone quality equivalent to 64-kb/s PCM has been achieved with delta coders at 32–40 kb/s. The technique is also relatively simple to implement.

On the other hand, CVSD coders do not deliver adequate quality at the 16-kb/s goal of today's mobile systems [53]. Indeed, difference-encoding techniques that work on inherent speech redundancies are apparently not sufficient to reach the desired low bit rates for the mobile radio application. To achieve lower rates, researchers have turned to more powerful techniques.

Adaptive Subband Coding

Delta-modulation techniques are not inherently adapted to the transmission of perceptually important characteristics of human speech. Subband techniques, on the other hand, *are* based upon a knowledge of the unique characteristics of human sound processing. By effectively weighting the different frequency components of the digitized signal to correspond to the perceptual weightings assigned by the ear, subband techniques are able to reduce the bit rate needed for perceived telephone quality.

The concept of subband coding is fairly simple. The input signal is first filtered into a number of strategically defined frequency subbands. Figure 12.25 shows an idealized set of subbands. In the simplest subband coders, each subband is digitized with a fairly conventional PCM-type waveform coder. (See Figure 12.26.)

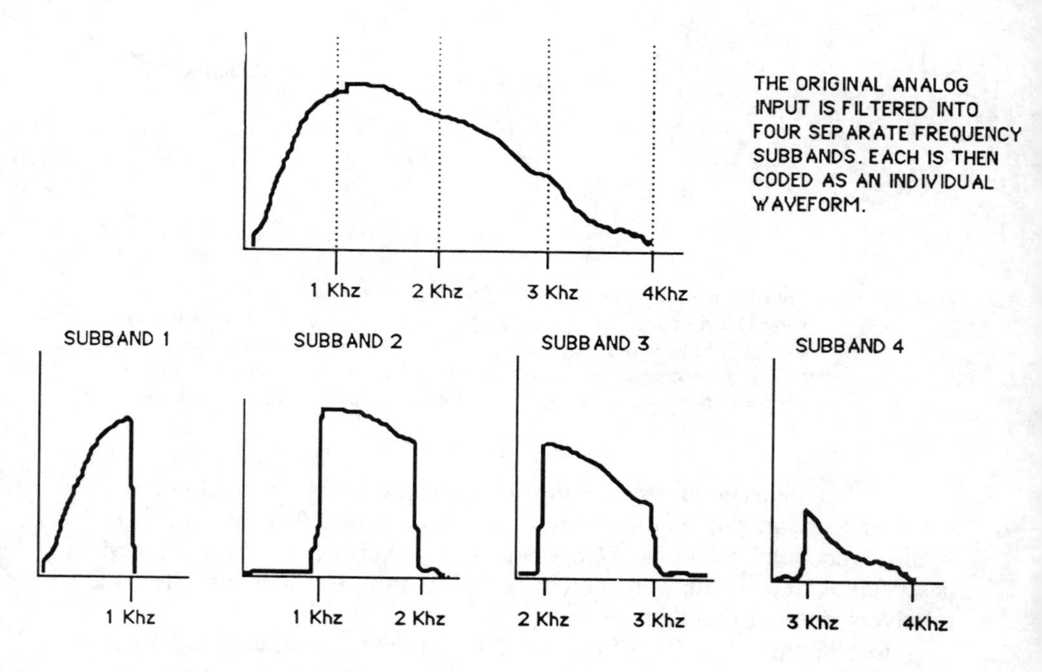

Figure 12.25 Hypothetical Subbands.

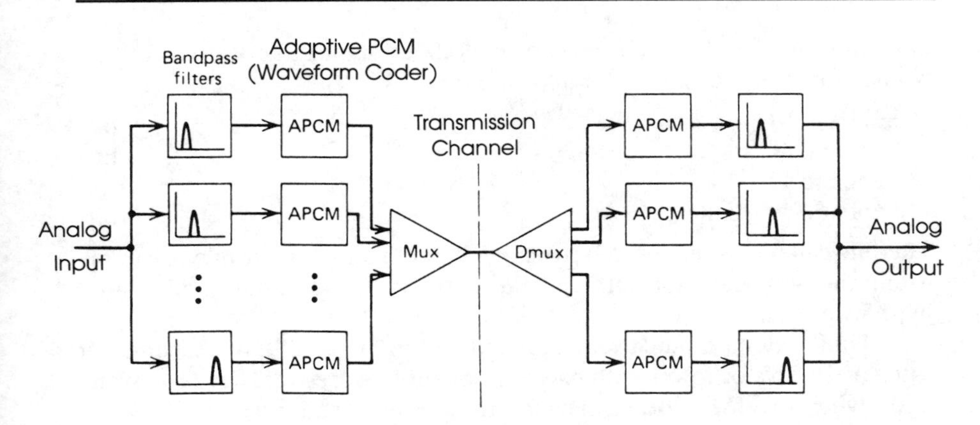

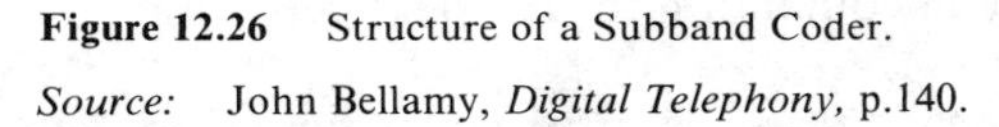

Figure 12.26 Structure of a Subband Coder.

Source: John Bellamy, *Digital Telephony,* p.140.

Per se, simply dividing the signal into subbands would not reduce the bit rate. The savings come about in several ways. First, instead of dividing the available bits in the sample equally across all bands, a relatively larger number of bits can be assigned to low-frequency bands below 1 KHz which tend to contain the most important tonal information that we associate aurally with quality, or fidelity, or speaker recognition. The higher bands tend to contain the hissing, noiselike sounds of certain consonants, such as "s" and "f," and adequate quality can be achieved in these subbands with much lower bit rates.

Another strategy, somewhat more adaptive to the changing input, adjusts the quantization step size for each band depending upon the energy level in that band. This allows the system to tolerate higher levels of quantization noise by masking it under high-amplitude bands.

More recently, more fully adaptive techniques have been applied. In standard subband coding, the bit allocations to each band are fixed. In adaptive subband coding, each sample is analyzed to determine in which band the greatest amount of speech energy, and information, lies. The available bits are then adaptively allocated to the various subbands. For example, in an "s" sound, which is unvoiced, most of the information will be in the higher frequency bands. If we assume an 8-bit sample and four subbands, the allocation of bits might be:

Subband A (lowest)	1 bit
Subband B	1 bit
Subband C	2 bits
Subband D (highest)	4 bits

On the other hand, a rich, voiced, melodious sound like the vowel "o" would generate a different bit allocation:

Subband A (lowest)	5 bits
Subband B	2 bits
Subband C	1 bit
Subband D (highest)	0 bits

The bits can be moved around from one band to another, from one sample to another. Overall, this strategy allows the coder to attain good quality with fewer bits.

Unlike delta modulation, subband coding is capable of reaching bit rates of 16 kb/s or lower with excellent quality. Moreover, because of the underlying waveform coding in each band, subband coders tend to be fairly robust. SBC coders are definitely in contention for the next generation of digital mobile systems.

On the other hand, subband coders are algorithmically more complex than delta-type coders, and it is only in the past few years that progress in digital signal processors has enabled the effective realization of many subband algorithms.

Residual-Excited Linear Predictive Coding

To understand the structure of a source coder, let us conceive of the vocal tract as similar to a musical horn instrument like a saxophone. At one end there is a source of sound energy, referred to as *excitation.* In the case of the saxophone it is a vibrating reed. The excitation source has two chief characteristics: pitch and loudness, both of which may vary, but tend to change slowly and may be conceived of as constant over short intervals, e.g., a few milliseconds. The sound produced by the excitation source is then channeled through a peculiar configuration of tubes and chambers — the body of the horn itself — which modulates the sound in rich and characteristic ways. From the standpoint of the acoustical engineer, the body of the horn is a kind of complex filter which attenuates, distorts, and amplifies the input sound to produce the throaty saxophone sound. (The same vibrating reed combined with a different "filter" will produce the clear notes of a clarinet.) The precise characteristics of this filter may vary from moment to moment as the saxophone player modifies the shape of the sound chambers by depressing different keys. With other types of horns, the player may modify the filter characteristics in other ways, using the hand in the bell for muting the French horn, or a hat over the mouth of a trumpet.

It is possible to construct a mathematical model of the saxophone in which the characteristics of the complex filter are represented in a polynomial type of equation with a number of coefficients that determine its precise value. In principle, if we knew the values of these coefficients, along with the pitch and the gain (loudness), we could describe the sound being produced at any given moment in terms of a very few bits of information. We could then transmit these bits electrically to a receiver which incorporates an identical mathematical model of the saxophone. Plugging in the coefficients and the pitch and gain factors, the receiver model could then electrically synthesize the original sound. This could be accomplished with far fewer bits than a conventional waveform coder.

Modeling the human voice is very similar. There is an excitation source modulated by a complex and variable filter corresponding to the vocal tract. The major structural difference is that in human speech there are two kinds of excitation: voiced speech, in which the excitation is generated by the vocal chords and exhibits a detectable pitch, and unvoiced speech, generated by certain consonants, like "s" and "f," which produce noiselike, relatively pitchless sound from turbulent air passing through constricted vocal passages. A source coder — such as a linear predictive coder (LPC) — must first determine whether the speech sample is voiced or unvoiced. If it is unvoiced, a noiselike excitation source will be indicated. If it is voiced, the pitch is extracted. The filter coefficients are determined by setting them in a feedback loop to compare with the input speech sample until the error, the difference between the modeled speech and the original sample, is minimized. (See Figure 12.27.) The only information that must be transmitted is:

1. The type of excitation, voiced or unvoiced;
2. The pitch, for voiced excitation;
3. The gain or loudness;
4. The filter coefficients (parameters of the vocal-tract model) [54].

A source coder can thus be constructed to yield intelligible speech at very low bit rates. In fact, conventional LPC has largely been employed at bit rates between 1.2 and 4.8 kb/s, much lower than waveform coders. Unfortunately, the quality achieved at such low bit rates has not been acceptable for telephonic communication. LPC often generates intelligible but machinelike speech and can distort some speakers' voices severely. LPC is also far more computationally intensive than waveform coders, and until relatively recently LPC coders required large computers for implementation in real time.

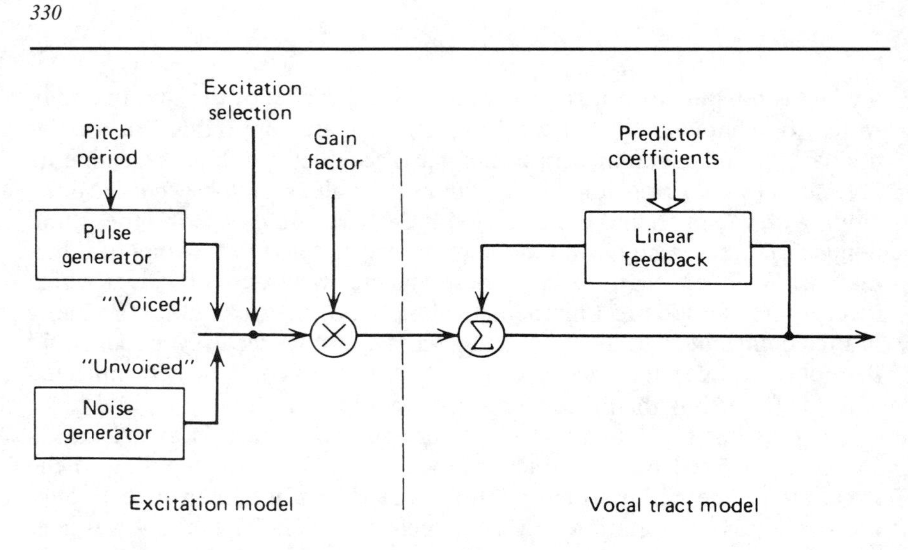

Figure 12.27 Structure of a Linear Predictive Coder.

Source: John Bellamy, *Digital Telephony,* p.146.

Thus there was a crucial gap in the region of greatest interest to digital mobile-radio engineers. (See Figure 12.28.) Traditionally, waveform coders, like subband coding, were unable to achieve good quality much below 16–20 kb/s. Source coders could reach much lower rates, but with a limit on speech quality below the telephony standards.

Recently, however, a new family of still more complex coders has emerged, starting to combine elements of both LPC and waveform coders to achieve good quality in the region of the gap. These techniques work on improving quality by better definition of the excitation source for the LPC-type coder. Instead of using a very simple measure of pitch, they invest a few more bits to create an excitation source that will more faithfully recreate the original sound. A number of techniques are currently being investigated, including regular pulse excited LPC (RPE-LPC), multipulse excited LPC (MPE-LPC), code excited LPC (CE-LPC), among others. Such enriched excitation-source coders are currently the most popular algorithms, along with subband coding, for mobile radio research. Bit rates of 8–16 kb/s are being tested in various implementations.

RELP illustrates the mixture of coding strategies. A speech sample of approximately 20 milliseconds is the input to a more or less conventional LPC vocal-tract modeler, which extracts pitch, filter coefficients, and gain, and uses a feedback loop to minimize the *residual,* or difference between the LPC output and the original. The residual, which is itself a waveform, is then encoded by means of a fairly conventional PCM-type coder, incorporating a few tricks to reduce the bandwidth of the residual, and is

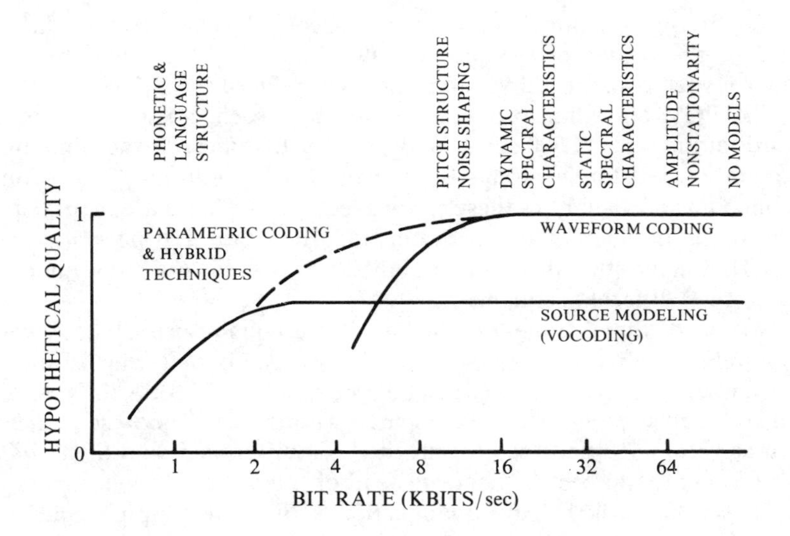

Figure 12.28 Relationship between Quality and Bit Rate.

Source: Crochiere and Flanagan, "Current Perspectives in Digital Speech," *IEEE Communications Magazine,* January 1983.

transmitted along with the coefficients of the LPC model. At the receiver, the residual is utilized along with the pitch and gain information to define the excitation source that is put through the LPC filters.

It is likely that the next generation of digital mobile-telephone systems will use RELP or similar coders, which seem, according to several studies, to be able to deliver better quality at lower bit rates than other algorithms that are currently implementable on a single microprocessor, more or less a requirement for low-cost mobile telephone units [55].

Vector Quantization

VQ is not an algorithm *per se,* but an idea that can be applied in conjunction with any coding algorithm to compress the data further. It is applied to the output of another coder in the same way that a turbo can boost the power of a conventional automobile engine. VQ is very powerful and computationally too intensive for today's one-chip signal processors; I believe, however, that it will become important within the next few years and could lead to bit rates as low as 4.8 kb/s for telephone quality by the early 1990s.

Vector quantization is effectively a second level of coding that is applied to the results of the primary algorithm [56]. To illustrate this, consider a vector-quantized version of ordinary PCM.

The PCM algorithm produces, as we have seen, a succession of 8-bit words at the rate of 8000 per second. Assume that these are accumulated in a memory register in groups of six words, 48 bits in all. The second level of coding now analyzes these 6-word sequences. There are in principle an enormous number of these sequences. There are 256 possible 8-bit words. That means that the number of different possible 6-word sequences is 256^6, or 2^{48} different sequences.

Not all of these sequences, however, are equally probable. A very large number probably never occur at all. If it can be determined that a small number of these sequences can be used as a kind of code book, every actual six-word sequence can be compared with the code-book sequences, known as *vectors,* to find the best fit, the least difference, and then *quantized* by sending the best-fit vector instead of the actual sequence. It is in principle similar to the original quantization of the analog input signal into 8-bit words. As with all quantization processes, a measure of irreducible *quantization error* is introduced. A high degree of data compression, however, can be achieved. For example, if we constructed a dictionary or code book of 65,536 possible vectors and quantized every 6-word sequence with the best fit among these, then, instead of sending a 6-word sequence, 48 bits, we could send a 16-bit vector — a compression of the data by a factor of three.

The principle of vector quantization can be applied to any coding technique. For example, an LPC coder which transmits 10 filter coefficients requires, say, 56 bits for each set of coefficients or vector. There are 2^{56} different possible sequences. If we could define 1024 most likely code-words, however, each of which is represented by a 10-bit sequence, we could perform the search and best-fit process, and then transmit only a 10-bit vector. The data compression factor would be 5.6.

The VQ designer faces a number of complex and interesting questions. How many code vectors should be used? How large should the code book be? The larger the code book, the smaller the average quantization error, the more accurate the code. Large code books mean that more bits must be used to transmit each vector, however, reducing the data-compression gains. Also, most importantly, the larger the code book, the longer the search to find the best-fit vector. If the search is exhaustive, i.e., the input sequence is compared with every single code vector, and a constant search rate is assumed, then it will take 256 times longer to search a code book of 16-bit code vectors, of which there are 65,536, than a code book of 8-bit code vectors. For a real-time process, the processing burden is a very significant constraint.

How should the code vectors be defined? The general answer is that the code vectors should be selected such that the average quantization error is minimized. This requires very intensive and skillful analysis of large numbers of speech samples.

Once the vectors are defined, how should the code book be organized, and how should the search-and-match process be carried out so as to speed processing and allow for the use of larger and more accurate code books? Procedures have been developed to allow for more rapid searching. For example, tree-structured code books can be constructed, where, instead of searching every code word, the processor goes through a series of binary choices. Is the input sequence more similar to vector A or vector B? If A, then is it more similar to A.1 or A.2? If A.2, then is it similar to A.2.1 or A.2.2? And so forth. Another expedited process is multistage vector quantization. A small code book is used, and a fast search quickly finds the best fit. The residual, or difference between the input sequence and the best-fit code vector, is measured, and then the residual is encoded with a second code book which is also small and amenable to rapid search. Several stages may be used. By keeping each codebook simple, the search process is rapid and better quality may be obtained more quickly. (See Figure 12.29.)

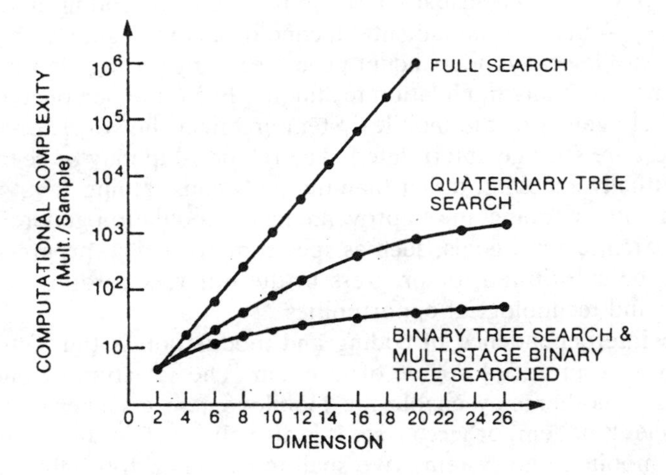

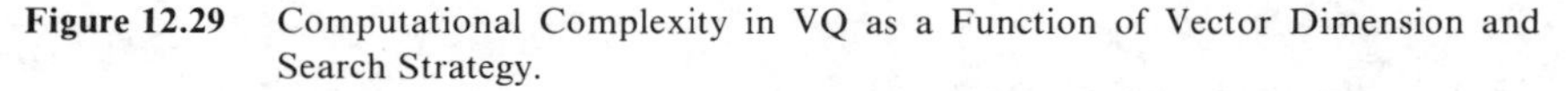

Figure 12.29 Computational Complexity in VQ as a Function of Vector Dimension and Search Strategy.

Source: Gersho and Cuperman, *IEEE Communications Magazine,* December 1983.

In all these versions, it is assumed that the code book has been constructed in advance and stored in read only memory (ROM) for access by the processor. (Memory limitations are another significant constraint for VQ systems.) This implies that the code book is, practically speaking, an average across a large number of speakers. If it were possible to construct an individually tailored code book for each speaker, the average quantization error could be further reduced for a given transmission rate. The ultimate VQ coder would be capable of re-tuning the codebook, optimizing it for each speaker. Who knows how many supercomputers it would take to accomplish this in real time.

The beauty of VQ is that it can be applied to the output of any algorithm. It can be applied to the degree desired, or permitted by the economic and processing constraints of the system. It is true that efficient VQ algorithms will require enormous processing power, compared to the underlying techniques they operate upon. As digital signal processors continue to increase in power and decrease in cost, however, vector quantization will come into play in low bit-rate coders appropriate for the bandwidth-limited mobile channel.

12.2.3 Interdependence of Coding and Modulation

We have stressed in earlier sections that voice coding and carrier modulation — the two fundamental technologies in a digital mobile-radio system — are logically independent processes. Any coding algorithm may be combined with any modulation technique. From the perspective of the higher-level agenda of the mobile-system architect, however, coding and modulation are strongly interrelated. The relationship may be seen as one that is mutually enabling, rather than mutually constraining. That is, since improvements in coding and improvements in modulation generally favor the same architectural goals, such as spectrum efficiency, progress in one area may be substituted for progress in the other, according to current economic and technological opportunities.

The interdependence of coding and modulation is thus felt on the *design* level as an added degree of freedom. The selection of one coder rather than another may enable the mobile-system designer to utilize a more efficient modem, or *vice versa*. It is also observed in the *performance* of digital mobile-radio systems. We shall touch on each briefly.

12.2.3.1 Interdependent Design

The higher-level design trade-off between coding and modulation becomes clearer if we analyze the goal of spectrum efficiency. (We can define *spectrum* efficiency for the moment as the number of telephone voice circuits per megahertz of spectrum; see Chapter 15.) Higher efficiencies — more circuits per megahertz — may be obtained by decreasing the bit rate of the voice coder, while maintaining quality, or by increasing the modulation level, the *spectral* efficiency. Figure 12.30 shows the relationship between all three variables: coding bit rate (per circuit), modulation level (spectral efficiency), and circuits per megahertz. The graph assumes a base transmission rate of one symbol per hertz.

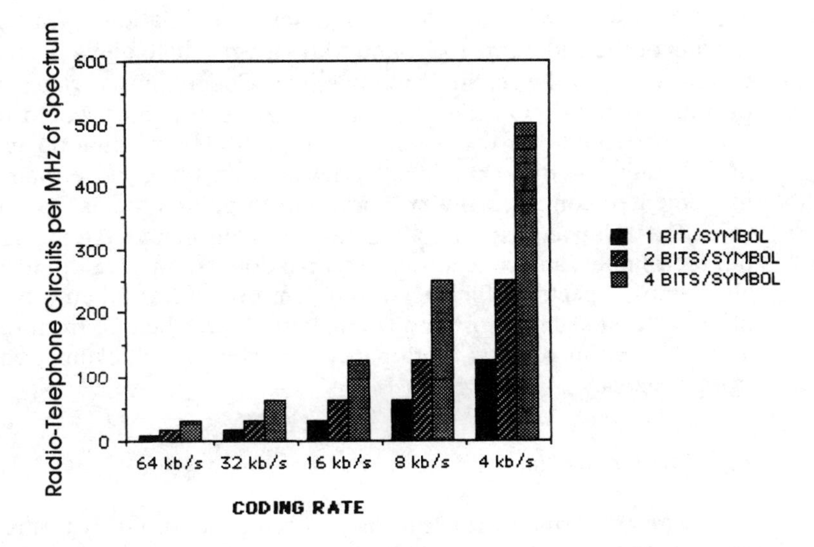

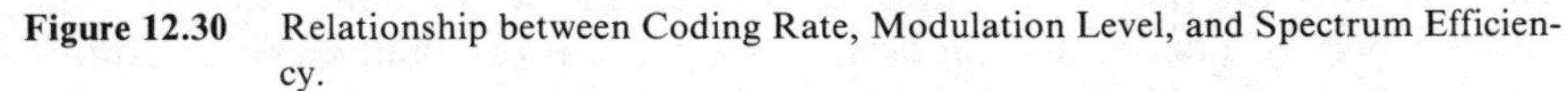
Figure 12.30 Relationship between Coding Rate, Modulation Level, and Spectrum Efficiency.

Note: This chart is an idealization that does not address adjacent channel or cochannel interference, guard-band requirements, or digital "overhead" for framing, synchronization, *et cetera.*

The available circuits per megaherz may be doubled either by halving the bit rate or doubling the modulation level. At any given historical moment, depending upon the specific application, e.g., fixed microwave *versus* mobile telephony, and upon the state of development of the underlying device technology, it may be that one of these two paths is easier to attack than the other. Through the 1960s and 1970s, modulation was the more promising avenue, at least for point-to-point microwave systems. Spectral efficiencies of 2, 3, even 6 bits/Hz were achieved. During this time, coding standards were stuck at 64 kb/s, essentially no improvement since the first T-carrier systems. In the past few years, however, there has been a revolution in coding, and bit rates have been driven down rapidly with the advent of modern digital signal-processor chips.

In the field of mobile radio, multilevel modulation is still in its infancy. Most mobile-radio architectures today are based on 2-level or 4-level modulation schemes. The only exception I am aware of is IMM's adaptive 4-level and 16-level DPSK system [57]. Voice coding gains, on the other hand, have been spectacular. Most researchers would probably say that halving the coder rate from today's standard of 16 kb/s would be about as easy, or as difficult, as doubling the spectral efficiency from a standard of 2 b/Hz to 4 b/Hz, from 4-level to 16-level modulation. While a further halving of the coder rate to around 4 kb/s is very challenging, it is, however, certainly within the realm of the possible. Coders in that range are being prototyped today. On the other hand, very few people would be willing even to speculate on the possibility of an 8-b/Hz modulation scheme for mobile radio — that would be 256 levels, four times more than has been implemented commercially even in point-to-point systems.

The relative ease of coding gains has led many system designers to pin their hopes, and concentrate research dollars, on developing new voice algorithms. Spectral efficiency in modem design has taken a backseat to other criteria such as cost and complexity. Nevertheless, from the system architect's point of view, both paths are open for achieving greater spectrum efficiency.

12.2.3.2 Interdependent Performance

The performance of the total system, measured in terms of voice quality, for example, is also affected by certain performance interdependencies between coder and modem. As a general rule and all other things such as transmitter power and average interference levels being equal, for example, the channel bit-error rate (BER) is determined in any given environment by the modulation technique. At a given signal-to-noise ratio (SNR), some modulation schemes will produce higher error rates than others. Or, to put it another way, some modulation techniques can deliver a given BER objective with a lower SNR, i.e., poorer channel conditions. For example, 16-ary QAM needs about 4 dB less than 16-ary PSK to achieve the same BER [58]. (See Figure 12.31.) In a given channel with an assumed SNR of 16 dB, a 16-ary PSK system will deliver a BER of about 10^{-4} — one bit in error for every ten thousand — while a 16-QAM modulator will deliver a BER of about 10^{-8} — one error per hundred million bits. (It is unfortunate that 16-QAM, with its amplitude reference, is not particularly suitable for the mobile environment.)

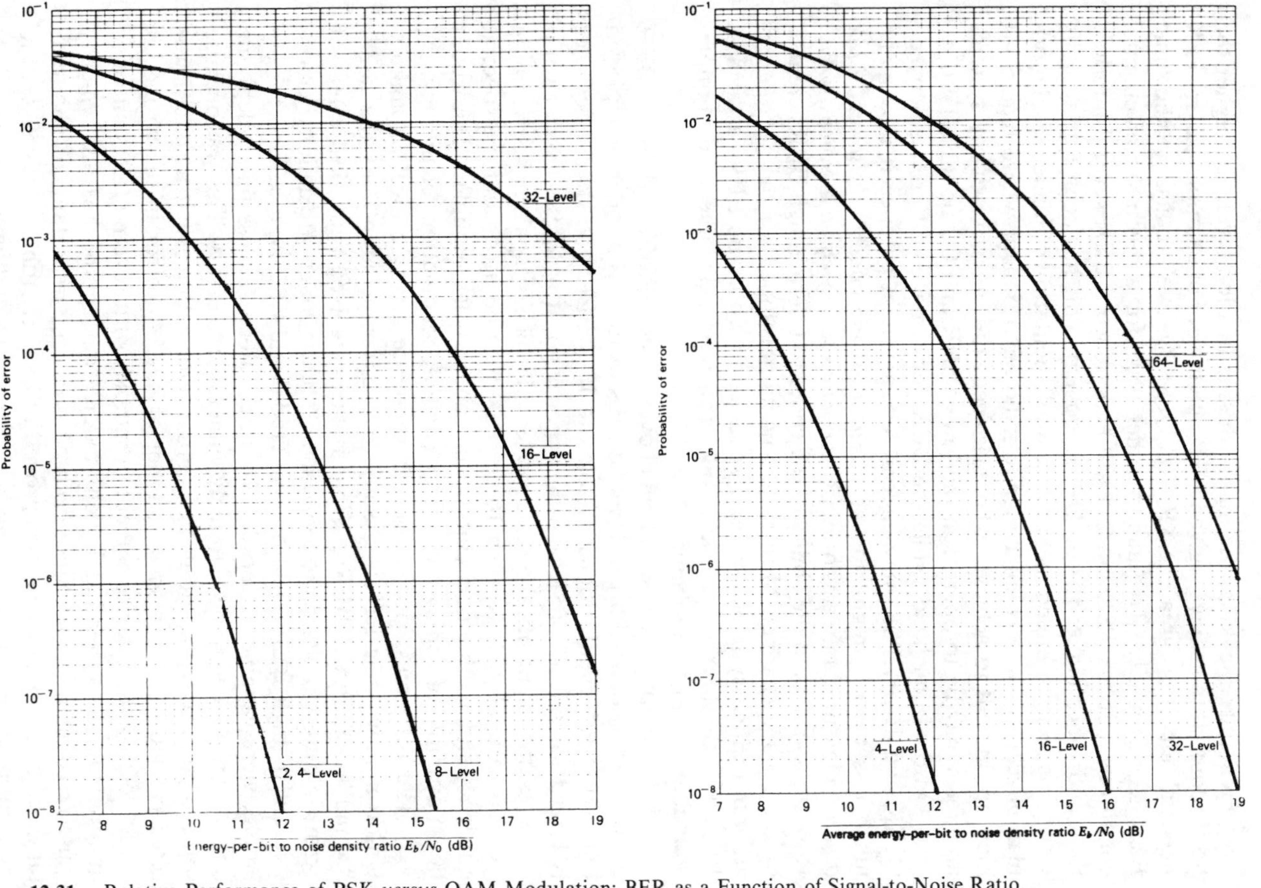

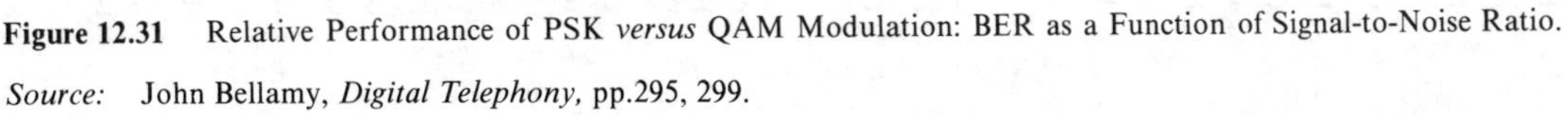

Figure 12.31 Relative Performance of PSK *versus* QAM Modulation: BER as a Function of Signal-to-Noise Ratio.

Source: John Bellamy, *Digital Telephony,* pp.295, 299.

Generally speaking, the higher the modulation level, the higher the SNR required to achieve a given BER objective. For example, 16-ary PSK needs about 13 dB greater SNR than QPSK (4-level) to achieve the same BER signal quality. To put it another way, for a given set of channel conditions a 16-ary PSK modem will experience a BER several hundred thousand times higher than a QPSK modem. This has been traditionally viewed as a major impediment to the implementation of higher-level modulation schemes in a cellular mobile-radio environment where high SNR may be difficult to achieve, where, indeed, the objective of the systems architects may be to operate at the lowest SNR and C/I possible [59].

On the other hand, while the BER performance difference between QPSK and 16-ary PSK may appear very great, in practice it may mean much less than it might appear, *because of the coder robustness. Link quality,* measured in terms of BER, *is not necessarily equal to voice quality.* Robustness refers to the ability of the coder to tolerate high BERs created by the mobile channel. Most coders being developed for the next generation of cellular systems are being specified to withstand BERs as high as 10^{-2}, 1% BER. A very robust coder can rescue an otherwise unimplementable modulation scheme.

For example, assume a channel over which QPSK will deliver a BER of 10^{-8}, while 16-ary PSK will yield a BER of 10^{-3}. If a coder is employed that is not sufficiently robust to tolerate 10^{-3} BER, the system designer is forced to choose QPSK. On the other hand, if a different coder with, let us assume, an identical bit rate is selected which is capable of performing well up to a 10^{-2} BER, then 16-ary PSK may be used, and the spectral efficiency and number of circuits per megahertz will be doubled. This would be an example of coding technology *enabling* modulation technology.

The robustness of the coder is therefore a very critical parameter. Unfortunately, it is largely a subjective criterion: how does the coder "sound" at different BERs? What is the perceived acoustical effect of errors in different bits in the coded signal? Some coders are definitely more tolerant of high BERs than others. Traditional LPC-type coders, for example, were capable of very low bit rates, albeit for less than telephone quality, but tended to be extremely vulnerable to errors. This was because some of the most important perceptual information, such as pitch, was often carried in a very small number of bits. An error in the wrong bit could destroy the pitch of the entire sample, with severe effects on voice quality. On the other hand, delta-modulation approaches are typically quite resistant to errors, since each bit is equally important, or unimportant. Subband coders have also proved to be fairly robust. Of course, the robustness of any technique can be improved by adding error correction, at least for crucial bits. This, however, increases the overhead.

Thus, the definition of the digital mobile-radio link, which encompasses the selection of coding and modulation techniques, involves a very rich, multidimensional decision set — even if we limit ourselves to the relatively conventional alternatives discussed in this section.

12.3 SPREAD-SPECTRUM TECHNIQUES

12.3.1 The Great Unknown

All conventional radio systems, including all those described in the preceding section, are based upon an architectural principle which seems so obvious as scarcely to require mention. Ever since the days of Marconi and De Forest — who started off battling one another to report the yachting race results off New York and often succeeded only in jamming each other's signals [60] — it has been a central tenet of radio design that each transmission must be confined to its own separate frequency, its own distinct channel, and carefully isolated from adjacent channels. Only one transmitter is allowed per channel at any given time. This is true of all FDM and TDM systems, analog and digital. It is true of broadcast radio and television. It is true of the current generation of cellular radio and it will be true of the next generation. The regulatory function of the FCC in the radio field is largely based upon the articulation and enforcement of this principle. Channels are defined, bandwidths specified, emissions masks promulgated, all in the name of controlling the interference problem, of keeping different transmitters from stepping on one another.

The fear of interference is rife in the radio community. A decades-long battle has been waged, is still being waged, between the broadcasters, especially TV, and the mobile-radio community over just how close, in space and frequency, the two types of system should be allowed to come together. Because of this fear, a vast conservatism has settled over the industry and huge blocks of spectrum have been set aside simply as guardbands between transmitters. For example, in the UHF TV band, the two channels adjacent to each broadcast channel in a given area must remain unoccupied. By definition, at least half of the UHF spectrum — equal to approximately 200 MHz or five times the spectrum originally allocated to cellular mobile radio! — must lie fallow forever in any given urban area.

Channelization is engrained in the thinking of radio engineers. They strive for better transmit filters to contain the transmission within the proper bounds. They strive for better receive filters to reject any interference that may assault their receiver. They strive for hyperstable frequency synthesizers to keep the carrier tuned as sharply as possible.

Channelization has been the keystone of mobile radio for fifty years. Because of the scarcity of spectrum, the channelization assumption has driven radio engineers to look for ways of reducing the bandwidth of the voice circuit, first by channel-splitting, more recently in digital systems by means of TDM techniques, lower bit-rate coders, more spectrally efficient modems. We have explored in considerable detail the logic and implementation of bandwidth-reduction methods in the preceding sections.

Viewed from this orthodox perspective, the vision of *spread-spectrum* transmission may seem so contrary, even perverse, that it might almost be taken for a jest upon the inflamed sensitivities of the interference-bedeviled radio community.

The logic of spread spectrum is based upon a negation of the channelization assumption and all its corollaries. Instead of carefully segregating each transmission in its own channel, spread-spectrum techniques heap circuits and transmitters on top of each other, by the dozens or hundreds, right in the same channel. Instead of trying to eliminate interference, it is welcomed; it is designed into the system. Instead of looking for narrower and narrower channels, which lead to more costly radios, tighter filters, more complex modems with sharper roll-offs, spread-spectrum techniques utilize very wide channels, transmitting each voice circuit over a bandwidth perhaps as wide as a conventional TV channel. The subscriber mobile radio is simpler, cheaper, lower power. The transmission is almost unbelievably robust, so robust, in fact, that spread-spectrum systems strive to achieve *negative* signal-to-noise ratios! There is no fixed limit on the number of simultaneous conversations that may be transmitted over the same spectrum band. There is no blocking, only a "graceful degradation" as traffic increases; the system is self-regulating, homeostatic. Even multipath, the demon of the mobile-radio designer's nightmares, may actually become a form of redundancy and diversity to strengthen the spread-spectrum signal.

Such, at any rate, is the vision of some spread-spectrum advocates. Whether or not it is completely realizable, it is such a radically different way of looking at radio that it can seem unsettling to those, especially in the regulatory community, who have spent a lifetime adhering to the "channelization orthodoxy." Spread-spectrum thinking breaks every conventional rule of spectrum management. Indeed, by controverting the rules thoroughly, spread spectrum promises to solve all the conventional problems — spectrum efficiency, mobile-radio cost, multipath, interference. It is as though a dieter who had lived his life on a severe and unappealing regime of calorie-counting were suddenly told that he could lose weight by eating several helpings of ice cream and apple pie with every meal!

Yet, we are perhaps even more surprised to learn, however, that spread spectrum not only works, but has been working for decades, almost as long as FM. Until the last ten years or so, however, it was cloaked in military secrecy. Papers were classified. Experiments were carried out in secrecy, inventions tested without the knowledge of the inventors. Projects were segregated into small jobs, mutually isolated from one another, to prevent those working on them from piecing together what it was all about. Patents were held up on grounds of national security for decades. The key patent in the United States, developed by a pair of ITT engineers shortly after World War II, was suppressed until 1978, thirty years after the inventors' conception, by which time the senior inventor had been dead for seven years [61].

The reason for this unprecedented policy of confidentiality lies in the enormous importance of spread-spectrum techniques for the military. Uniquely among radio techniques, spread spectrum can withstand active jamming that would occur on the battlefield in wartime. It can provide excellent security against interception for sensitive communication. Moreover it is not only technically secure, it is also *operationally* secure; if the enemy possesses the technology, or even the equipment, and knows exactly which spectrum band is being used, he is still unable either to jam or to decode the transmission. Spread spectrum also can inhibit an enemy's ability to locate military units through their radio transmissions. With spread-spectrum techniques, the transmitter can conceal itself in a shroud of noise, blending into the background, electromagnetically invisible except to a receiver with the right decoding key. Spread spectrum has many applications to radar and navigation systems.

In short, it is a truism that modern military systems and operations are crucially dependent upon wireless communication, and spread-spectrum techniques are the backbone of modern Milcom. This military genealogy lends a further mystique to spread spectrum. It suggests the image of a "super-technology" that could solve all the problems of the commercial mobile-telephone systems engineer, yet which has been sequestered for esoteric military purposes.

Is this a true picture? Is spread spectrum a path to a radically different alternative for the next generation of cellular radio? Can it really be all the things it is sometimes claimed to be?

Unfortunately, the answers to these questions are not clear. There are still many fundamental parameters of spread-spectrum performance that are open to dispute among engineers in good standing [62]. Opinions differ as to whether spread spectrum can really function in a commercial mobile-radio environment, whether it can actually attain some of the forecast spectrum efficiencies, whether the subscriber mobile radios in a spread-spectrum system would be dramatically cheaper, or considerably more expensive.

The European standards group evaluating alternative proposals for the next-generation digital cellular standard in Europe recently voted in favor of conventional narrowband architectures (TDMA) and against two wideband spread spectrum–type proposals. The vote was 13–2. The two supporters of the wideband approach, however, were the French and the West Germans, who have been strong supporters of this technology. In many discussions of mobile telephony, there is a tendency to rule out spread spectrum right at the start, under the assumption that it cannot really be made to perform up to expectations in a mobile-telephone system, or that it is so complex that it is still many years, or several generations, in the future.

I believe that it is probably impossible for spread spectrum to receive fair consideration at this time, for the reason that most commercial radiotelephone engineers, not to speak of government regulators, are so utterly unfamiliar with it. It is no wonder: no large-scale commercial spread-spectrum system has ever been fielded. There are simply too many unknowns. For this reason, I personally believe that the next generation of cellular radio will not be a spread-spectrum system.

This, however, makes me somewhat apprehensive. I remember quite vividly the conventional wisdom of the mobile-radio industry back in, say, 1980. At that time, analog cellular was being heralded as the ultimate solution. Plans were being laid to sell enormous numbers of cellular phones, and there was talk of urban systems like Chicago growing to a million subscribers or more within the AMPS framework. The view of digital technology was simply that it was not needed; moreover, it was too embryonic, too far in the future. Many breakthroughs would be required, it was felt, to bring a digital mobile-radio system into being, and it was not forecast until sometime well into the twenty-first century. That view was based upon ignorance. Digital techniques were still *terra incognita* for the world of mobile radio. The very idea of what a digital mobile-radio system encompassed was unclear.

We, this author included, are more or less in the same situation today with respect to spread-spectrum techniques. Yet while it is true that spread spectrum has never yet been adapted to commercial application, there is a tremendous base of experience in the military sphere with spread-spectrum communication. Some of the results are provocative, to say the least, in the sense that if we could project currently available military spread-spectrum capabilities into commercial applications, our thinking about mobile telephony would be revolutionized. Also, in most cases the obstacle to such a projection is economic, the high cost of military hardware. Yet if there is one constant in the modern world of digital communication, it is the inconstancy of cost barriers.

This is unsettling. I believe that if there is one "macro-risk" associated with the decision concerning the next generation of digital cellular radio, it is the possibility that spread-spectrum techniques could advance far more rapidly than we currently expect, and we could find ourselves a few years from now laying the groundwork for yet one more generation of mobile-radio technology based on conventional techniques that will be obsolete before they can be deployed.

12.3.2 The Background of Spread Spectrum

So what is this protean system?

Spread spectrum is not a single technique, not even a technology. It is almost a *philosophy* of communication, linking under one nomenclature several different methods of implementing communication systems which arose separately and were not even recognized as similar for quite a while [63]. It is the communication theory equivalent of, say, quantum physics; it is the first communication system based upon a thoroughly modern theoretical underpinning. Even the very advanced digital TDMA systems described in the previous section are still essentially linear derivatives of early telephone architectures. They comprise a set of circuits which are made available to individual users, dedicated to that user for the duration of the call. There is an intuitive line of descent from early channelized FM systems and the much more advanced digital channelized systems being proposed for the next generation. The fundamental communicative goals and principles are the same.

Spread spectrum is rooted theoretically in a very different, more modern, less intuitive, view of the communication process that was first articulated by a number of people, but especially by Claude Shannon, during and shortly after World War II [64]. Several of the key breakthroughs were apparently stimulated directly by the revolutionary theoretical implications of Shannon's writings [65]. It is not clear whether spread-spectrum techniques would have emerged intuitively from tinkering with existing radio techniques, although some point to the roots of spread spectrum–type thinking in Armstrong's work on FM, which did, after all, involve an expansion of the signal bandwidth in return for much improved performance [66]. I believe, however, that, like quantum physics or the Copernican revolution, it required virtually a subversion of conventional intuitions to bring about the critical breakthroughs for spread spectrum.

It required what has begun indeed to be called the "Copernican System of Communications": Shannon theory [67]. Indeed, the implications of spread-spectrum techniques were not immediately obvious. It was some time before the early inventors or, for that matter, their future military sponsors and censors fully realized what they had spawned.

Actually, spread-spectrum techniques first arose in connection not so much with communication, as with what could be called *anticommunication.* The practical spur to the development of spread-spectrum techniques was the problem of *preventing* communication: in particular, preventing an enemy from receiving and decoding, or from detecting and jamming, a military radio communication. It was only later that the positive potential of spread spectrum to *improve* communication was recognized.

The best way to approach an understanding of spread spectrum — and it would be far beyond the scope of this book to do much more than point the way toward that approach — is to look at the early evolution of the two chief spread-spectrum techniques. *Frequency-hopping spread spectrum* was developed as an antijam strategy. *Direct-sequence* or *noise-modulated spread spectrum* first emerged as a promising approach to secure voice communication. Both techniques first emerged in tantalizingly incomplete form in analog implementations — there is nothing inherently digital about spread spectrum, although digital techniques vastly enhance the ability of system designers to optimize specific spread-spectrum techniques.

12.3.3 Frequency-Hopping Spread Spectrum (FH/SS)

The problem of jamming is sometimes analyzed as a game, not in the sense of an amusement, but according to the formalism of game theory, a branch of modern decision theory. The jamming game is played by a communicator, e.g, a battlefield headquarters, who wants to communicate with a receiver, such as a specific tank group, and an enemy jammer who wants to prevent that communication from taking place by blasting out his own transmitter to obliterate the communicator's signal [68].

Let us assume that both the communicator and the jammer have a number of different courses of action they can take: specifically both the communicator and the jammer can choose to transmit on a number of different frequencies. In this game, the receiver does not transmit; he only listens for the communicator's message. Let us also assume that the jammer has the ability to scan the spectrum to find out what frequency the communicator is transmitting on. To make the game more interesting, let us also further assume that the jammer has a transmitter one hundred times more powerful than the communicator.

During World War II, Allied and Axis forces engaged in this game constantly throughout the European theater. It rapidly became clear that ordinary fixed-channel radio transmission, whether AM or FM, was highly vulnerable to jamming; the jammer could easily locate the transmission, tune his jamming transmitter to the same frequency, and let loose. Communicators began to develop a rather obvious, *ad-hoc* solution: when jamming was detected, both transmitter and receiver changed to a new frequency, and the game began anew. The communicator was able to transmit on the new frequency for as long as it took the jammer to locate and jam him once again.

We can now formulate the jamming game, diagrammed in Figure 12.32. The communicator's only course of action in this game is to change frequencies, which he can do at any time. He can place his message chip on any row in the game matrix. The jammer does the same thing, placing his jamming chip on any column, except that the jammer has higher power and can put down more chips. Assume that he has one hundred times the power and can place ten jamming chips on ten different columns, and still be confident that any one of his jamming chips will be ten times more powerful than the communicator's signal. (See Figure 12.33.)

The communicator's goal is to make sure that he is not transmitting on the same frequency as the jammer. If he can achieve that for a specified percentage of the total time available — enough to get his messages through — then he wins. The jammer wins if he can keep his transmitter on the communicator's frequency most of the time.

Obviously, the communicator and the receiver must both possess an agreed-upon sequence of frequency changes. Otherwise it would take the receiver at least as long as the jammer to find the new frequency, and the jammer will get there at the same time and win the game. The only advantage the communicator and the receiver can use is their foreknowledge of the correct sequence of frequency changes.

Historically, the frequency change by both parties was manual at first, as was the jammer's search process. Since the frequency sequence was known to the communicator and the receiver, but not to the jammer, and since the jammer's ability to find the new frequency was rather slow, the communicator generally won the game.

Let us make the game even more interesting. Let us assume that the jammer has the ability to scan all available channels instantly and find which one the communicator is operating on within a very short period of time, say, a few milliseconds. Now the jammer will win every time over a communicator who is still changing frequencies manually. So the communicator must automate and speed up his frequency change process. Both his transmitting and receiving equipment must step through a precise sequence of frequency hops. On each hop, only a short burst of information is transmitted; then the frequency is changed and the next burst is transmitted, and so forth. As the jammer becomes quicker at finding the transmission, the hopping must become faster and faster.

Let us now assume that both the jammer and the communicator can react instantaneously; that is, the communicator can change frequencies as fast as he wishes, and the jammer can detect and jam the communicator's signal as fast as he wishes. Both are able to move and react as fast as the laws of physics will allow. Who wins?

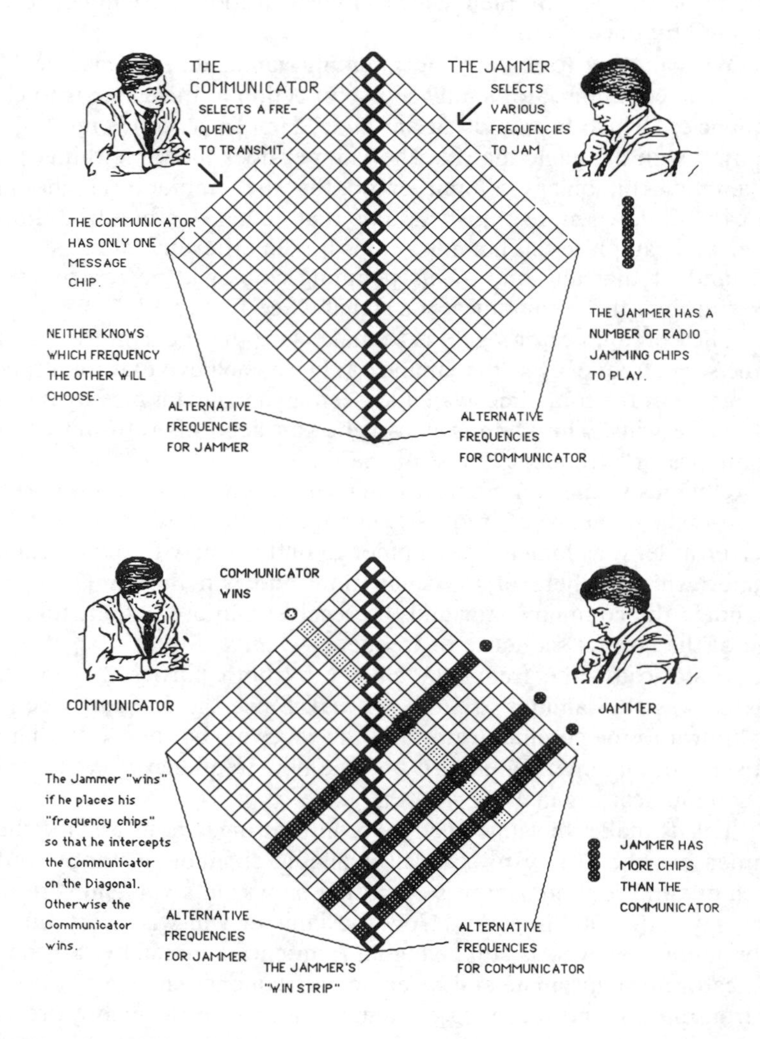

Figure 12.32 The Jamming Game.

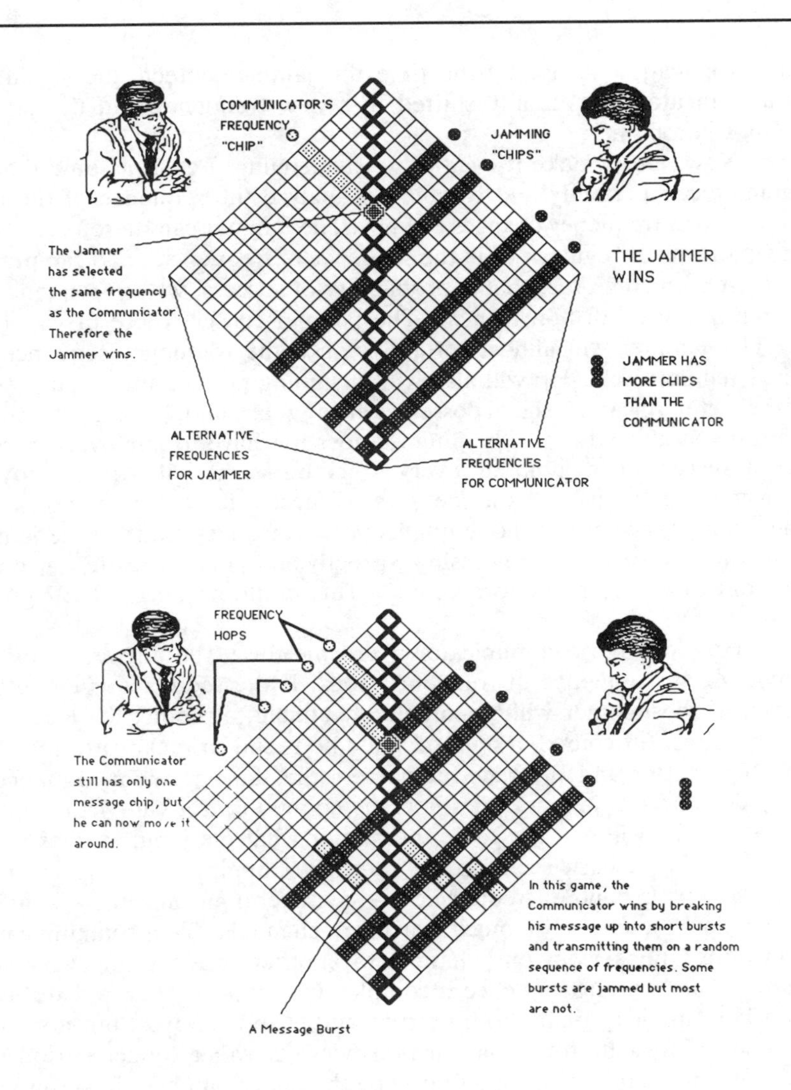

Figure 12.33 Jamming Game (continued).

There is a strategy which ensures that the communicator can win. It is known today as fast-frequency hopping [69]. The communicator reduces the length of each burst to be equal to or less than the propagation time — the time it takes for the radio wave to travel from the communicator's transmitter to the receiver, which may typically be on the order of 10–100 microseconds. He may transmit only a single bit, or a very few bits, assuming a digital signal, before hopping to the next frequency. Because of the propagation delay, by the time the jammer detects the signal, the communicator has already shifted to a new frequency and the jammer cannot jam him.

Now, let us make it even more challenging. Let us assume that the jammer can perfectly and completely analyze the sequence of the communicator's frequency changes or hops. The jammer can therefore identify any patterns or regularities in the sequence of frequencies and can predict, at least with some reasonable probability of success, which frequency the communicator will hop to next. The jammer can get there first and jam it. The jammer will almost certainly win if the frequency sequence is a fixed regular cycle. He will also win, assuming perfect analytical powers, if the communicator's signal possesses any regular, and, hence, predictable, patterns at all. After all, the jammer does not have to jam every burst; if he is successful in jamming every other burst, he will still destroy the transmission. In fact, if the measure of success for the communicator is the maintenance of telephone-quality voice transmission, then the jammer will win if he succeeds in guessing correctly and jamming as few as one or two out of every one hundred bursts. This would produce a BER greater than 10^{-2}.

How can the communicator overcome this? The answer is: by *randomizing* his sequence of frequency hops. If the sequence is completely random, the jammer will be completely unable to predict the location of future hops. Of course, a method must be found to make sure that both the transmitter and the receiver possess the same random sequence. If they do, and the jammer does not, the communicator can virtually guarantee that he will win the game, no matter what the jammer does.

This is the basic scenario for frequency-hopping spread spectrum. A relatively wide band of frequencies — say several megahertz — is divided into a large number of much narrower channels. The communicator's transmitter hops from one channel to another, transmitting very short bursts. The hopping sequence is completely random, generated according to a key that is available to both transmitter and receiver, but not to the jammer. Thus, the total transmission, viewed over a longer period such as a second, actually appears to occupy the entire bandwidth — this is the "spreading" of the spectrum — although at any one moment, for any one burst, it occupies only a very small percentage of the channel.

Now let us rename the jammer. Let us call him Noise. (Another of his names is Multipath.) Noise plays the same game as the jammer, except that he places his noise chips on every single frequency. (Multipath does not place his multipath chips on every frequency — recall that deep multipath fades are uncorrelated on different frequencies and occur only from time to time on any one frequency; he places his multipath chips on a few frequencies and constantly shifts them around.) The communicator, whom we may now assume to be a mobile-telephone user, continues to place his message chip on only one frequency at a time, shifting it rapidly from frequency to frequency in a completely random series of hops. (See Figure 12.34.)

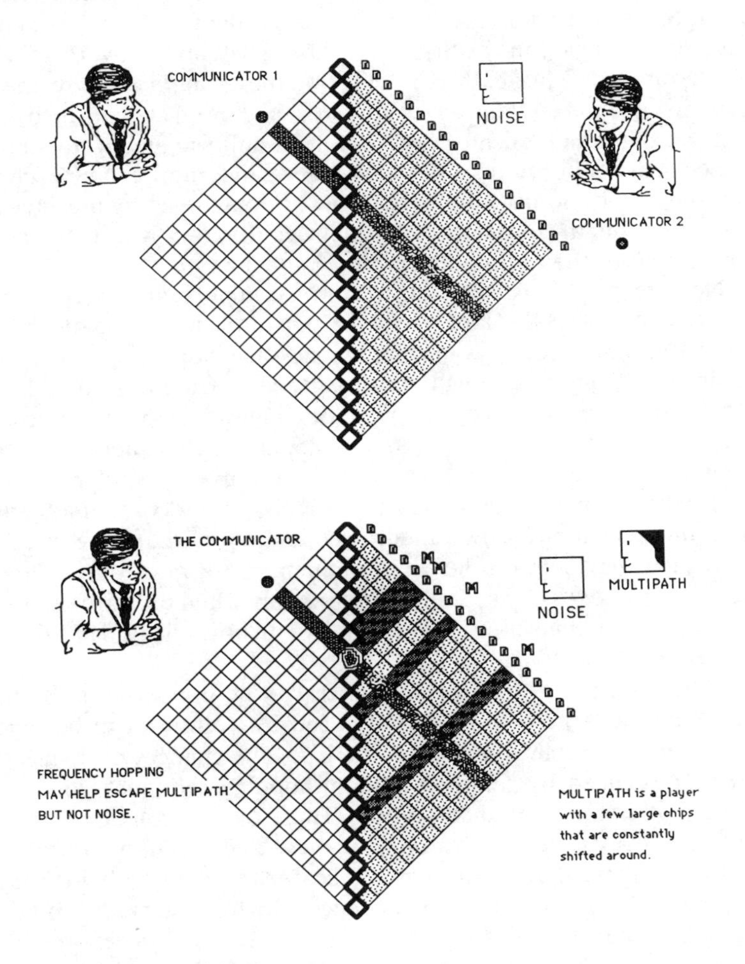

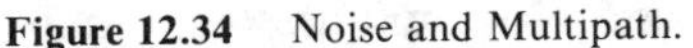

Figure 12.34 Noise and Multipath.

How can the communicator win, since now there are noise chips on every frequency? Fortunately for the mobile communicator, he can still get through, because, although the total power of the noise in the entire channel is very large, it is spread over the entire band and it is lower than the mobile communicator's signal in any particular frequency at any particular moment. Thus, even though the signal power in the entire band may be well below the noise power — in other words, there may be a signal-to-noise ratio of less than one! — the mobile communicator will succeed. Since there will also be multipath problems on only a small percentage of the hops, multipath, too, will be substantially overcome. Also, neither Noise nor Multipath is intelligent; they cannot analyze the communicator's frequency-hopping sequences and predict where to attack. Therefore, the communicator does not need quite so elaborate a hopping sequence. His sequence does not have to be random; it can be cyclic. Nor does it have to be so fast; the pace of the hopping is set by the speed with which the multipath chips change frequencies. In other words, the hopping rate is related to the fading rate in the channel.

Now for the most amazing result. Assume that a second mobile communicator enters the game — on the side of the noise and the multipath players! Communicator 2 is placing his message chips in his own random sequence of frequencies along the columns, while Communicator 1 is moving through his sequence of the rows. (See Figure 12.35.) In effect, Communicator 2 has assumed the jammer's role, although he adopts a different strategy for a different purpose. We can call him the interferer. He does not intend to jam the communicator, but he may do so, inadvertently. Both communicators follow random sequences of frequency hops; from time to time they land on the same frequency, but most of the time they avoid one another. They actually interfere only when they are on the same frequency on the same hop. If this happens with only one hop in a thousand, the BER would be 10^{-3}.

If the number of frequencies which can be hopped onto within the total channel is large enough, the chance of a collision can be made arbitrarily small to achieve any desired BER. The chances of collisions can be further reduced by selected random-hopping sequences that are *orthogonal,* or completely uncorrelated. Thus, *two communicators can occupy the same total bandwidth, and yet coexist with minimal interference.*

Now assume that a third communicator is added, going through his own random frequency-hopping sequence which is selected to be orthogonal to the other two. The possibility of collisions and interference for all three will increase slightly. In fact, as more and more mobile communicators are added to the frequency-hopping system, each with his own orthogonal sequence of hops, the level of interference gradually rises. In

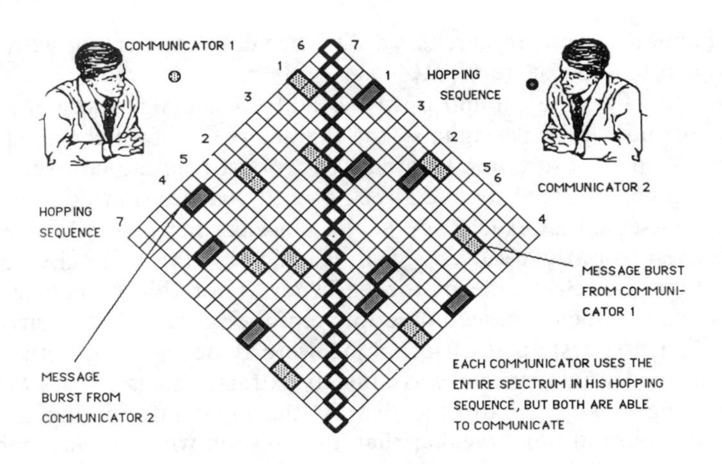

Figure 12.35 Two Communicators Playing the Game against One Another.

fact, there is no fixed limit to the number of mobile communicators who may share the same channel in an FH/SS system. New ones can keep coming in, the interference levels will rise gradually, until at some point the process will become self-regulating: the quality of the voice link will deteriorate slowly to the point where callers begin to cut short or to refrain from undertaking new calls. But no one is ever blocked in the conventional sense, as they are when all the channels or slots in an FDMA or TDMA system are occupied.

How many users can occupy the same channel simultaneously while maintaining telephone-quality voice communication? Alas, the answer to this crucial question is not entirely clear at present. There has simply not been enough applicable experience to project to a large-scale commercial system for mobile telephony, and opinions vary [70]. Before delving further into the remarkable performance characteristics of FH/SS, we should review another, entirely different, approach to spread spectrum, which is, if anything, even stranger to commonsense intuitions.

12.3.4 Direct-Sequence Spread Spectrum (DS/SS)

A second, and equally urgent, wartime communication requirement was the development of reliable secure-voice communication. We have already seen that the search for an encryptable voice transmission stimulated the development of pulse code modulation, which opened the door for all digital communication. A few years after the war, spurred directly by Shannon's work, researchers at ITT began to experiment with another approach to voice secrecy [71].

The ITT group found a way to add a voice signal together with a much stronger noiselike signal — the characteristics of which were precisely known — and in so doing create a very robust transmission vehicle. The resulting mixed signal sounded like noise. It was transmitted to a receiver which possessed an exact copy of the noiselike input — which was then subtracted from the received signal, leaving only the original voice input. (See Figure 12.36.) One of the early terms for this process was *noise modulation,* which is indeed descriptive of the process that is carried out.

The first tests of this technique were done in 1950, with stunning results. The crude system was able to extract signals 35 dB *below* the interfering noise [72]! In other words, the input information signal was several thousand times weaker than the noise in which it was embedded, and yet it could be transmitted successfully and extracted from the noise background with remarkable clarity. The implications were revolutionary. Although most of the early test results are still classified, one of the participants has recalled that during the tests a large solar magnetic storm happened to occur, knocking out most of the long-distance radio communication systems then in operation. A 50-kW conventional transmitter was unable to make contact with its receiver; the noise-modulated signal, however, was successfully transmitted from coast to coast, at the height of the storm, with only 25 W of power — a factor of 2000 times less power [73].

Conventional radio people had grown up with the idea that noise is a bad thing, that it hinders transmission. The idea that adding noise, huge amounts of it, could actually strengthen a communication link to the point where it could operate at very small SNRs, was startling, to put it mildly. Noise modulation was a very strange idea indeed.

Today, this approach is exemplified by a digital version of noise modulation which is generally referred to as *direct-sequence* or *direct-coding* spread spectrum. The idea is to add together two digital signals, two bit streams, to create a third bit stream which will be the one actually transmitted. The first signal is the information signal, such as the output of a digitized voice circuit. Let us assume a bit rate of 10 kb/s. The second signal is produced by a random-sequence generator. It is simply a stream of random bits, flowing at an enormously faster rate, say 100 Mb/s. When the two are added together, the result is a third bit stream with the same bit rate as the second signal, but which now contains the information of the first signal as well. (See Figure 12.37.) At the receiver, an identical random-sequence generator produces a random bit stream which is exactly the same as the original random sequence added at the transmitter, bit stream number 2. By subtracting this random sequence from the received

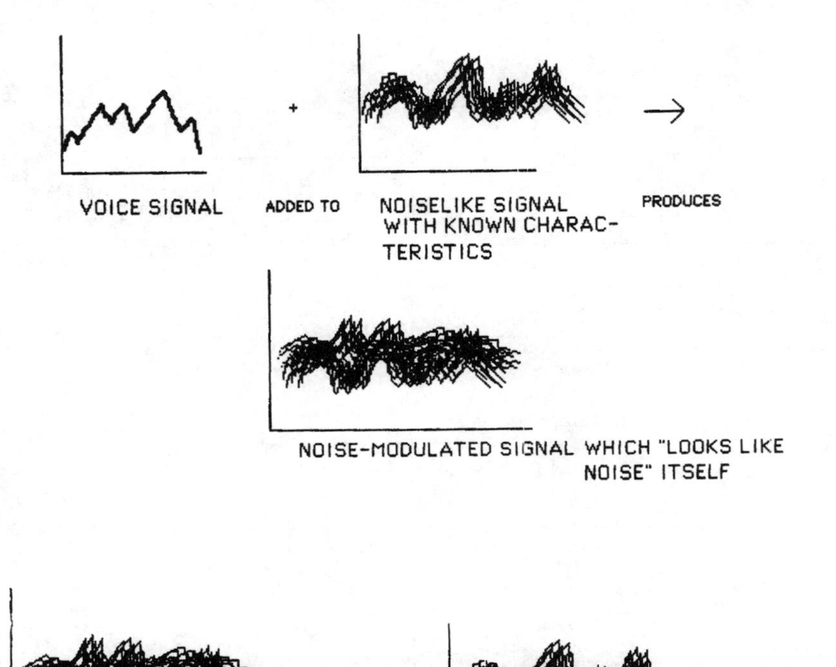

Figure 12.36 Noise Modulation.

signal, the difference is obtained — which is exactly equal to the original 10 kb/s voice input.

Because the transmitted signal has a bit rate of 100 megabits, the transmission has expanded to, say, around 100 MHz. This is the spectrum-spreading function. Unlike FH/SS, however, direct-sequence spread spectrum transmissions occupy the entire bandwidth all of the time. Just as

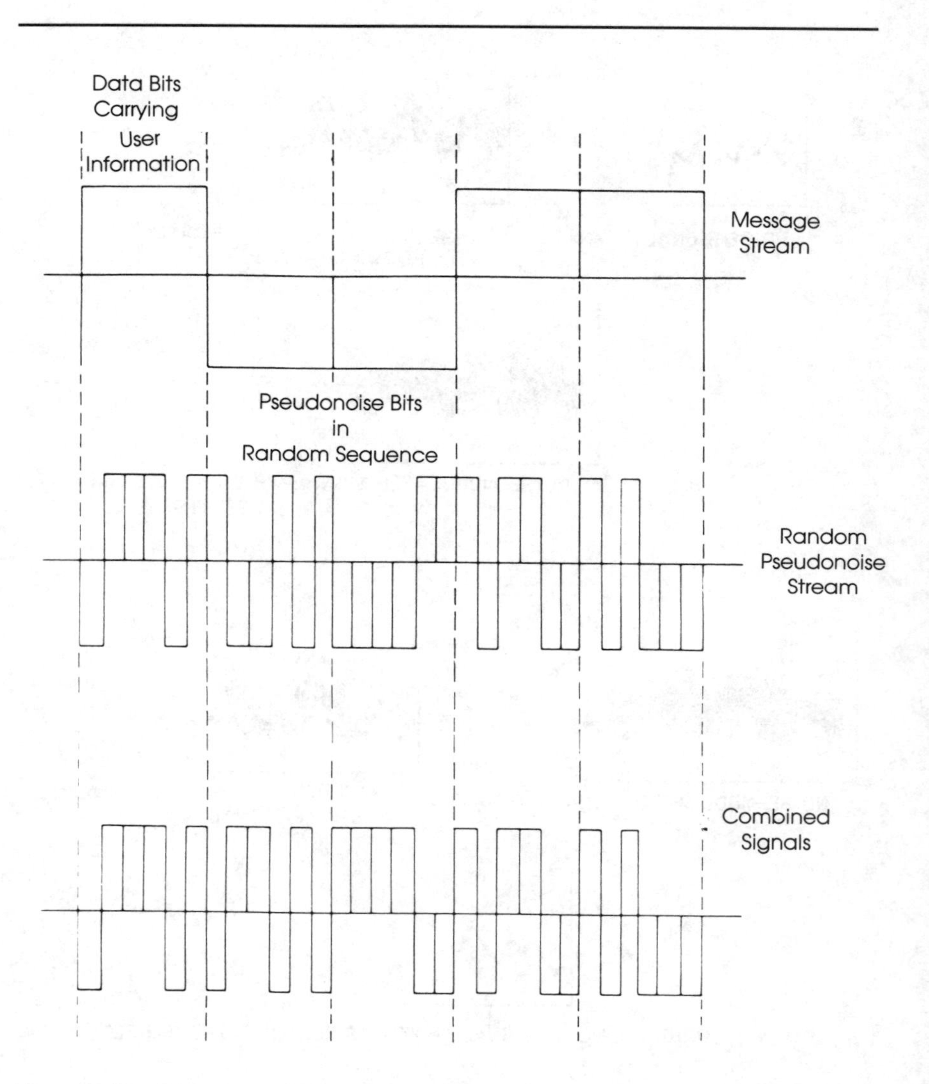

Figure 12.37 Direct Sequence Spread Spectrum.

Source: Marvin K. Simon, et al., *Spread Spectrum Communications,* Vol. 1, p. 144.

with FH/SS, it is possible to operate many DS/SS transmitters simultaneously over the same channel. The dynamics of the channel are mathematically identical to the FH/SS example, except that instead of having each communicator follow an orthogonal random sequence of frequency hops, each user now employs an orthogonal random sequence for the noiselike signal that is mixed with the information signal. As more and more users transmit over the channel, the effect is the same as in FH/SS: the level of interference gradually rises. A large number of users can share the same spectrum, even though each one of them is occupying the entire channel at exactly the same time!

12.3.5 Performance of Spread-Spectrum Techniques

There are other techniques which have the same spread-spectrum effect, including time-hopped systems in which bursts are randomly sequenced in time [74] and Chirp systems in which each burst has a narrow-frequency bandwidth that rapidly sweeps the entire channel bandwidth [75]. All these systems utilize conventional voice-coding and modulation techniques actually to encode the bursts of the random bit stream. PSK, for example, is commonly the modulation technique of choice [76]. Spread-spectrum systems differ in the way they manage the transmission.

The essential shared characteristics of these very different techniques are:

1. The transmission of a voice circuit over a bandwidth much wider than would be normally required in a conventional channelized radio system;
2. The coding of the transmission by means of a random sequence that is shared by both transmitter and receiver;
3. The assignment of different random sequences to distinguish different users.

There are several interesting and valuable characteristics of spread-spectrum performance, regardless of the technique employed. The most important is the *processing gain* [77]. Measured in terms of decibels, the processing gain is approximately equal to the logarithm of the ratio of the channel bandwidth (after the signal has been "spread") to the information bandwidth (before the signal is spread). In a direct sequence spread-spectrum approach, this is equivalent to the ratio of the channel bit rate to the information or voice-coding bit rate. In a frequency hopping spread-spectrum approach, the processing gain roughly corresponds to the ratio of the total channel width (in megahertz) to the width of the frequency occupied during each hop.

For example, in a DS/SS approach, if we assume that the circuit rate, the information rate, is 10 kb/s, which is embedded in a channel rate of 10 Mb/s, the ratio is 1000:1, or 30 dB. An FH/SS system would achieve the same processing gain with a 10-MHz channel divided into 1000 hops each 10 KHz wide. This processing gain can be added directly to the numerator of the conventional signal-to-noise ratio. As noted above, a large processing gain can allow the recovery of a weak signal from a much stronger noise background. For example, if an SNR of 10 dB is normally needed to achieve good circuit quality, then an SS system with 30 dB processing gain can deal with a received SNR of −20 dB, where the desired signal is 100 times weaker than the noise [78].

Another advantage of spread-spectrum techniques is a tremendous reduction in multipath fading. Deep Rayleigh fades tend to be frequency selective. In an FH/SS system, if there is a fade on the specific frequency of a particular hop, it is unlikely that there will be a fade on the next hop, especially if the hops are separated by a certain distance, known as the coherence bandwidth — about 600 khz at these frequencies. Whereas in a conventional mobile system the mobile unit is assigned to a given frequency for the duration of the call and is vulnerable to very deep and possibly long-lived fades on that frequency, the FH/SS mobile unit skips through the whole frequency range very rapidly and never spends more than a brief hop in any one fade. Thus, the average fade is much less. In fact, the maximum fades experienced by SS systems should be the equivalent of about 2–3 dB, instead of the 20 dB or more in conventional systems [79]. This means that the fade margin can be reduced, bringing down mobile power levels, which further aids the cochannel interference situation [80].

12.3.6 Application to Mobile-Telephone Systems

There is substantial disagreement over just how well an SS system would perform in a cellular mobile-telephone system and over which type of SS technique would work best; the imperfect consensus seems to favor FH/SS systems [80]. Some of the proposals for the European standard make use of slow frequency hopping in order to combat fading [81]. The now-rejected CD 900 proposal for the European standard incorporates "moderate spectrum spreading" and elements of a spread-spectrum code scheme [82]. More recently, proposals for a full spread-spectrum approach to portable indoor communication have also been discussed.

The era of spread-spectrum techniques has not yet arrived for commercial mobile telephony. It is likely, however, that with major continuing advances in digital-processor technology over the next ten years spread spectrum will command increasing interest from mobile-systems architects. It is likely, in my opinion, that spread-spectrum concepts will be important in the third-generation systems that may begin to emerge in the latter part of the next decade.

REFERENCES

[1] Thomas L. Jones and William A. Kissick, "ACSB: What is Adequate Performance?," *Proceedings of the 37th IEEE Vehicular Technology Conference,* Tampa, June 1–3, 1987, pp. 492–497; see also *IEEE Communications Magazine,* "Single-Sideband Mobile Radio: A Review & Update," March 1979, pp. 25–28.

[2] S. C. Gupta, R. Viswanathan, and R. Muammar, "Land Mobile Radio Systems-A Tutorial Exposition," *IEEE Communications Magazine,* June 1985, pp. 34–45.

[3] *Ibid.*

[4] F. de Jager and C. B. Dekker, "Tamed Frequency Modulation, a Novel Method to Achieve Spectrum Economy in Digital Transmission," *IEEE Transactions on Communications,* Vol. COM-26, May 1978, pp. 534–542; see also Flavio Muratore and Valerio Palestini, "Features and Performance of 12PM3 Modulation Methods for Digital Land Mobile Radio," *IEEE Journal on Selected Areas in Communications,* Vol. SAC-5, June 1987, pp. 906–914.

[5] K. S. Chung and L. W. Zegers, "Generalized Tamed Frequency Modulation," *Philips Journal of Research,* Vol. 37, 1982, pp. 165–177; see also Muratore and Palestini, *op. cit.*

[6] Sirikiat Ariyavisitakul and Tsutomo Takenchi, "A Novel Anti-Multipath Modulation Technique—DSK," *IEEE Transactions on Communications,* December 1987, pp. 1252–1258.

[7] Subbarayan Pasupathy, "Minimum Shift Keying: Spectrally Efficient Modulation," *IEEE Communications Magazine,* July 1979, pp. 14–22; see also Muratore and Palestini, *op. cit.*

[8] W. J. Weber, "A Bandwidth Compressive Modulation System Using Multiamplitude Minimum Shift Keying (MAMSK)," *IEEE Transactions on Communications,* Vol. COM-26, May 1978, pp. 543–551.

[9] K. Murota and K. Hirade, "GMSK Modulation for Digital Mobile Radio Telephony," *IEEE Transactions on Communications,* Vol. COM-29, July 1981, pp. 1044–1050; see also Muratore and Palestini, *op. cit.*

[10] C.-E. Sundberg, "Continuous Phase Modulation in Cellular Digital Mobile Radio Systems," *Bell System Technical Journal,* September 1983, pp. 2067–2086.

[11] Hiroshi Suzuki and Kenkichi Hirade, "System Considerations of Mary PSK Land Mobile Radio for Efficient Spectrum Utilization," *The Transactions of the IECE of Japan, Vol. E 65, No. 3, March 1982, pp. 159–165; Hiroshi Suzuki, "Canonic Receiver Analysis for M-ary Angle Modulations in Rayleigh Fading Environment," IEEE Transactions on Vehicular Technology,* Vol. VT-31, February 1982, pp. 7–14; Hiroshi Suzuki and Kenkichi Hirade, "Spectrum Efficiency of M-ary PSK Land Mobile Radio," *IEEE Transactions on Communications,* Vol. COM-30, July 1982, pp. 1803–1805.

[12] Yoshihiko Akaiwa and Yoshinori Nagata, "Highly Efficient Digital Mobile Communications with a Linear Modulation Method," *IEEE Journal on Selected Areas in Communications,* Vol. SAC-5, June 1987, pp. 890–895; S. A. Gronemeyer and A. L. McBride, "MSK and Offset QPSK Modulation," *IEEE Transactions on Communications,* Vol. COM-24, August 1976, pp. 809–820.

[13] Akaiwa and Nagata, *op. cit.*

[14] Sundberg, *op. cit.;* see also Justin C. I. Chuang, "The Effects of Time Delay Spread on Portable Radio Communications Channels with Digital Modulation," *IEEE Journal on Selected Areas in Communications,* Vol. SAC-5, June 1987, pp. 879–889.

[15] S. Asakawa and F. Sugiyama, "A Compact Spectrum Constant Envelope Digital Phase Modulation," *IEEE Transactions on Vehicular Technology,* Vol. VT-30, August 1981, pp. 102–111.

[16] Justin C. I. Chuang, "The Effects of Time-Delay Spread on QAM with Non-Linearly Switched Filters in a Portable Radio Communications Channel," *Proceedings of the International Conference on Digital Land Mobile Radio Communications,* Venice, June 30–July 3, 1987, pp. 104–113.

[17] James Mikulski, Remarks at an FCC Seminar on "The Future of Cellular Radio," September 2, 1987.

[18] Minoru Kuramoto and Masaaki Shinji, "Second Generation Mobile Radio Telephone System in Japan," *IEEE Communications Magazine,* February 1986, pp. 16–21.

[19] James J. Mikulski, "Dyna*T*A*C Cellular Portable Radiotelephone System Experience in the U.S. and the U.K." *IEEE Communications Magazine,* February 1986, pp. 40–46.

[20] Mikulski, Remarks, *op. cit.*

[21] *Ibid.*

[22] William C. Y. Lee, Remarks at an FCC Seminar on "The Future of Cellular Radio," September 2, 1987.

[23] E. F. O'Neill, ed., *A History of Engineering and Science in the Bell System: Transmission Technology (1925–1975),* AT&T Bell Laboratories, 1985, p. 79.
[24] *Ibid.,* pp. 322 ff.
[25] *Ibid.,* p. 334
[26] R. M. Wilmotte and B. B. Lusignan, *Spectrum Efficient Technology for Voice Communications,* UHF Task Force Report, FCC/OPP UTF 78-01, Federal Communications Commission, February 1978.
[27] William C. Jakes, ed., *Microwave Mobile Communications,* New York: Wiley, 1974, p. 201.
[28] For a good short summary of ACSB and a good bibliography of the FCC's efforts to promote it, see Thomas L. Jones and William A. Kissick, "ACSB: What is Adequate Performance?," *Proceedings of the 37th IEEE Vehicular Technology Conference,* Tampa, June 1–3, 1987, pp. 492–497.
[29] *IEEE Communications Magazine,* "Single-Sideband Mobile Radio: A Review & Update," March 1979, pp. 25–28.
[30] William C. Y. Lee, "Comparing FM and SSB in Cellular Systems," *Communications,* November 1985, p. 98.
[31] Jim Hendershot, "ACSB vs. FM in Cellular," *Personal Communications Technology,* March 1986, p. 16.
[32] D. Ridgley Bolgiano, "Spectrally Efficient Digital UHF Mobile System," *Proceedings of the 38th IEEE Vehicular Technology Conference,* Philadelphia, June 1988.
[33] John C. Bellamy, *Digital Telephony,* New York: Wiley, 1982, p. 273.
[34] *Ibid.,* p. 273.
[35] M. H. Meyers & V. K. Prabhu, "Future Trends in Microwave Digital Radio: A View from North America," *IEEE Communications Magazine,* February 1987, pp. 46–49; see also T. Noguchi, Y. Daido, & J. A. Nossek, "Modulation Techniques for Microwave Digital Radio," *IEEE Communications Magazine,* October 1986, pp. 21–30.
[36] William C. Y. Lee, *Mobile Communications Engineering,* New York: McGraw-Hill, 1982, p. 42; see also D. Berthoumieux, P. Huish, R. Hulthen, and A. Mawira, "Wideband Mobile Radio Channel Characterization," *Proceedings of the International Conference on Digital Land Mobile Radio Communications,* Venice, June 30–July 3, 1987, pp. 171–179.
[37] Donald C. Cox, Hamilton W. Arnold, and Philip T. Porter, "Universal Digital Portable Communications: A Systems Perspective," *IEEE Journal on Selected Areas in Communications,* Vol. SAC-5, June 1987, p. 765.
[38] Lee, *Mobile Communications Engineering, op. cit.,* p. 42.
[39] Cox *et al., op. cit.,* p. 767.

[40] Justin C. I. Chuang, "The Effects of Time Delay Spread on Portable Radio Communications Channels with Digital Modulation," *IEEE Journal on Selected Areas in Communications,* Vol. SAC-5, June 1987, p. 888.
[41] A. W. D. Watson and P. H. Waters, "An Experimental Evaluation of Narrowband TDMA Digital Mobile Radio Multipath Propagation Compensation Techniques for City and Rural Environments," *Proceedings of the International Conference on Digital Land Mobile Radio Communications,* Venice, June 30–July 3, 1987, p. 350.
[42] Justin C. I. Chuang, "The Effects of Time-Delay Spread on QAM with Nonlinearly Switched Filters in a Portable Radio Communications Channel," *Proceedings of the International Conference on Digital Land Mobile Radio Communications,* Venice, June 30–July 3, 1987, pp. 104–113.
[43] Jan Uddenfelt and Bengt Persson, "A Narrowband TDMA System for a New Generation Cellular Radio," *Proceedings of the 37th IEEE Vehicular Technology Conference,* Tampa, June 1–3, 1987, pp. 286–292.
[44] Joseph Tarallo and George I. Zysman, "A Digital Narrowband Cellular System," *Proceedings of the 37th IEEE Vehicular Technology Conference,* Tampa, June 1–3, 1987, p. 279.
[45] Jakes, *op. cit.,* p. 219.
[46] Robert D. Hoyle and David D. Falconer, "A Comparison of Digital Speech Coding Methods for Mobile Radio Systems," *IEEE Journal on Selected Areas in Communications,* Vol. SAC-5, June 1987, pp. 915–920.
[47] Bellamy, *op. cit.,* p. 113.
[48] S. Millman, ed., *A History of Engineering and Science in the Bell System: Communications Sciences (1925-1980),* AT&T Bell Laboratories, 1984, Chapter 2.
[49] Robert V. Bruce, *Bell: Alexander Graham Bell and the Conquest of Solitude,* Boston: Little, Brown, 1973, p. 36.
[50] *Ibid.,* p. 5.
[51] Millman, *Communications Sciences, op. cit.,* pp. 99 ff.
[52] R. E. Crochiere and J. L. Flanagan, "Current Perspectives in Digital Speech," *IEEE Communications Magazine,* January 1973, pp. 32–40; Nuggehally S. Jayant, "Coding at Low Bit Rates," *IEEE Spectrum,* August 1986, pp. 58–63; Nobuhiko Kitawaki, Masaaki Honda, and Kenzo Itoh, "Speech-Quality Assessment Methods for Speech-Coding Systems," *IEEE Communications Magazine,* October 1984, pp. 26–33; Manfred R. Schroeder, "Linear Predictive Coding of Speech: Review & Current Directions," *IEEE Communications Magazine,* August 1985, pp. 54–61; Hoyle and Falconer, *op. cit.;* Bellamy, *op.*

cit., Chapter 3; David R. Smith, *Digital Transmission Systems,* New York: Van Nostrand Reinhold, 1985.

[53] Hoyle and Falconer, *op. cit.*

[54] Bellamy, *op. cit.,* p. 147.

[55] Hoyle and Falconer, *op. cit.*

[56] Allen Gersho and Vladimir Cuperman, "Vector Quantization: A Pattern-Matching Technique for Speech Coding," *IEEE Communications Magazine,* December 1983, pp. 15–21.

[57] Bolgiano, *op. cit.*

[58] Bellamy, *op. cit.,* p. 295.

[59] *Ibid.,* p. 295.

[60] Lee de Forest, *Father of Radio: The Autobiography of Lee de Forest,* Chicago: Wilcox & Follett, 1950, p. 124.

[61] Marvin K. Simon, Jim K. Omura, Robert A. Scholtz, and Barry K. Levitt, *Spread Spectrum Communications: Volume I,* Rockville, Maryland: Computer Science Press, 1985, p. 80.

[62] Andrew J. Viterbi, "Spread Spectrum Communications: Myths & Realities," *IEEE Communications Magazine,* May 1979, pp. 11–18.

[63] Probably the definitive work on spread spectrum at this time is Simon *et al., Spread Spectrum Communications,* cited in Note 61. This work is projected for three volumes and will almost certainly constitute the most complete and up-to-date treatment of a complex and fascinating subject. The extensive historical chapter in Volume I is a real treat. A sample of shorter references that are reasonably accessible to the nonspecialist would include: George R. Cooper, Ray W. Nettleton, and David P. Grybos, "Cellular Land-Mobile Radio: Why Spread Spectrum?," *IEEE Communications Magazine,* March 1979, pp. 17–23; William Utlaut, "Spread Spectrum: Principles & Possible Application to Spectrum Utilization and Allocation," *IEEE Communications Magazine,* September 1978, pp. 21–30; Viterbi, "Spread Spectrum Communications: Myths & Realities," *op. cit.;* Robert Eckert and Peter M. Kelly, "Implementing Spread Spectrum Technology in the Land Mobile Radio Services," *IEEE Transactions on Communications,* Vol. COM-25, August 1977, pp. 867–869; Charles E. Cook and Howard S. Marsh, "An Introduction to Spread Spectrum," *IEEE Communications Magazine,* March 1983, pp. 8–16; Robert A. Scholtz, "The Spread Spectrum Concept," *IEEE Transactions on Communications,* Vol. COM-25, August 1977, pp. 748–755; Marlin P. Ristenbatt and James L. Daws, "Performance Criteria for Spread Spectrum Communications," *IEEE Transactions on Communications,* Vol. COM-25, August 1977, pp. 756–763; Raymond L. Pickholtz, Donald L. Schilling, and Laurence B. Milstein, "Theory of Spread Spectrum Communications: A Tutorial," *IEEE*

Transactions on Communications, Vol. COM-30, May 1982, pp. 855 ff.

[64] Simon *et al., Spread Spectrum Communications,* Vol. I, pp. 44 ff.

[65] *Ibid.,* p. 76.

[66] *Ibid.,* p. 63.

[67] James L. Massey, "Information Theory: The Copernican System of Communications," *IEEE Communications Magazine,* December 1984, pp. 26–28.

[68] This discussion was inspired by the presentation in Simon *et al., Spread Spectrum Communications, op. cit.,* Chapter 1.

[69] Cook and Marsh, *op. cit.*

[70] Cooper *et al., op. cit.,* "Editor's Note," p. 17.

[71] Simon *et al., Spread Spectrum Communications, op. cit.,* p. 76.

[72] *Ibid.,* p. 78.

[73] *Ibid.,* p. 79.

[74] Ristenbatt and Daws, *op. cit.*

[75] *Ibid.;* see also Utlaut, *op. cit.*

[76] Ristenbatt and Daws, *op. cit.*

[77] Cook and Marsh, *op. cit.*

[78] Utlaut, *op. cit.,* p. 26.

[79] Cooper *et al., op. cit.,* p. 20.

[80] *Ibid.*

[81] Uddenfeldt and Persson, *op. cit.,* p. 287.

[82] K. D. Eckert, "Conception & Performance of the Cellular Digital Mobile Radio Communication System CD 900," *Proceedings of the 37th IEEE Vehicular Technology Conference,* Tampa, June 1–3, 1987, p. 373.

Chapter 13
ALTERNATIVE SYSTEM ARCHITECTURES: CELL LEVEL

In the previous chapter, we dealt with the alternatives for the transmission link between the cell-site and the mobile unit. This may be referred to as the *circuit-level architecture.* The next higher level in the cellular architecture encompasses the creation of a *cell-level* operating system. The cell-site manages the communication traffic with a large number of mobile units operating more or less independently of one another. There are two broad questions to be addressed at this level:

1. *Spectrum management.* How is the total spectrum allocation divided into individual mobile-telephone circuits?
2. *Access.* How are individual mobile-telephone circuits assigned on demand to specific users?

The first question is of greater interest because there are several viable alternatives. The techniques of access *per se* — call setup, circuit assignment, exclusion of other users from occupied channels — are less controversial, primarily because today's cellular systems have established certain operating standards which are not likely to change.

All of the cell-level architectures we shall consider in this chapter are *multiple-access* systems, which means that every radio-telephone circuit is a *trunk* [1]. In other words, it is shared among all users and may be accessed by any user for any given telephone call. Full trunking is an essential requirement for a modern mobile-telephone system.

Older radio-telephone systems did not employ multiple access. Instead, specific users were preassigned to specific channels; originally, each mobile unit was tuned to one and only one frequency, which was shared like a party line among a number of mobiles. Obviously, like party lines, such systems had their limitations. Later, tuning elements were incorporated into mobile units and manual scanning of a set of channels was allowed. This brought the question of access to the fore.

Older systems also approached the access question in a very makeshift fashion, by comparison at least with the operating procedures in the wireline network. In the precellular era, access was generally under the control of the mobile unit. The basic technique was simply *contention:* mobile units would attempt to transmit on an unoccupied channel without any preclearance from the base station, and would continue transmitting unless interference from another user was detected. In the event of such a collision, certain back-off routines were instituted, designed to stagger the new access attempts by different users so as to allow them all to get through, ultimately. This is sometimes called *carrier-sense multiple access* (CSMA) — a rather fancy term for what one author defines as follows: "Each terminal must listen to the channel first before transmission. If the channel is idle, it sends; otherwise, it waits [2]." In older systems, the carrier sensing was done by the human user; the user could tune his unit to the various channels, like turning the channels of a TV set, until he found an empty channel on which to transmit. Later the process was automated. One type of automatic scanning involved the transmission by the base station of a tone on all idle channels; the mobile units would home on this tone to find a usable circuit. The advent of such relatively primitive multiple-access systems is recent; as late as the late 1970s, trunking was being heralded as a major new advance in spectrum efficiency.

The benefit of such crude access control was that it did not require very much *system*-level control. The penalties included: reduced throughput, potential lack of privacy, and potential for mobile units to transmit on unauthorized channels, thereby creating uncontrolled interference.

Analog cellular radio brought out what may be termed the first modern multiple-access system. Separate control channels were used to initiate the call requests to and from the mobile units, greatly improving system efficiencies and reducing collisions, attempts by two or more users to access the same channel at the same time which occur frequently in heavily loaded — i.e., efficient — systems. Contention was limited to the call setup request, which is a much shorter data burst; access to voice circuits is strictly controlled by the base station. A key feature of modern multiple-access systems is that all mobile units are under the direct and continuous control of the cell-site base station. Their transmissions are under system control, which makes possible system-level optimizations, like hand-offs, that cannot be easily accomplished in less sophisticated architectures. It also means that the system can enforce privacy and police unauthorized emissions with much greater effectiveness.

The various proposals for the next generation of cellular telephone systems will all retain these access features. Where they differ is in the area of spectrum management. In general, there are three types of cell-level architecture, based on three different types of spectrum management:

1. Frequency-division multiple access (FDMA);
2. Time-division multiple access (TDMA);
3. Code-division multiple access (CDMA).

Two terminological notes: First, TDMA systems may also encompass FDMA simultaneously and are, therefore, sometimes referred to as TDMA-FDMA. It is the TDMA characteristic, however, that differentiates them from conventional FDMA architecture. Second, some writers persist in applying the terms TDMA and FDMA in their original context only, where they referred to systems in which specific frequencies or time slots were in fact preallocated to individual users [3]. This is now an obsolete usage. TDMA and FDMA as used here refer to fully trunked systems, where there is no preassignment of frequency or time slots.

13.1 FDMA ARCHITECTURE

In an FDMA architecture, the total number of channels available within a particular cell are assigned on demand, on a first-come, first-served basis, to users asking to initiate a call, or to those for whom an incoming call has been received. Each frequency carries one single call at any one time. Any given mobile user may be assigned to any of the available voice channels for any call. Over time, any given user may expect to use all of the available frequencies at one time or another. (See Figure 13.1.)

In modern FDMA systems, including analog cellular radio, one or more of the available channels is set aside as a *control channel*. In the case of a mobile-initiated call, the call setup request, including transmission of dialed digits and sometimes other initiation information, is transmitted over this channel from the mobile to the base. The base transmits instructions to the mobile unit to move to a particular voice frequency to carry out the call. This frequency assignment is carried out rapidly and automatically; it is transparent to the mobile user.

At this time, several of the more important industry players in the United States, including AT&T and Motorola, are apparently backing digital FDMA architectures for the next generation of mobile telephony [4]. Some of the important features of digital FDMA are:

Single circuit per carrier. Each FDMA channel is designed to carry only one telephone circuit.

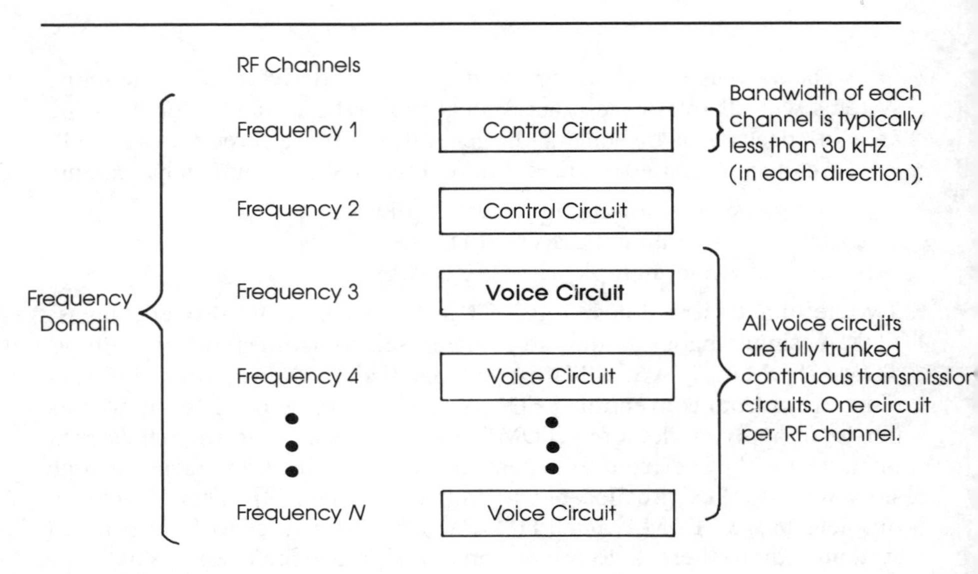

Figure 13.1 FDMA Architecture.

Continuous transmission. Once the voice channel has been assigned, both the mobile unit and the base station transmit continuously and simultaneously.

Bandwidth. FDMA channels are relatively narrow, usually 30 kHz or less, since they carry only one circuit per carrier. In fact, the trend is toward narrower bandwidths: some recently proposed digital FDMA systems have called for channels of 10 or 15 kHz [5].

Bit rates and symbol durations. Using constant-envelope digital modulation, bit rates equal to about 1 b/Hz are anticipated in most FDMA proposals [6]. In a 25-kHz channel, with 25 kb/s transmission and one bit per symbol, the symbol time would be about 40 microseconds. This is quite a long symbol time compared to the average delay spread, typically a few microseconds at most, depending upon the environment. (See Figure 13.2.) In the traditional view, this is a very important advantage for FDMA systems, because it means that the amount of intersymbol interference is likely to be quite low and little or no adaptive equalization may be required [7]. In actuality, however, the processing overhead associated with adaptive equalization is likely to loom less and less as digital signal processors become faster and cheaper.

Lower mobile-subscriber unit complexity. Along the same lines, it is sometimes claimed that FDMA mobile units will be less complex and,

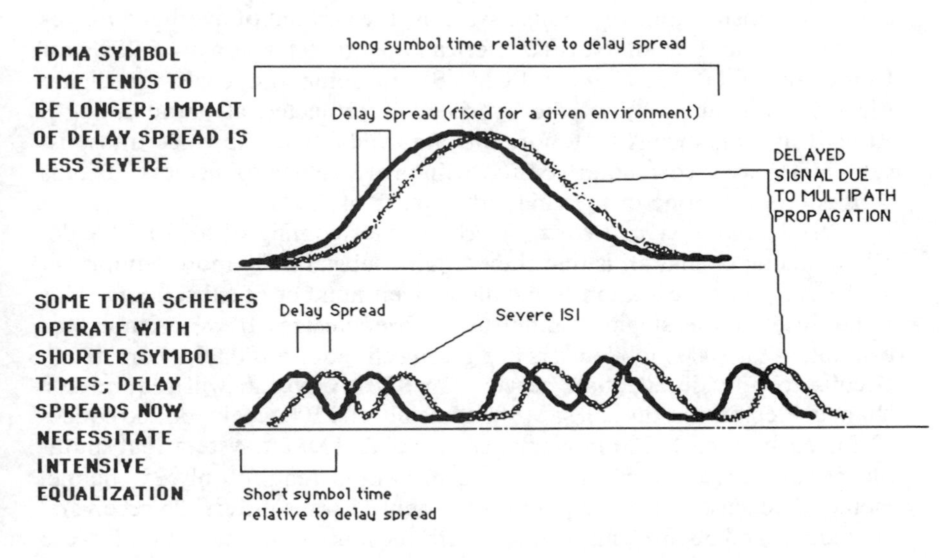

Figure 13.2 Delay Spread in an FDMA Channel *versus* Some TDMA Channels.

therefore, perhaps less costly than TDMA units, since they may not require equalization or the complex framing and synchronization associated with burst transmission in TDMA systems. I believe this is a transitory consideration; TDMA subscriber units are being designed today to operate on a single digital signal-processor chip, and the fact that the chip may be operating at 90% capacity, in a hypothetical TDMA implementation, instead of at 60% capacity, in a hypothetical FDMA implementation, should have virtually no impact on recurring product cost. The real mobile-unit cost drivers lie elsewhere, mainly in the radio subsystem of the unit, as opposed to the digital baseband processors.

Low transmission overhead. In an FDMA channel, the transmission to or from the mobile unit is continuous, much like T-carrier transmission on wireline or in point-to-point microwave radio. As in T-carrier, a few overhead bits may be inserted into the bit stream to allow for synchronization, framing, and certain control information, such as hand-off instructions, which have to be transmitted in the voice channel, since they occur when the call is already in progress. Because the transmission is continuous, however, the FDMA system should maintain good synchronization fairly easily with fewer bits devoted to overhead than in TDMA systems which operate in the burst mode and are constantly starting and stopping transmission. In wireline systems, the amount of overhead ranges from less than 1% for North American T-1 transmission to 6.25% for European CCITT 32-channel PCM [8]. In some proposals for digital FDMA for cellular radio, the overhead is projected at about 2% [9]. Effectively, this means that more bits are available to the voice transmission or to error-correction coding to improve signal robustness. This is considered to be one of the chief advantages of FDMA.

High shared-system costs. A critical shortcoming of all FDMA designs, analog or digital, is that they require substantially more equipment at the cell-site base station to handle a given number of subscribers. This is because of the single channel per carrier design. If we assume, for example, that 1000 mobile users in a given cell require 100 telephone trunk circuits to provide adequate service, an FDMA system will require 100 channel elements at the cell site, 100 transmitters, 100 receivers, 200 codecs (2 for each circuit), 200 modems, *et cetera*. A TDMA system that mutliplexes four circuits per carrier, however, would require only 25 channel elements to serve the same population, only 25 transmitters, 25 receivers, 50 codecs, and 50 modems — one-fourth the amount of base-station radio equipment compared to the FDMA system. The TDMA equipment would probably be somewhat more expensive, but the cost of shared-system equipment per subscriber would be much less. (The economics of this very important point will be discussed in greater detail in Chapter 15.)

Duplexer required. Because the transmitter and the receiver must operate at the same time — both are on continuously during an FDMA call — the FDMA mobile unit must incorporate a duplexer to prevent the mobile transmitter from overwhelming the mobile receiver. The duplexer circuitry, which may be taken here to include more stringent filter requirements, imposes a cost penalty of perhaps as much as 10% upon the FDMA mobile unit compared to a TDMA mobile unit, which does not need a duplexer for reasons discussed in the following section. This is a moderate disadvantage for FDMA systems.

Hand-off complexity. Because the FDMA transmission is continuous, the accomplishment of the hand-off to another channel in another cell is somewhat more difficult than in a TDMA system, where the shift can be made during an idle time slot [10]. In today's cellular systems, this has led to the awkward "blank and burst" technique, which creates difficulties for superimposed data transmissions [11].

13.2 TDMA ARCHITECTURE

In a TDMA system, each radio channel carries a number of telephone trunk circuits which are time division multiplexed. The structure of the TDMA channel is considerably more complex than an FDMA channel. Figure 13.3 portrays a typical TDMA channel structure. This TDMA system creates four time slots on each channel. A mobile unit that has accessed one of these channels transmits in a buffer-and-burst fashion; transmission from or to an individual mobile unit is not continuous. The mobile transmits on slot 1; holds for slot 2; receives on slot 3; holds for slot 4; and then transmits again on slot 1, and so repeats the cycle. This burst-transmission mode has a number of implications.

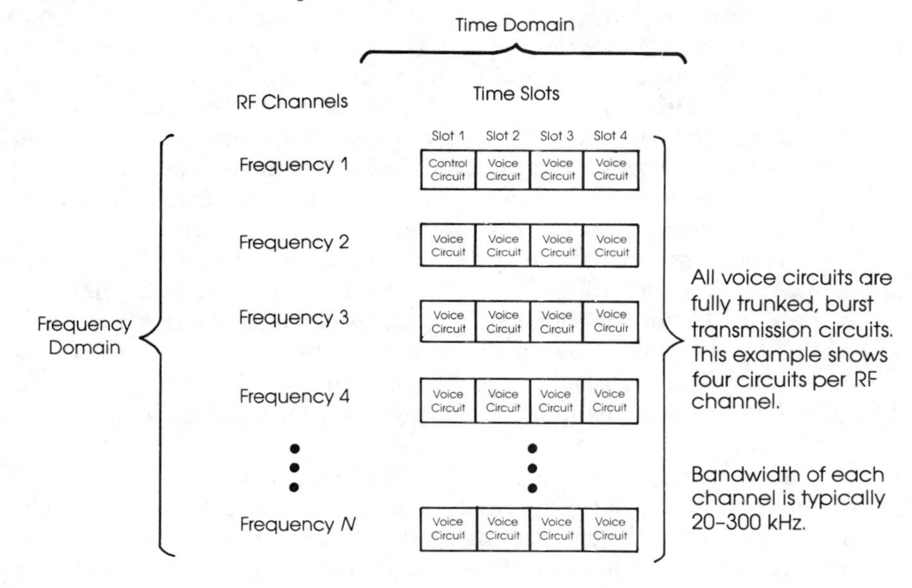

Figure 13.3 TDMA Architecture.

For example, the channel-transmission rate is no longer equal to the coding rate, or even the coding-plus-overhead rate. The channel rate is faster by a factor equal to the number of time slots in the frame. For example, in the system described above, the voice-coding rate, including overhead, is 16 kb/s. Voice coding is continuous. Since the radio transmitter only burst-transmits during one time slot out of four, however, the transmission rate during that burst must be *four times faster,* or 64 kb/s. If there were eight time slots per frame, as contemplated by the European standard, the transmission rate during the slot-burst would have to be eight times the coding rate.

How is multiple access achieved in TDMA? The TDMA system creates a frequency-time matrix: on each frequency it creates four time slots, or eight, or whatever number may be chosen by the system architects. Each of the cells in this matrix is a telephone trunked voice circuit, which can be accessed by any mobile user on any given call. As in FDMA, the assignment of users to circuits takes place on a call-by-call basis under the control of the base station. Call setup is performed over a separate *control slot,* which corresponds to the control channel or channels in an FDMA system. While the mobile unit is on hook, it monitors the control slot continuously, listening for a page indicating that it has an incoming call. If a page is detected, the mobile unit acknowledges the page to the cell-site base station over the control slot, and then the base station transmits another message over the control slot to instruct the mobile unit to move to a given frequency and a given time slot to take the call. (The mobile-initiated sequence is virtually the same.)

TDMA proposals have gained considerable support, especially in Europe, where the GSM standard has focused upon an eight-circuit per carrier TDMA format. In the United States, TDMA has also been championed by some [12]. The significant features of TDMA systems include:

Multiple circuits per carrier. As indicated, all TDMA formats seek to multiplex at least two, and more often four, or eight, or more, circuits per carrier.

Burst transmission. As noted, the transmission from the mobile units is not continuous, but occurs during specified time slots only. This has many implications for circuit design and system control. It may also impact, positively, the cochannel interference equation, since at any given moment only a percentage of the mobile units in operation are actually transmitting. (See Chapter 15.)

Bandwidth. The bandwidths of proposed TDMA systems range from around 20–30 KHz, equivalent to today's analog channels [13], to more than ten times that wide. Bandwidth is determined in part by the choice of modulation technique [14]. Spectrally inefficient modulation approaches require a wider bandwidth to transmit the same bit stream. For example, the European standard appears to be heading toward a 200-KHz channel, multiplexing eight circuits of about 30 kb/s each, for a total channel rate of about 240 kb/s. This is a little more than 1 b/Hz, about the same as

FDMA approaches. By comparison, the IMM TDMA architecture multiplexes four circuits per carrier, but by using higher-level modulation is able to transmit the required 64 kb/s, for 16-kb/s voice circuits, within a 20-KHz occupied bandwidth, equal to about 3.2 b/Hz [15].

Bit rates and symbol durations. The higher channel rates proposed for some TDMA systems — up to 300–400 kilosymbols per second, one bit per symbol — can create a much more severe intersymbol-interference problem than FDMA systems. For a 300-kilosymbol rate, for example, the symbol time is 3.33 μs, which is approximately the same as the delay spread that might be expected in dense urban centers. Even using four-level modulation to attain 2 b/Hz would only increase the symbol time to 6.67 μs. Adaptive equalization would be absolutely necessary. For TDMA systems with lower channel rates, however, the equalization requirement may be no more severe than for FDMA systems. The IMM system, for example, with a 16-kilosymbol rate, four bits per symbol, has a symbol time of 62.5 μs, and is relatively impervious to the average delay spread. It is important to remember that TDMA is *not inherently more vulnerable to the delay spread;* the degree of vulnerability in any digital system depends not upon the form of access, but upon the symbol rate in the channel, which in turn rests upon the degree of spectral efficiency of the modulation scheme.

Higher mobile-unit complexity. The TDMA mobile unit has more to do than the FDMA unit, at least on the digital-processing side. Whether this is a crucial difference depends upon the processing-device technology that is assumed. As time passes, the added processing for TDMA grows less and less significant in the total picture. Some TDMA units may also be required to do slow-frequency hopping, to improve resistance to multipath fading [16]. Slow FH circuitry would complicate the mobile design. Other TDMA proposals, however, do not call for frequency hopping; this can be treated as a separate design requirement.

Higher transmission overhead. A TDMA slotted transmission forces the receiver to reacquire synchronization on each burst. Also, guardbands may be necessary to separate one slot from another, in case unequalized propagation delay causes a far user to slip into the adjacent slot of a nearby user. Because of this, TDMA systems usually need much more overhead than FDMA systems. The conventional view is that TDMA overhead requirements can run to 20–30% of the total bits transmitted, which is indeed a significant penalty [17]. With proper design, however, this can be lessened. The IMM TDMA system today utilizes less than 9% for overhead. Nevertheless, it is true that TDMA overhead will always exceed

FDMA overhead, which means that fewer bits will be available at a given bit rate for voice coding or error correction. The significance of this inefficiency depends upon the scope of the TDMA overhead requirement, and upon the choice of coding schemes.

Lower shared-system costs. As already indicated, the major advantage of TDMA systems over FDMA arises from the fact that each radio channel is effectively shared by a much larger number of subscribers. The cost of the central-site equipment is reduced dramatically. A more detailed analysis of these savings is presented in Chapter 15.

No duplexer required. An additional cost advantage for TDMA over FDMA arises from the fact that by transmitting and receiving on different slots it is possible to eliminate the duplexer circuitry entirely, replacing it with a fast switching circuit to turn the transmitter and receiver on and off at the appropriate times. This should lead to a reduction in the cost of the mobile unit [18].

Hand-off complexity. Because the TDMA transmitter is turned off during idle slots, TDMA units have the opportunity to carry out a more efficient hand-off procedure [19]. In particular, the blank-and-burst problem can be avoided to maintain data-transmission integrity.

Openness to technological change. TDMA systems have another edge over FDMA which may outweigh everything else, at least in the eyes of the system architect or regulator who looks to the long-term technological viability and flexibility of the system. As bit rates fall for coding algorithms, a TDMA channel is more easily reconfigurable to accept new techniques. Within the existing channel rate, the slot structure can be redefined to support lower bit rates or variable bit rates. For example, if the channel rate is 64 kb/s, divided into four 16-kb/s slots, an 8-kb/s coder may be accommodated by redividing the channel into eight slots instead of four. This modification, if carried out with careful attention to other architectural constraints, should be implementable by means of ROM *(read only memory)* changes in the digital circuitry. That is, existing radio hardware can most likely be utilized at the base station. Retrofitting existing mobile units may or may not make economic sense; a TDMA format, however, can be designed to accommodate different bit rates, different slot lengths. Since the channel rate, bandwidth, and other features of the radio transmission remain the same, such changes can be introduced without disrupting the cellular network frequency plan.

Although not as straightforward, modifications in modulation techniques can also potentially be accommodated in TDMA systems through software upgrades or options. For example, the IMM system previously referred to possesses the ability to adapt modulation levels from 4-level PSK to 16-level PSK, depending upon channel conditions [20]. The same hardware is used; the option is built into the system and mobile-unit software. It is possible to envision future modulation adaptations that could be implemented through stored program changes to digital circuitry without affecting radio-channel characteristics significantly.

By comparison, incorporating new technology into an FDMA would appear far less straightforward. To realize the potential spectrum-efficiency gain from, say, a reduction in coding rates, it would be necessary to redesign the FDMA to operate on still narrower channels. Aside from the system-wide implications, including changed carrier-to-interference ratios and reuse patterns which could be disrupted by rechannelization, much of the radio hardware at the base station would have to be substantially altered or replaced. This would involve a considerable investment.

In short, TDMA systems leave the door open wider to continuing technological improvements. Since they operate fully in the digital domain, many upgrades can be implemented "in software" — which is to say, with minimal impact upon expensive radio hardware. FDMA systems, which can only be upgraded through overhauling the radio channel itself, are much less flexible.

13.3 CDMA ARCHITECTURE

Spread-spectrum techniques have been described in detail in Section 12.3. The characteristic form of multiple access for spread-spectrum systems is known as code division multiple access. Each mobile unit is assigned a unique randomized code sequence, different from and orthogonal to, i.e., uncorrelated with, all other codes. In an FH/SS system, this code is used to generate a unique sequence of frequency hops. In a DS/SS system, the code is used to generate the randomized noiselike high-bit-rate signal that is mixed with the information signal to spread the spectrum.

"Unlimited" circuits per carrier. SS systems utilize single, or very few, wideband carriers to transmit a great many individual telephone circuits. Where FDMA systems transmit only one circuit per carrier, and TDMA a handful, CDMA systems could potentially transmit hundreds. In fact, as noted in the discussion of spread-spectrum methods, there is no hard limit on the number of circuits that may be transmitted.

Bandwidth. Proposed SS channels are very wide — typically 1–10 MHz — compared to TDMA and FDMA systems.

Bit rates and symbol durations. Because of the very high bit rates in the SS channel, the symbol times are very short. In a DS/SS system, with a one-megabit channel rate, each symbol would be only one microsecond in length, assuming BPSK or some other 2-level modulation. This is less than the average delay spread. Intensive equalization would be required.

Lower mobile complexity (?). Some writers have suggested that the mobile units in a CDMA system would be very simple, compared to TDMA and FDMA units. If so, this simplicity would appear to benefit chiefly the radio portion of the subscriber unit by relaxing or eliminating filter requirements. The digital portion of the mobile unit would likely be as complex as TDMA units, if not more so. Certainly the requirement for frequency hopping in FH/SS systems is no simple task.

Overhead (?). I am not sure that it is meaningful to discuss overhead for CDMA SS systems in the same way we have discussed overhead for FDMA and TDMA. The concept of overhead as bits expended for channel administration, bits which cannot be used for voice communication, is difficult to apply here. The entire SS concept is in effect a way of building in a massive overhead which, paradoxically, improves the ability of the signal to coexist with other transmissions and with environmental noise and multipath.

Shared-system costs. Presumably, an SS system ought to enjoy relatively low system costs per subscriber, since a very large number of subscribers are sharing the cost of the base-station channel equipment. This indeed may prove to be a decisive future advantage for SS, once the costs are better known.

Unfortunately, as noted in Section 12.3, there are still too many unknowns for most people, this author included, concerning the actual implementation of SS techniques and their performance in the field. It is still difficult to draw conclusions about CDMA SS architectures in comparison to better understood alternatives like FDMA and TDMA.

REFERENCES

[1] Victor O. K. Li, "Multiple Access Communications Networks," *IEEE Communications Magazine,* June 1987, pp. 41–48.

[2] *Ibid.,* p. 43.

[3] *Ibid.*

[4] Joseph Tarallo and George I. Zysman, "A Digital Narrowband Cellular System," *Proceedings of the 37th IEEE Vehicular Technology Conference,* Tampa, June 1–3, 1987, pp. 279–280.

[5] *Ibid.*

[6] Jan Uddenfelt and Bengt Persson, "A Narrowband TDMA System for a New Generation Cellular Radio," *Proceedings of the 37th IEEE Vehicular Technology Conference,* Tampa, June 1–3, 1987, pp. 286–292.

[7] *Ibid.*

[8] John C. Bellamy, *Digital Telephony,* New York: John Wiley and Sons, 1982, pp. 206–208.
[9] FCC Seminar on "The Future of Cellular Radio," Washington, D. C., September 2, 1987.
[10] Uddenfelt and Persson, *op. cit.*
[11] Z. C. Fluhr and P. T. Porter, "AMPS: Control Architecture," *Bell System Technical Journal,* January 1979, pp. 43–69.
[12] D. Ridgley Bolgiano, "Spectrally Efficient Digital UHF Mobile System," *Proceedings of the 38th IEEE Vehicular Technology Conference,* Philadelphia, June 1988.
[13] *Ibid.*
[14] Uddenfelt and Persson, *op. cit.*
[15] Bolgiano, *op. cit.*
[16] Uddenfelt and Persson, *op. cit.*
[17] James Mikulski, Remarks at the FCC Seminar on "The Future of Cellular Radio," Washington, D. C., September 2, 1987.
[18] Uddenfelt and Persson, *op. cit.*
[19] *Ibid.*
[20] Bolgiano, *op. cit.*

Chapter 14
ALTERNATIVE SYSTEM ARCHITECTURES: NETWORK LEVEL

The telephone business grew for many years before there was a network. In the very early years of telephony, local exchanges developed independently; each engineer followed his own rules, learning what he could from others, but untrammeled by any sense of the emerging whole. The limitations of early long-distance voice transmission isolated cities from one another. It was not until 1892 that service from the East Coast to Chicago began, after a fashion. California was isolated until World War I. Slowly, however, local systems began to interconnect. With greater numbers of subscribers, greater circuit densities along cable routes and in switchboards, and more widespread interexchange links, a qualitatively new set of problems began to emerge which could only be addressed by greater coordination and standardization of materials and procedures. The concept of the "Network" began to assume its present prominence in the minds of telephone engineers, planners, and managers, until, in the words of a recent historian, they "held the network in almost mystical awe [1]." This implied, and corresponded to, a very strong belief in the need for rationalized, uniform standards, promulgated from a central technical resource, e.g., Bell Labs, and enforced from the top down.

Today, we tend to assume that network planning is automatically a top-down process. This is not always the case, however, especially in a nascent field like cellular radio. The cellular business in the United States has grown largely without network standards as such. While the air interface is standardized on FM, almost nothing else is. There is considerable diversity among different system operators on most of the issues addressed in this chapter. The network-level architecture of cellular radio is still fluid. This is not necessarily a problem at the current stage of development of mobile telephony. By comparison, the long-distance operations of AT&T

were not established until Year 9 (1885) of the telephone, and the goals of network standardization were not really articulated until the late 1880s and 1890s [2]. As mobile systems achieve larger capacities and higher user densities, and as customer expectations begin to rise, cellular network issues will come to the fore. To a considerable degree, however, network-level problems are a bridge that can be crossed when we get there.

The network-level issues that next-generation cellular architects may confront will include:

1. Frequency plans, including reuse patterns and different cell geometries;
2. Growth plans, including cell-splitting;
3. Network-wide measures to reduce or control cochannel interference;
4. Network-wide traffic and service objectives, such as the mix of portable and mobile subscribers;
5. Roaming and related issues such as numbering plans;
6. Definition of interface to the public switched (wireline) telephone network;
7. Distribution of system control (centralized *versus* distributed architectures).

A full discussion of alternative network-level designs is beyond the scope of this book. In keeping with our focus on the impact of digital technology, we shall not delve deeply into these issues. Most of these network-level questions are, by definition, little affected by the choice of radio-link technology. *Per se,* the digital revolution does not greatly change the agenda for the network-level planner. The digital communicator's ability to manage control information more efficiently will tend to make many of these issues somewhat easier to deal with, but the selection of one or another voice coder, modulation scheme, or access technique will not alter the decision set.

14.1 FREQUENCY PLANS

The network planner is concerned with frequency management across a wide area, encompassing a large number of coverage zones or cells. Initially, at low densities, with large cells and substantial excess capacity, the operator does not have to cut things too finely. As the system grows, however, there are at least three questions the operator must be concerned with:

1. Reuse patterns;
2. Cell-site geometry and antenna directionality;
3. Dynamic channel allocations.

Reuse is of course a central concern to the cellular system operator. Based chiefly on cochannel-interference criteria (see Chapter 9), a *reuse pattern* is defined. Initially, 12-cell patterns were deployed [3]; later, 7-cell patterns and even 4-cell patterns were advocated [4]. In a large system, a lower reuse number will generate much more capacity — more circuits per megahertz per square mile. As noted in Chapter 9, however, there is uncertainty and considerable disagreement over the validity of many cellular geometries on paper. For example, a fundamental question such as the relationship between capacity (reuse) and channel bandwidth is still subject to wide disagreement among cellular proponents [5].

Cell-site coverage characteristics also influence the cellular geometry. Initial cellular architectures called for omnidirectional antennas, which were conceived of as being located in the center of the cell [6]. Subsequently, a *sectorized* cell architecture was advocated, where 120° sectors were defined and the antennas were conceived as being at the corners of the cells [7]. Proponents of sectorizing argued that it gained capacity without multiplying cell-sites. Recently, it has been suggested that sectorized cells actually *lose* capacity because the trunking efficiencies in each sector are worse due to a smaller pool of available channels per sector. Figure 14.1 shows an example of a frequency plan for (idealized) sectorized cells. Today there are a very large number of proposals for exotic cellular geometries, many based on advanced sectorized or ring concepts [8].

Dynamic channel allocation involves the idea that channels can be temporarily "borrowed" by a cell that has reached saturation from another cell that temporarily enjoys unused capacity. Traditional frequency plans assigned an equal number of channels to all cells. Obviously the traffic is rarely divided as equally; downtown cells will tend to have a surfeit of traffic during the business day, while suburban cells are underutilized. During the evening rush hour this pattern may tend to reverse itself. In principle, it is possible to conceive of a base station that could, under system control, make use of a greater or lesser number of the total available channels, depending upon its need at a particular moment.

Some cellular planners have envisioned a cellular network architecture that would be capable of "breathing" with the daily traffic flow: channels would tend to flow inward toward the city center during the morning rush hour, remain there throughout the business day, and then follow the commuters home toward the periphery in the evening [9]. Claims for very significant improvements in system capacity have been made; indeed, the potential gains are fairly obvious, on paper.

There are, however, a number of obstacles in the way of effective dynamic channel allocation. For one thing, the packing of additional channels into the standard frequency plan for a given cell would reduce the

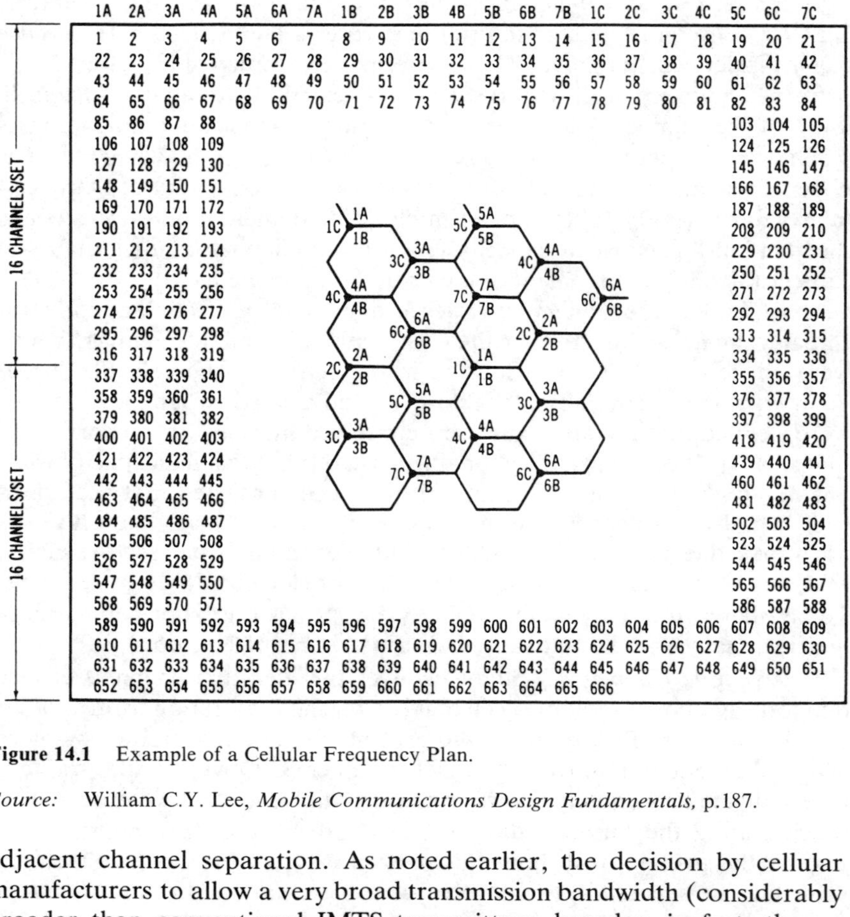

16 CHANNELS/SET

16 CHANNELS/SET

1A	2A	3A	4A	5A	6A	7A	1B	2B	3B	4B	5B	6B	7B	1C	2C	3C	4C	5C	6C	7C
1	2	3	4	5	6	7	8	9	10	11	12	13	14	15	16	17	18	19	20	21
22	23	24	25	26	27	28	29	30	31	32	33	34	35	36	37	38	39	40	41	42
43	44	45	46	47	48	49	50	51	52	53	54	55	56	57	58	59	60	61	62	63
64	65	66	67	68	69	70	71	72	73	74	75	76	77	78	79	80	81	82	83	84
85	86	87	88															103	104	105
106	107	108	109															124	125	126
127	128	129	130															145	146	147
148	149	150	151															166	167	168
169	170	171	172															187	188	189
190	191	192	193															208	209	210
211	212	213	214															229	230	231
232	233	234	235															250	251	252
253	254	255	256															271	272	273
274	275	276	277															292	293	294
295	296	297	298															313	314	315
316	317	318	319															334	335	336
337	338	339	340															355	356	357
358	359	360	361															376	377	378
379	380	381	382															397	398	399
400	401	402	403															418	419	420
421	422	423	424															439	440	441
442	443	444	445															460	461	462
463	464	465	466															481	482	483
484	485	486	487															502	503	504
505	506	507	508															523	524	525
526	527	528	529															544	545	546
547	548	549	550															565	566	567
568	569	570	571															586	587	588
589	590	591	592	593	594	595	596	597	598	599	600	601	602	603	604	605	606	607	608	609
610	611	612	613	614	615	616	617	618	619	620	621	622	623	624	625	626	627	628	629	630
631	632	633	634	635	636	637	638	639	640	641	642	643	644	645	646	647	648	649	650	651
652	653	654	655	656	657	658	659	660	661	662	663	664	665	666						

Figure 14.1 Example of a Cellular Frequency Plan.

Source: William C.Y. Lee, *Mobile Communications Design Fundamentals,* p.187.

adjacent channel separation. As noted earlier, the decision by cellular manufacturers to allow a very broad transmission bandwidth (considerably broader than conventional IMTS transmitters, broader, in fact, than a single cellular channel) in order to keep down mobile-unit costs creates a potential problem for dynamic allocation schemes. Perhaps of greater long-range concern is the complexity that dynamic channel-allocation schemes would introduce into the cochannel-interference situation, which is, as we have seen, the foundation of the cellular geometry. One advantage of fixed, equal-allocation frequency plans is that the cochannel-interference environment for any given cell is also fixed, except, of course, for relatively small-scale environmental fluctuations. Moreover, the cochannel-interference situation is the same, or very nearly so, for every channel in the set allocated to that cell. If channels begin moving around among the cells, it will no longer be easy to ensure this. It is likely, given the complexity of cellular geometries, that the cochannel-interference characteristics of a given frequency in a given cell will change over time, and it is also likely that not all the channels in a given cell will exhibit the same interference characteristics.

14.2 GROWTH PLANS

A cellular network must have some plan for coping with growth in the subscriber base without additional spectrum. The original cellular concept of *unlimited* cell-splitting has proven doubtful. Within limits, cell-splitting is still part of the answer. Cellular engineers are discovering the practical lower limits of cell size; again, however, there is disagreement over what these limits are. Some operators claim that cells of less than 1 kilometer in radius are readily achievable; others believe that cells smaller than 1–2 miles radius are not practical. When the poor economics of very small cells — the proliferation of cell-sites and the rising cost per circuit (see Chapter 4) — are factored in, it may be that *sectorized* transmission schemes will ultimately prove to be a better way of gaining capacity than cell-splitting.

Another approach that has recently received support is the *microcell.*

> The concept is to provide coverage of high density teletraffic areas with a number of very small cells created by low power radio links between the base station antennae, located at street lamp elevations, and the mobile units. The wide coverage area of a conventional cell [in this system referred to as the macrocell] is used to overlay the microcells . . . Areas suitable for covering with microcells are expected to consist of city center streets and motorways [10].

The microcells are applied like patches to the cellular network. (See Figure 14.2.) Microcell base stations are controlled by the macrocells within which they reside.

This may represent a more feasible growth strategy than cell-splitting, since the boundaries between individual microcells may not have to be engineered as precisely as between macrocells. If a user falls off the edge of a microcell, he is caught in the safety net of the macrocell. For this to work, however, the hand-off algorithms will become more complex, as the proponents of microcells admit [11].

14.3 COCHANNEL-INTERFERENCE REDUCTION MEASURES

As the centrality of cochannel interference has become recognized, a number of proposals have emerged for the alleviation of such interference at the network level. Some proposals have focused on ways of reducing

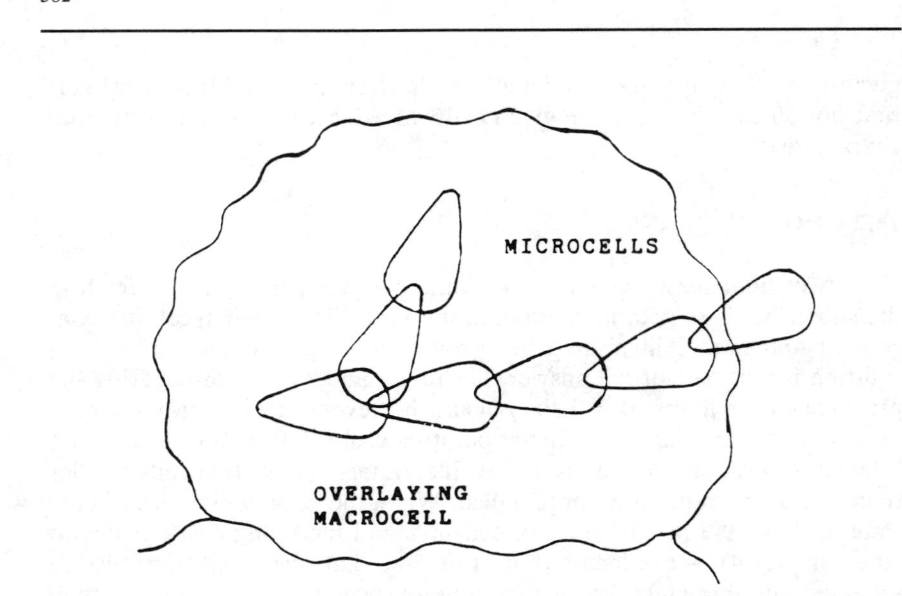

Figure 14.2 The Idea of Microcells.

Source: A.R. Potter, et al., 1987.

the emissions of the mobile units. For example, it has been suggested that transmission could be suppressed during intervals of silence (so-called "speech silents" or pauses). In a typical telephone conversation, only one party is talking at any given moment. Actual measurements indicate that the "activity factor" (measured by what are called "talkspurts") ranges between 0.4 and 0.6 — one party is talking only 40–60% of the time, the rest of the time he is listening. Yet the transmitter of the listening party continues to transmit, even though there is no information to communicate. The principle of "pause-encoding" (or "activity detection") has been used elsewhere to gain transmission efficiencies; for example, time-assigned speech interpolation (TASI) techniques have been used on undersea cables, satellites, and other expensive circuits for many years to increase effective capacity. The idea of turning off the radio transmitters when the user is not actually speaking could, in principle, improve the cochannel-interference situation:

> It is possible to use speech silents to reduce overall interference by suppressing the carrier transmission during those silents . . . . Simulations have been made with a mean activity factor of 0.6 . . . [which] show an improvement in performance of about 2.5 dB [12].

In other words, if the C/I criterion for a given modulation technique is set at 16 dB, the suppression of the carrier during speech pauses could reduce the threshold to 13.5 dB, reducing the D/R ratio by 10–15% (according to Lee's equation, see Chapter 9), which could allow a better reuse factor.

Similar results could be obtained by TDMA architectures, where the transmitters operate in burst mode, reducing overall emissions compared to FDMA systems, and by means of unusual cell geometries, such as the doughnut-shaped cells now being tested in some cellular systems. In fact, the power-control schemes already used in cellular radio, whereby the transmitter power of the mobile units are continuously adjusted under base-station control, are designed in part for this purpose.

Another approach to possibly reducing cochannel-interference levels is based on the use of *frequency offsets*. This technique has been evaluated in certain noncellular land mobile services as an alternative to other capacity-boosting measures such as channel-splitting [13]. It involves modifying the nominal carrier frequency of the channel so that, for example, the actual center frequency of the channel might differ by a few kilohertz from one cell to another. On paper, and in initial studies outside the cellular bands, frequency offsets appear to offer considerable promise for cochannel-interference reduction. The use of offsets, however, would create an additional dimension in the complexity of the frequency plan.

14.4 TRAFFIC AND SERVICE OBJECTIVES

Another set of network-level decisions involves the types of traffic and the classes of service to be provided by cellular systems in the future. Perhaps the most important question is whether portable telephony will be integrated with mobile telephony. The assumption in the first generation of cellular systems was that mobile and portable subscriber units would and could operate in the same network. In practice, however, the two services have rather different requirements and performance characteristics. Mobile units operate at higher power, unconstrained by battery limitations. Mobiles also operate outdoors most of the time. Other than the occasional parking garage, mobile telephones generally will not be taken inside man-made structures. Building penetration is probably much more important for portable users. On the other hand, mobiles normally operate at ground level; portables may climb hundreds of feet in the air as their owners go to work in modern skyscrapers. Also, mobiles operate at speeds where multipath fading is normally very rapid and can be handled "statistically." Portables, on the other hand, are probably operated most often from stationary or very slow-moving positions; the frequency of fades may be less, but the duration of deep fades could be longer, creating higher hand-off uncertainties.

Recognizing these and other differences between mobiles and portables, there is discussion as to whether the next generation of architecture should attempt to continue to integrate both types of user into a single network. Many have come to believe that portable communication will ultimately have to develop in a separate network, with a different architecture, at a different frequency band [14].

The question of the service objectives of cellular radio is very broad (see Chapter 16). Many network-level decisions are going to be determined by whether cellular is viewed as an urban car-telephone service for businessmen, or whether it is envisioned as a broad-based "wireless access technology" encompassing fixed telephony as well as mobile data, paging, and other services. In other words, is the *ultimate* design goal for cellular radio to serve 3% of the population, or 30%? Shall we assume that the average busy-hour traffic per user will be characteristic of mobile usage, say 0.02 Erlangs, or will cellular usage approximate, or even surpass, wireline traffic levels of 0.1 Erlangs or more?

14.5 ROAMING AND RELATED ISSUES

As cellular networks become ubiquitous, users will begin to rely upon mobile telephony not only for local mobile communication but also for travel beyond their local areas. This raises the issue of *roaming*.

Roaming — that is, the operation of mobile units in areas outside their home territory — is an important concern today for many operators, because it has the potential for generating substantial additional revenues. Roaming rates are normally much higher than usage charges for local service; an operator therefore wants to ensure that network-level standards are in place to allow easy roaming. In fact, some cellular operations are highly dependent today upon roaming traffic along heavily traveled interstate corridors. Whereas roaming may constitute a small percentage of the traffic in a large urban system, it may be the dominant source of revenue in some smaller markets. Indeed, the opportunity to capture roaming revenues may spell the difference between success and failure for an operator in a small market.

Unfortunately, the haphazard development of cellular systems in the United States (exacerbated by the lottery process for allocating cellular licenses, which made it difficult for operators to develop integrated regional networks) has created difficulties for roamers that frustrate users and operators today. Hand-offs from one system to another often cannot be made; the call is simply dropped and must be redialed, as in the precellular era. Moreover, many systems apparently did not have provisions initially for capturing the necessary billing information from roamers. On the one hand, lost revenue and fraud have apparently been more serious in the case of roamers than for the locals; on the other hand, some systems initially had difficulties recognizing and serving roamers at all.

The roaming issue blends into the more far-reaching question of the *numbering plan.* Some cellular architects are inspired by a vision of a wireless world where telephone numbers are assigned at birth like social security numbers, where dialing an individual's number will reach him wherever he may be, even in another city or state. Of course, to design a network — even on paper — that would be capable of locating an individual anywhere in the country on the basis of a simple ID number is a tall order. Moreover, many people are not sure that we would entirely welcome such an all-seeing network peering into our lives, tracking our movements. Yet as mobile services become more pervasive, numbering and location plans will have to evolve considerably.

14.6 INTERFACE TO THE WIRELINE NETWORK

The most critical network-level decisions facing cellular architects have to do with the definition of the interface between the mobile network and the existing wireline network. The nature of the interface is intimately related to the control architecture of the cellular system, the degree of centralization or decentralization of system control — see Section 14.7.

The problem of interface definition arises in part from the unique control requirements of cellular systems (Section 14.7) and in part from the fact that the front-line, traffic-collection points in the wireline network (the local switches) and the cellular network (the cell-sites) do not match up geographically, and likely never will. Wire-center locations reflect historical settlement patterns and are sited to minimize, approximately, total copper costs in the loop plant. Cell-sites are located for purposes of good radio propagation within an overall frequency reuse plan. Cell-coverage areas will not tend to map onto existing exchange boundaries. Therefore, it is necessary to design a special interface.

There are basically two opposing visions of the cellular network of the future. The first is what may be called the *overlay architecture,* characteristic of most of today's cellular systems [15]. A separate network is created to link individual cell-sites together, under the control of a central mobile-switching office. The cell-sites do not interface directly to the wireline network, but communicate through the mobile-switching office. (See Figure 14.3.) In the original Chicago AMPS test in the late 1970s:

> The MTSO occupies a position in the switching hierarchy below a class 5, or local office . . . The MTSO can be interconnected with one or

more local offices over standard trunk facilities. Directory numbers for mobile telephones are assigned from within the local exchanges that are served by those interconnected offices. The MTSO interconnection arrangement is similar to that used with a Private Branch Exchange (PBX), and it makes use of existing capabilities in ESS local offices [16].

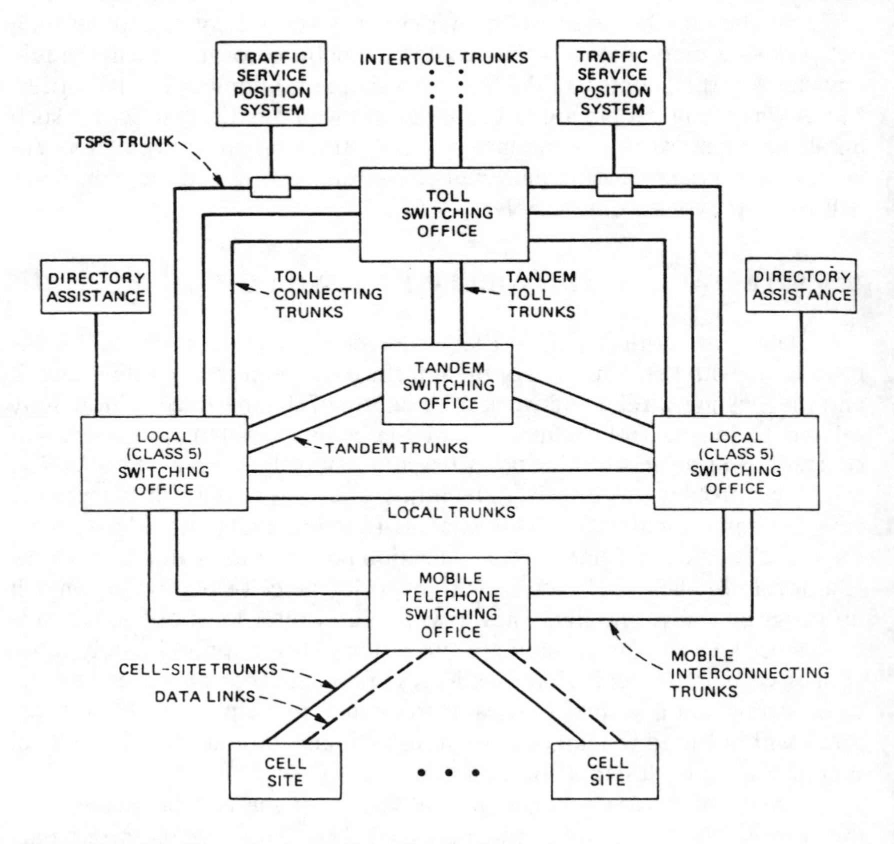

Figure 14.3 The Hierarchy in a Cellular Overlay Network.

Source: K.J.S. Chadha, D.F. Hunnicutt, S.R. Peck, and J. Tebes, *Bell System Technical Journal,* September 1979. Reprinted by special permission. Copyright 1979 AT&T.

The chief implication of an overlay network is that the cellular system must duplicate a great deal of the interconnective tissue of the wireline network. All mobile calls must be routed back to the mobile-switching office for processing. This means that there must be an extensive, and expensive, intercell trunking system to funnel the voice traffic back to the center where all connections are made. (See Figure 14.4.) This backhauling can add significant costs. For example, a call originating from a mobile unit in Cell A, destined for another mobile unit in the same cell, must still be routed back to the mobile-switching office for processing, and then back out to Cell A. (See Figure 14.5.) Two exchange trunks are occupied unnecessarily. The cost of leasing trunk lines from local exchange carriers has become a major component of the operating costs of many cellular operators (see Chapter 4).

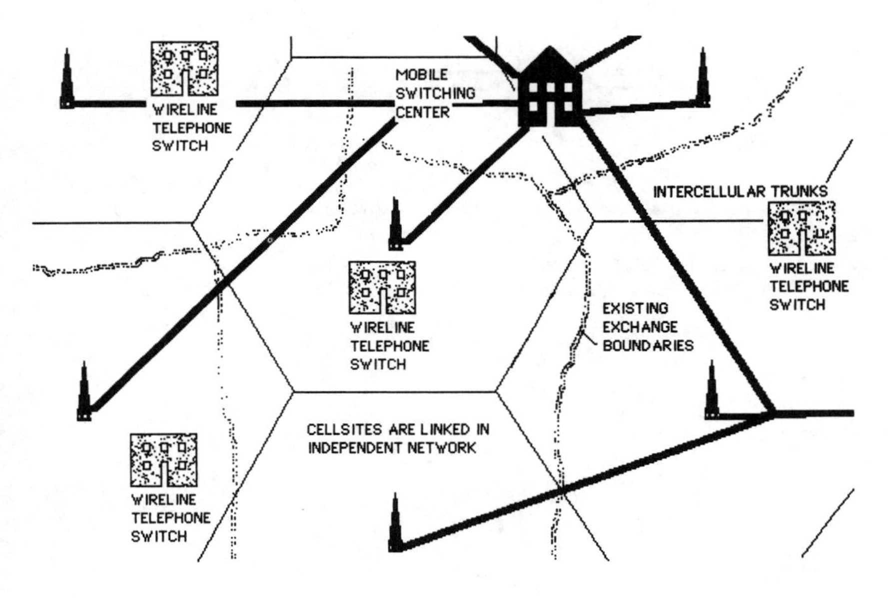

Figure 14.4 Standard Cellular Overlay Network.

A different structure is represented by what we may call the *integrated network.* (See Figure 14.6.) The cellular system is interfaced to the wireline network at each cell-site, or perhaps at satellite offices that collect traffic from only a few cell-sites. In effect, instead of one interface-point per system (city), there may be many interface-points, perhaps even one per cell, or one per Class 5 central office. Such a system naturally involves decentralization of control, since each interface point must be capable of making call-connection decisions.

The purported advantage of integrated networks is, of course, the avoidance of backhauling charges. The chief disadvantage is the higher cost of local switching and interconnection facilities. This advantage, however, is correlated with the degree of intra-calling, calls placed between

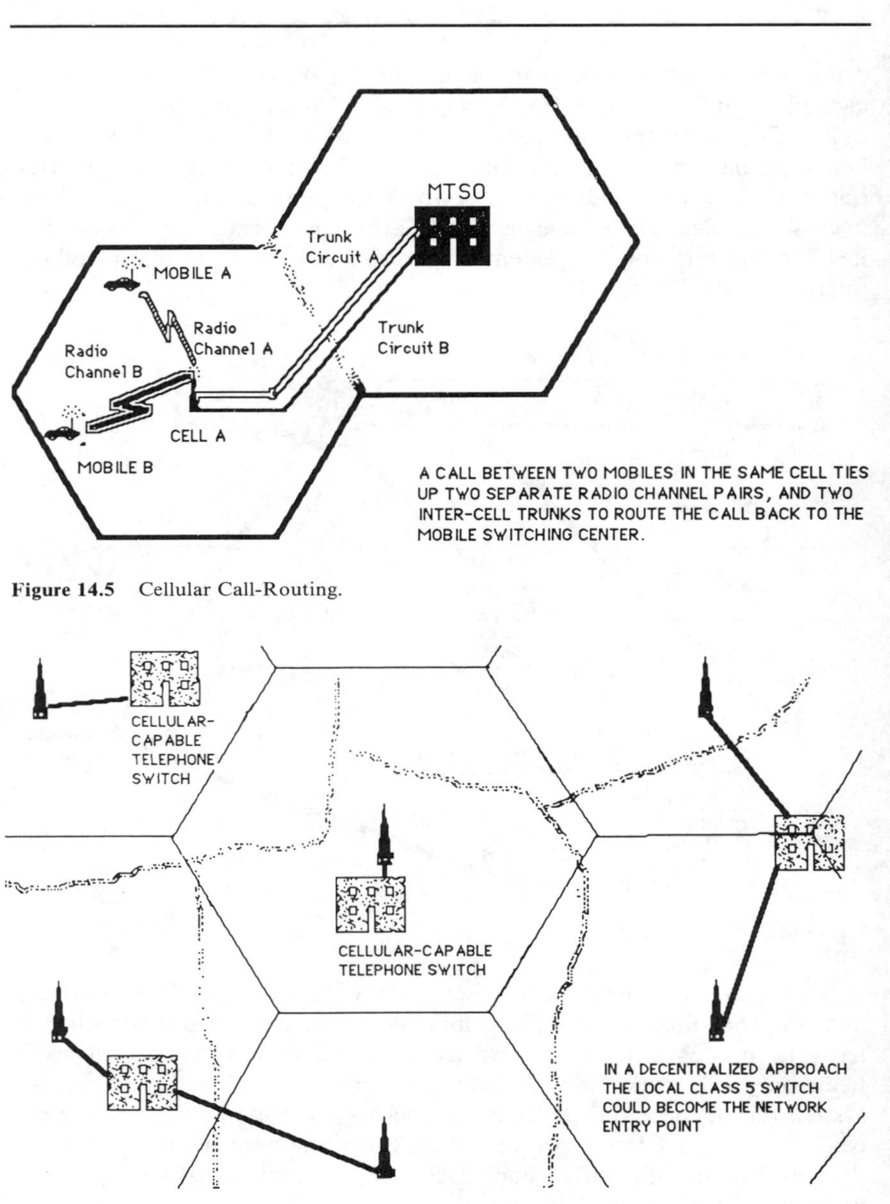

Figure 14.5 Cellular Call-Routing.

Figure 14.6 Idealized Integrated Cellular Network.

users in the same cell or in the same few cells, including landline callers within the geographical exchange areas covered by those cells. In other words, an integrated network makes most sense if there is a strong *community of interest* [17]. If a subset of the network can be identified within which there is a high level of intra-calling — where people will tend to call one another, rather than calling outside — then it makes economic sense to place the switching capability locally and avoid the backhauling charges.

This logic is straightforward and conventional: it is the same logic that underlies, in part, the placement of central offices and the definition of exchanges in the wireline network. It fosters the deployment of remote switching architectures in the wireline network today. In fact, the overlay approach is not qualitatively different from the integrated approach; the overlay architecture simply represents an implicit judgment that the relevant community of interest is the entire city, whereas the integrated concept is based on the assumption that smaller communities of interest exist within the city so that the cellular system can be optimized by designing multiple interface and control points within the network.

The economics of overlay architectures appear to be more favorable than the economics of integrated architectures at the present time. The higher costs of cell-site interconnection and switching facilities are not offset sufficiently by savings in backhaul costs. The analysis is not entirely clear, however, since it involves a comparison of capital costs *versus* operating costs which can only be made based on the often tenuous assumptions regarding both the level and the speed with which "full capacity" is achieved. It would seem likely that as cellular operations grow and penetrate more deeply into the communication mainstream, communities of interest will emerge that will dictate multiple interface points within a single urban network.

14.7 CONTROL: CENTRALIZED *versus* DISTRIBUTED

When cellular radio was planned out in the late 1960s and early 1970s, the microprocessor did not exist. The model for much of the thinking behind cellular architecture was the "big machine": the central mainframe controlling a large number of "dumb terminals." In today's systems, most of the decisions are referred up the hierarchy to the mobile-telephone switching office. The MTSO receives call requests from the cell-sites, identifies and validates users, locates users for incoming calls, determines channel selections, controls hand-offs, keeps all records of call information for billing and administration, monitors system performance including diagnostics and alarms for cell-site equipment [18].

Once again, over the past decade or so there has emerged an alternative vision of the control structure, propelled by the evolution of distributed architectures in computing networks, based on intelligent nodes, cells that are capable of considerable call processing, decision making, and record keeping. In fact, substantial computing power can be integrated directly into individual cell-sites without a significant cost impact.

This is not the place to rehash the virtues and trade-offs of distributed *versus* centralized architectures, which have become a broad subject in computer and communication network design. Suffice it to say that in a cellular network, distributed systems appear to promise greater flexibility, greater robustness (and survivability — the loss of the MTSO need not take the system down), and perhaps lower operating costs. (By concentrating information and decisions at lower levels it may be possible to reduce backhauling requirements further). In the next generation the issue of distributed control architecture will occupy a central position in the debate. It seems safe to say that a higher degree of decentralization will likely be implemented in future networks.

A more interesting question may be whether the push for distributed control will extend to the mobile units themselves. The next generation of digital cellular mobile units will incorporate extremely powerful processors; indeed, from the standpoint of today's dumb-terminal orientation, the digital mobile units will look like special-purpose microcomputers optimized for real-time signal processing. The mobile units will also possess, in principle, at least, the ability to make many of the decisions that are today strictly system decisions. It is not inconceivable, for example, that the mobile units themselves might initiate hand-off decisions, or even determine channel assignments, although none of the current proposals envisions this degree of decentralization.

The question of direct mobile-to-mobile operations is another matter of interest for the evolution of the network. Today the vast majority of mobile calls terminate to a landline telephone; mobile-to-mobile calls are rare. The patterns of usage of CB radio, however, would tend to indicate, for example, that as mobile telephony becomes more widespread, mobile-to-mobile traffic could become a very important part of the total cellular service. Some mobile architectures can allow for direct mobile-to-mobile links, not mediated through a central base-station, except perhaps for call setup. This may improve spectrum efficiency; a mobile-to-mobile call uses two radio channels if mediated through a base station, whereas only one might be necessary in a direct link. It may also open the way to radio nets, similar to today's fleet dispatch services, where a large number of users can be in communication at one time in a private net. None of the proposals that are currently being discussed for the next generation of cellular systems, however, addresses the issue of net operations. It seems likely that, as with many of the other questions raised in this chapter, net operations as a network-level architectural alternative will not emerge until a later stage in the development of mobile telephony.

REFERENCES

[1] Peter Temin, *The Fall of the Bell System,* New York: Cambridge University Press, 1987, p. 44.

[2] M. D. Fagen, ed., *A History of Engineering and Science in the Bell System: The Early Years (1875-1925),* AT&T Bell Laboratories, 1975.

[3] V. H. McDonald, "The Cellular Concept," *Bell System Technical Journal,* January 1979, pp. 15–41.

[4] William C. Y. Lee, *Mobile Communications Design Fundamentals,* Indianapolis: Howard W. Sams, 1986.

[5] "The Future of Cellular Radio," seminar sponsored by the Federal Communications Commission, September 2, 1987, Washington, D. C.

[6] Nathan Ehrlich, "The Advanced Mobile Phone Service," *IEEE Communications Magazine,* March 1979, pp. 9–15.

[7] Lee, *op. cit.,* p. 187.

[8] V. H. McDonald, *op.cit.*

[9] James F. Whitehead, "Cellular System Design: An Emerging Engineering Discipline," *IEEE Communications Magazine,* February 1986, pp. 8–15.

[10] A. R. Potter, E. Green, and A. Baran, "Increasing the Capacity of a Digital Cellular Radio System by Using Microcellular Techniques," *Proceedings of the International Conference on Digital Land Mobile Radio Communications,* Venice, June 30–July 3, 1987, p. 393.

[11] *Ibid.,* p. 394.

[12] D. Berthoumieux and M. Mouly, "Spectrum Efficiency Evaluation Methods," *Proceedings of the International Conference on Digital Land Mobile Radio Communications,* Venice, June 30–July 3, 1987, p. 248.

[13] Philip Y. Byrd and Gene A. Buzzi, "A Method for Improving Radio Frequency Spectrum Efficiency in the UHF and 800 MHz Bands," Tallahassee: Omnicom, August 1986.

[14] "Future Public Land Mobile Telecommunications Systems," *CCITT Interim Working Party 8/13, Report of the 2nd Meeting,* Melbourne, March 11–18, 1987.

[15] K. J. S. Chadha, C. F. Hunnicutt, S. R. Peck, and J. Tebes, "Mobile Telephone Switching Offices," *Bell System Technical Journal,* Vol. 58, No. 1, January 1979, pp. 71–95.
[16] *Ibid.,* pp. 73–74.
[17] Paul Clements, Director of Canadian Sales & Marketing, IMM, personal communication.
[18] Chadha *et al., op. cit.,* pp. 77–78.

Chapter 15

CHOOSING THE FUTURE: EVALUATING THE ALTERNATIVES

One's assessment of alternative architectures for the next generation of cellular radio depends upon what one thinks mobile telephony will, or should, become. Is cellular radio ultimately only a niche service for wealthy urban commuters, or is it a mainstream wireless access technology? Is it optimized for voice only, or should it be capable of bearing new digital services as well? One architect may view the ease of transitioning from today's analog systems as the primary consideration; another may be most concerned about compatibility with *future* generations of digital technology. If cost of the mobile unit is the chief concern, as it was with the first generation of architecture, different solutions will emerge than if system capacity is to be optimized.

At the very least, a digital generation of cellular technology must be capable of providing high quality (voice) telephone communication in an environment dominated by the severe problems of multipath propagation (see Chapter 8) and strong cochannel interference (see Chapter 9). It must alleviate the two chief shortcomings of today's analog cellular systems: capacity and cost (see Chapter 4). Also, it should address the other fundamental concerns identified in Chapter 10 and elsewhere throughout this book: modularity and geographical flexibility, digital network compatibility, and openness to new technology. These are the issues we shall touch upon in this chapter.

There are other questions that must be left to one side. Shall mobile telephony encompass portable telephony? Shall mobile services include nontelephone-type services such as paging or local area networks? Should cellular networks accommodate unrestricted national roaming? These questions go beyond the scope of the fundamental dilemmas of cellular radio today, and probably lie over the visible horizon of service and technology alternatives.

15.1 SPECTRUM EFFICIENCY

The impact of new technology on systems capacity is tied up with the vexing question of *spectrum efficiency* — a notorious portmanteau concept. Clearly, if mobile telephony is to penetrate beyond the niche to which it now seems confined, the most critical of all requirements for any new technology is the ability to generate substantial additional capacity from the existing spectrum allocations. But there are many different ways to view and define system capacity. "Spectrum efficiency" is not a simple, unidimensional measure, as is often stated or implied in the cellular literature. There are at least six different approximations, all of which are, to some degree, relevant to the evaluation of alternative mobile-telephony technologies.

1. Spectral efficiency;
2. Circuit spectrum efficiency;
3. Geographical spectrum efficiency;
4. Economic spectrum efficiency;
5. Communication efficiency;
6. Allocation efficiency.

The first four are "narrow" measures of efficiency, in that a definite, objective index can be stated. The last two are "broad" measures of efficency; they require considerable judgment and interpretation.

15.1.1 Spectral Efficiency (Information Density)

A very common and useful measure of the efficiency of any digital radio transmission is the *spectral efficiency* or *information density,* as defined in Chapter 12. Mathematically, it is obtained by dividing the bit rate of the transmission (in bits per second) by the bandwidth of the digital signal in hertz [1]. Sometimes referred to, loosely, as "bits per hertz," spectral efficiency is a measure of the modulation efficiency. For example, the IMM Ultraphone system referred to in earlier chapters achieves a bit rate of 64,000 b/s in a 20-kHz channel, for an information density of approximately 3.2 bits per second per hertz [2]. A recent proposal by Ericsson for a "narrowband TDMA system" indicates a bit rate of 340,000 b/s in a 300-kHz channel, or approximately 1.1 bits per second per hertz [3]. The emerging European digital cellular standard is apparently similar, by this measure, to the Ericsson approach.

The usefulness of this first-order approximation is clouded, however, by a number of modulation trade-offs that make straightforward comparisons difficult. For example, a 16-level PSK system should be capable, in principle, of achieving close to 4 bits per second per hertz. A 4-level PSK modulator can do only about half as well, or 2 bits per second per hertz. The 4-level system, however, can achieve a given bit-error criterion with about 12–13 dB *lower* signal-to-noise ratio than the 16-level system. Another consideration is that some modulators can achieve a given information density only by heavily filtering the transmitted signal to eliminate sidebands and "out-of-band" emissions; another scheme might reach the same spectral efficiency without such filtering. Thus, information density *per se* may not always be terribly meaningful. But it is at least a readily definable first approximation.

15.1.2 Circuit Spectrum Efficiency: Circuits per Megahertz

Information density tells us nothing about the way in which the bit stream is divided into telephone-grade voice circuits. The next, more general approximation, *circuit spectrum efficiency,* is defined as the number of telephone voice circuits per megahertz of spectrum. It measures coder efficiency as well as modulation efficiency. For example, a 64-kb/s channel rate can be used to carry a single PCM voice circuit, or two 32-kb/s ADPCM voice circuits, or four 16-kb/s RELP circuits, and so forth. Obviously, we would want to say that a system which creates four telephone circuits per 20-kHz channel is more efficient than another coder that creates only one.

Guardbands and adjacent channel restrictions must be taken into account. For example, in today's analog cellular systems it is not possible to utilize adjacent channels in a given cell; in fact, it is necessary to skip at least one or two frequencies between occupied channels to accommodate the relaxed filtering characteristics of most mobile units. This reduces the number of circuits per megahertz.

A simple calculation of circuit spectrum efficiency can be based upon the following equation:

$$\frac{C}{B \cdot (RA + 1)} \cdot \frac{1000\ (\text{MHz})}{2} = \text{Circuits/MHz}$$

where C = Number of Circuits per Carrier
B = Channel Bandwidth (including guardbands) in kHz
RA = Number of "restricted" adjacent channels

The final division by two is necessary to account for the doubling of transmit and receive frequencies.

Thus, today's analog cellular systems achieve about 5.5 circuits per megahertz (assuming that C = 1, B = 30, RA = 2 — in other words, two adjacent channels must be skipped).

By comparison, the proposed European standard should achieve about 20 circuits per megahertz (C = 8, B = 200, and assuming that RA = 0 — i.e., that adjacent channels can be used, which is not entirely clear). The IMM system currently achieves 80 circuits per megahertz in a fixed application (C = 4, B = 25, and RA = 0), and has demonstrated 40 circuits per megahertz in certain mobile applications [4].

Circuit spectrum efficiency is more useful than information density. It is in some ways the cleanest baseline comparison for alternative technical proposals. Some cellular analysts, however, would argue that it is not really relevant for cellular systems because of its emphasis on adjacent-channel restrictions, which may be a moot issue in cellular reuse schemes that tend to force each cell to utilize only a subset of the total available channels anyway. Also, it does not capture the aspect of spectrum efficiency that is claimed to be unique to cellular architecture: frequency reuse.

15.1.3 Geographical Spectrum Efficiency: Circuits per Square Mile

In a single-cell IMTS–type of system, circuit spectrum efficiency is the appropriate narrow measure of capacity. In a cellular system employing frequency reuse, the better measure is the number of circuits created per megahertz of spectrum *per square mile.*

Geographical spectrum efficiency relates in part to the average size of the cells and the degree to which cell-splitting *can be and has been* applied. For example, consider a cellular system covering a metropolitan area in a roughly circular area 60 miles across. The coverage area is nominally about 2800 square miles. Assume that the area is covered by cells with a radius of 5 miles, with a 7-cell reuse pattern. Each cell covers 78 square miles; assuming optimum packing and no terrain difficulties, the city can be covered by 36 cells. Each cell has 1/7 of the available channels, about 47 channels in a standard 20-MHz allocation. Each frequency is used about five times. Assuming that adjacent channel restrictions do not enter into the calculation, i.e., that the number of restricted adjacent channels is less than the number of cells in the reuse pattern, the equation for geographical spectrum efficiency is

[Number of Cells · Number of Circuits per Cell] / Total Area Covered

The number of circuits per cell is equal to

$$\frac{TA}{(B/C) \cdot R \cdot 2}$$

where TA = Total Spectrum Allocation (20 MHz)
B = Channel Bandwidth (30 kHz)
C = Circuits per Carrier (1)
R = Reuse Pattern (7)

The complete calculation is therefore

$$\frac{TA \cdot N}{(B/C) \cdot R \cdot 2 \cdot AR} = \text{Circuits/Megahertz/Square Mile}$$

where TA = Total Spectrum Allocation (in kHz)
N = Number of Cells
B = Channel Bandwidth (in kHz)
C = Circuits per Carrier
R = Reuse Pattern
AR = Total Area Covered

In this example, therefore, the cellular configuration will create about 0.6 circuits per megahertz per square mile.

The situation changes dramatically, however, if smaller cells are employed. With one-mile radius cells, there would be 900 cells covering the city. The 7-cell reuse pattern remains unchanged, so the calculation now yields 15.2 circuits per megahertz per square mile — a substantial improvement.

In part, as the equation indicates, geographical spectrum efficiency is determined by the reuse factor, which is largely determined by the cochannel-interference criterion. The 7-cell pattern is based, typically, upon a C/I criterion of 17–18 dB. If, hypothetically, digital techniques could permit a reduction in C/I to 10–12 dB for the same signal quality (see Chapter 9), it might be possible to implement a 3-cell reuse pattern. (The precise relationship between C/I criteria and reuse patterns is subject to considerable dispute; the numbers used here are illustrative only.) Applying a 3-cell reuse pattern to the first example given above would more than double capacity to about 1.4 circuits per megahertz per square mile. In the second example, a 3-cell pattern would improve geographical spectrum efficiency to 35 circuits per megahertz per square mile.

15.1.4 Economic Spectrum Efficiency: Cost per Circuit (per Megahertz per Square Mile)

The fourth approximation to a useful measure of system capacity introduces the question of cost. Obviously, a very expensive modulation scheme, operating at extremely high information densities, might achieve greater spectral efficiency than a simpler, less expensive system. Nevertheless, there is really no such thing as "pure capacity" for analytical purposes. There is always an implicit cost criterion: how much money has to be invested by the operator to add another circuit?

This observation is obviously quite relevant to cellular's current dilemma; cell-splitting is one method of gaining capacity, but it is apparently too expensive, aside from questionable performance, because of real-estate and other site costs that grow as small cells proliferate. For this reason — cost, or poor economic spectrum efficiency — operators are searching for other solutions to their capacity problems, even if this involves an upheaval in the technological base. In Chapter 4, we developed a hypothetical example comparing the cost per circuit for cell-splitting *versus* the cost per circuit for expanding into additional spectrum; at least in this model, the economic spectrum efficiency of cellular architecture appears to be substantially lower than originally predicted.

15.1.5 Communication Efficiency: Improved Individual Communication Utility

In the previous approximations of the elusive and multidimensional idea of spectrum efficiency we have concentrated on the creation of *circuits,* rather than on measures of telephone *traffic.* Many discussions of spectrum efficiency in the literature focus on traffic — how many Erlangs per megahertz or per square mile. In the narrow measures, this is unnecessarily obfuscatory; a circuit is a circuit, and trunking efficiencies apply equally to all schemes. (In effect, trunking efficiency — which is described in Chapter 4 — is an added plus for systems that generate more circuits per megahertz.)

The issue of traffic does become important, however, in a broader analysis. Traffic raises the issue of grade of service, which in turn opens the Pandora's box of service objectives. In particular, we come to the question of whether the uniform subdivision of the spectrum into telephone-grade voice circuits is actually the most efficient way to manage the scarce resource. Let us take a simplified example: consider a salesman who periodically calls his office from his cellular phone to pick up any messages that have come in for him, presumably from people who do not have his car-phone number. It may take him two or three minutes to reach his secretary, exchange amenities, and pick up half a dozen phone numbers. Hypothetically, if the future-generation cellular system were capable of operating in a data mode somewhat like today's alphanumeric pagers, the desired information — a thousand bits or so — could be dumped to his cellular terminal in a matter of seconds, stored for his recall, and the circuit cleared for the next user. The utility for the individual would be the same; the circuit capacity required to satisfy that user would be drastically reduced. (Of course, if the exchange of amenities is desired, a voice link could still be provided.)

This example may strike some as frivolous; I am convinced that the operators of future mobile-communication systems will not think so. They will be vitally interested in finding ways to improve the *communication efficiency* of the cellular system. We may define communication efficiency perhaps as messages per megahertz per hour (per square mile, per dollar invested). There are at least four ways in which technological and architectural decisions may influence communication efficiency:

1. Interchangeability of data and voice communication;
2. Increased use of queuing and sophisticated circuit-allocation techniques;
3. Shifting traffic away from the busy hour with techniques such as voice messaging and voice mailbox;
4. Flexible bandwidth allocations.

The possibilities for substituting data transmission for voice transmission in some circumstances hold considerable promise for increasing messages per megahertz per hour. The example given above is only one illustration. At projected data rates for the next generation of cellular systems, a page of text can be transmitted in the time it takes to say "Hi, how are you?"

Queuing is a technique that has already been applied in some mobile systems with some effectiveness. It is likely that a large percentage of callers would be happy to place a large percentage of their calls into a short queue, a maximum of 30–90 seconds, say, if (1) they had the opportunity to override when necessary and (2) they received some financial incentive for queuing. Reasonable, short queues such as this could potentially increase traffic throughout the system by a significant factor without diminishing the personal utility of the service. In effect, we have already built such measures into our *nonelectronic* message-transmission systems: most mail carriers, whether UPS or the US Post Office, offer several different grades of service that correlate price with speed of delivery.

Future cellular architects will undoubtedly come around in the same general direction.

More sophisticated call-setup procedures could also increase circuit throughput. For example, it is estimated that 20–30% of all calls do not connect, i.e., terminate as *busy* or *no answer.* Moreover, the average ringing time for calls that are answered is on the order of 10–20 seconds, 2–3 rings. Cellular systems have already been designed to move the time wasted during dialing off the shared radio circuit by using a *send* button. It should be possible to design a cellular call-setup procedure to allocate a radio circuit only once the called party has gone off hook. Certain statistical protections against the possibility of not finding a circuit available would have to be instituted; cursory analysis, however, has indicated a potential savings of 10–25% capacity in typical mobile-communication systems from this measure alone!

In traditional telephone thinking, every call is *urgent,* and all calls are equally urgent. In fact, a great many calls, viewed from a message perspective, are not especially urgent. This is particularly true of many one-way messages (the typical cellular morning call to the office, "I'm stuck in traffic and I won't be in until . . ."). A third approach to increasing communication efficiency is to flatten the peaks in the telephone traffic by allowing the system, rather than the user, to determine, within limits, the timing of certain messages. I believe that voice mail and similar techniques already being extensively applied to one-way communication (paging) would, if made available in a flexible digital architecture, allow much of today's "urgent" traffic to be downgraded (and billed at a lower rate).

Finally, the concept of *flexible bandwidth,* or *bandwidth-on-demand,* has lately begun to stimulate interest. Digital systems can be designed to be inherently reconfigurable; if a particular caller needs only one-half the bandwidth of a standard channel, for data transmission or for user-selected, less than toll-quality voice, which may be fine for a local telephone connection, the system can allow him to utilize only what he needs, instead of "wasting" bandwidth by offering only a standard fixed-circuit configuration.

Communication efficiency is admittedly something of a novel concept. Traditional telephone thinking has been conditioned by the existence in the traditional wireline network of *dedicated local circuits.* In the wire loop, there was always a tremendous excess of bandwidth, and no particular economic incentive to utilize it efficiently. In the bandwidth-constrained, *shared-circuit* radio environment, measures such as those described here will, I predict, become simply commonplace in the not too distant future. This trend will be driven by economics. Once the transition to digital architecture has been achieved, it will be much cheaper to gain capacity by such techniques, rather than by constantly reengineering the basic transmission link.

15.1.6 Allocation Efficiency: Improved Social Utility of Communication

The radio spectrum is a limited resource for which there are competing uses and competing users. The determination of the proper mix of uses is difficult, because both law and tradition have prevented the exercise of conventional market mechanisms. The FCC has repeatedly expressed the desire to move toward flexible spectrum-allocation procedures, even to the point of undertaking spectrum auctions; much as offshore oil leases are sold or auctioned to the highest bidder, ensuring at least some degree of rationality in the allocations process, many observers believe that fiascoes like the cellular lottery could be avoided by offering licenses to the highest bidder. Unfortunately for these advocates, the courts have ruled that the auctioning of spectrum exceeds the statutory authority of the Commission.

The FCC has, therefore, attempted to act as the rational allocator of the radio spectrum, assessing the various needs for radio services and trying to match blocks of spectrum to the perceived demand. Inevitably, since prime spectrum is scarce, there are judgments of the *social utility* of various services. For example, as we have seen in Chapter 2, mobile-radio development was held back in the postwar years by the implicit, and sometimes explicit, judgments on the part of the Commission that the relative social utility of television was greater than the utility of car telephones.

If the market mechanism were able to work, we would have, according to economic theorists, at least, a visible measure of the social utility of different uses of the spectrum: *revenue (or profit) per megahertz*. An examination of revenue and earnings from broadcasters and mobile-telephone operators leads to some interesting conclusions. For example, a television station in a major market reportedly generated revenues of approximately $100,000,000 in 1986 [5]. That is equal to approximately $16 million per megahertz. To generate the same revenue per megahertz, the cellular operator would have to pull in some $400 million in annual revenue, assuming 25 MHz. That would be equal to about 275,000 cellular users at the current average rate of $120 per month — about a 6% penetration of the market (population 4,770,000 according to Shosteck [6]). Shosteck's projections for actual demand in the market call for about 40,000 users by 1990, with the growth rate slowing. That would suggest that by the yardstick of the market mechanism, television is approximately seven times more "spectrum efficient" than cellular radio. Or, to put it another way, the current cellular technology must be improved by a factor of at least seven before it will equal the "revenue/MHz" of television.

Of course, this suggests one reason that policy-makers have not entrusted spectrum management to the market mechanism: the gross inefficiency of conventional mobile telephony, including analog cellular radio, means that it would not stand a chance in open competition with other services such as broadcasting. If businessmen were free to determine the uses to which the spectrum would be put, nobody would waste his spectrum earning one dollar on cellular phone service when he could earn seven dollars with TV.

This is a fundamental lesson for cellular architects. The traditional tendency of those in the mobile-telephone industry has been to view the advantage of the broadcasters as purely political. Actually, the real political power of the broadcasting interests is based on their vastly more efficient utilization of the spectrum. With a mere 6 MHz a broadcaster can reach millions of households, very inexpensively, on a per household per minute basis, and quite profitably. By comparison, a mobile-phone operator can serve perhaps 2000 or 3000 mobile-telephone users with the same spectrum. The imbalance in social utility provided is striking. (We are not making a judgment here about the value of TV *versus* car phones; we are simply observing the enormous disparity in the number of people who can be served with the same spectrum.)

Herein lies the real, chronic disadvantage of today's mobile telephony technologies. In fact, the "usurpation" by the FCC of the market's allocation function is primarily designed to *protect* less efficient services like mobile telephony from being swept away by far more lucrative services, including not only broadcasting, but paging as well.

Nevertheless, the administrative mechanism was, and is, sticky and inefficient. Spectrum allocations, once fixed by the Commission, tend to remain so for long periods. As conditions change and flaws in the original judgments are exposed, inefficiencies are created: some services are desperately crowded, to the point of crippling the industries that depend upon them, e.g., CB radio, while others harbor enormous blocks of spectrum that languish unused for decades, such as UHF television. Because small-stepped, regular change is prevented, pent-up market forces build like the pressure in the San Andreas fault, until a cataclysm finally breaks through the political process. Then, as with cellular radio in the late 1960s, the allocations are violently realigned, and frozen once again. The stresses begin building anew.

This has proved a very frustrating problem for almost everyone associated with it. The FCC is beseiged with specious and self-serving industry positions, all clamoring for spectrum. Like the robin with one worm and many gaping mouths to feed, the Commission finds itself embroiled in the politics of Grab. The strongest chick gets the meal, and all are hungry again when the next worm comes. Yet from almost every industry's point of view, except for the fattened few, there is a genuine and pressing need.

This type of spectrum inefficiency has several dimensions, which interact to some degree with the choices facing the next generation of cellular radio. First, there are *geographical inefficiencies.* For many services, including mobile telephony, the spectrum shortages are felt most acutely in the major cities. As noted in earlier chapters, the shortage in conventional IMTS mobile channels for the major metropolitan areas prior to the cellular era was extreme: only two dozen channels for all of New York City! This engendered terrible blocking, long waiting lists, and an unusable service. Yet even today, in the late 1980s, there is considerable unused radio spectrum in the IMTS channels lying fallow in almost every rural area. Cellular operators in Los Angeles and New York are worried that even 800 channels may not meet the demand for cellular service in their cities; on the other hand, it is hard to see how the operators in smaller cities and rural areas will ever be able to use even a small fraction of those channels.

Another dimension of the allocation problem is *service inefficiency.* The classic example is again the perennial struggle between the broadcasters and the mobile-radio community. This may be perceived as a problem of regulatory policy, rather than technology. Yet there is an emerging technological dimension to the allocation problem: one of the tremendous *potential* advantages of digital communication technologies is the opportunity for much greater flexibility and integration of different services. As noted in Chapter 7, the developing concept of an *integrated-services digital network* is predicated broadly upon the idea that as all forms of telecommunication, including voice, data, and video, are digitized, it becomes increasingly possible to share the same facilities for the delivery of such services. Indeed, this is one of the driving ideas behind the fiber-optics revolution: the vision of a single fiber link providing voice telephony, flexible bandwidth data, as well as television, over the same facilities.

15.1.7 Improving Spectrum Efficiency

How do the technological alternatives compare on this multidimensional criterion? A few general comments can be offered.

1. *Higher-level modulation.* Higher-level modulation schemes provide greater information density. Accepting the caveat that there is almost always a price to be paid in terms of power levels, SNR required for a

given BER, or the shape of the unfiltered power spectrum, the benefits of higher information density are impressive. In the past year or so, there has been a distinct reversal of the consensus among digital cellular planners: initially, many felt that constant-envelope modulation techniques were preferable because of (1) their narrow power spectrum and (2) the avoidance of the need for expensive linear power amplifiers; recently the importance of capacity has convinced many that the so-called linear-modulation techniques, such as PSK, are preferable, in spite of the potential cost penalty, because they are so much more spectrally efficient, given today's implementations.

2. *Lower bit-rate voice coding.* Improved voice-coding algorithms are an unmitigated blessing for the digital cellular architect from the standpoint of spectrum efficiency. A few years ago 64 kb/s seemed a fairly firm requirement; digital cellular radio was out of the question because of the bandwidth this would have devoured. Today, the rate of contemporary coding stands at around 16 kb/s for telephone quality; more than any other single technological development, it has been penetration of the 16 kb/s level that has kindled interest in digital cellular. Moreover, it is a fairly safe prediction that within five years telephone-quality voice coding at 8 kb/s or less will be readily available in forms appropriate for cellular implementation. I believe that by the mid-1990s, bit rates of 4–6 kb/s will be reached.

Even these forecasts may prove to be too conservative. Until now there has never really been a driving need in the telecommunication network for low bit-rate voice coding. Wireline transmission techniques were, and, now, fiber optics are, so bandwidth-rich, at least relative to the 3–4 kHz analog voice signal, that ingenious but computationally intensive techniques that achieved low rates have been prized less than economical implementations of moderate bit-rate coders, especially the PCM standard. The advent of digital cellular radio will change that. For the first time we will have a large-scale telecommunication system operating in a situation of severe bandwidth scarcity. Low bit-rate voice coders will assume a much greater importance. For example, in almost any of the digital architectures under discussion at the present time, a reduction in the bit rate from, say, 16 kb/s to 8 kb/s can be relatively easily translated into a *doubling* of system capacity. From 8 kb/s to 4 kb/s is of course another doubling, and from 4 kb/s to 2 kb/s still another. The *value* of such coding technologies to cellular operators will be enormous. If someone possessed today an algorithm capable of providing telephone-quality voice at 2 kb/s, implementable on a single processor, that would constitute a technology of great market value. It is quite possible that as research funding flows toward this lucrative opportunity we shall see coding breakthroughs coming more quickly than we have been conditioned to expect.

3. *Less overhead.* Every digital transmission system requires certain additional signaling or supervisory bits to be transmitted in addition to the voice bits, to permit the system to acquire and maintain synchronization, pass control messages, and monitor its performance. Continuous BER measurement typically requires the transmission of parity bits, for example. This overhead occupies bandwidth in the channel that cannot be used for voice; in short, it reduces the usable capacity of a bandwidth-limited channel.

In wireline transmission systems, again, the available bandwidth is usually so large that the overhead issue may not greatly affect voice transmission. In the radio channel, however, where bandwidth is precious, a system that requires 30% overhead is at a distinct disadvantage compared to one that requires 3%. Either voice quality or system capacity will be reduced.

The percentage of the total bit stream that is devoted to overhead bits varies considerably. Typically, digital transmission systems that transmit continuously, such as FDMA-SCPC, need fewer overhead bits than systems which operate in a burst-transmission mode, such as TDMA. In a TDMA system, the subscriber transmits and receives only during defined time slots; synchronization must normally be reaquired at the beginning of each slot, which means that certain bits in the header of each slot must be set aside for this purpose. In addition, there may need to be certain "guardbits" at the beginning and end of each slot to allow for the potential slight overlap of transmissions from two subscribers on adjacent slots, operating at different distances from the base station. (The overlap is created by different propagation delays.) In FDMA-SCPC architectures this overhead is not necessary because transmission is continuous, the receiver can maintain synchronization from the data stream itself, and there is only one user per channel. This is sometimes adduced as a one of the few clear advantages of FDMA architectures over TDMA architectures.

In some TDMA architectures, it is true, a relatively large percentage of the transmission must be devoted to framing overhead. In an architecture proposed by Ericsson, the framing and synchronization overhead constitutes 24% of the channel [7]. The difference in overhead rates is diminishing, however, as TDMA architectures are improved. FDMA-SCPC architectures will still require perhaps 2–3% overhead; the IMM TDMA architecture today requires overhead of about 9% [8]. Systemwide, I believe this will prove to be a small difference, more than offset by the other advantages of TDMA architecture.

Another form of overhead is the capacity that must be set aside for call setup. Today's analog cellular systems are surprisingly inefficient: of

333 channels originally allocated to each mobile operator, about 21 channels were required for call setup — more than 6%. In digital systems this should be greatly reduced. The IMM digital TDMA system today requires something well under 1% of the available circuits to be set aside for call-setup purposes. Presumably, digital FDMA and TDMA architectures should not differ much on this score.

4. *Robust voice coding and modulation mean less channel coding.* Most proposals for a digital mobile-radio system involve some form of channel coding, or error-correction capabilities, which are designed to enhance transmission integrity in the difficult mobile environment. Channel coding is another form of overhead in the sense that it increases the number of bits that must be transmitted for a single voice circuit. In a typical proposal, for example, a voice coder operating at a rate of 16 kb/s is processed through a channel-coding procedure that results in a 50% expansion of the signal to 24 kb/s [9]. The resulting signal is reportedly capable of withstanding a 25% BER.

As we noted in Chapter 12, however, some coders are inherently much more susceptible than others to bit errors; or, to state the issue more precisely, the *effects* of bit errors are much more perceptible for some coders than they are for others. A PCM-type coder should ordinarily show greater signal degradation at a moderate error rate than a delta modulation–type coder, for example. Some coders work well at 10^{-5} BER and deteriorate rapidly at 10^{-3} or higher BER. Other coders show perhaps lower quality at 10^{-5} but can survive down to 10^{-2} or worse.

The same is likely to prove true of modulation techniques. Some techniques, such as QAM, are apparently not able to survive well under typical mobile-channel conditions compared to others, such as PSK.

The point here is that more robust coders and modulation schemes, which can inherently withstand higher channel bit-error rates, will require less channel coding, and should be more spectrum efficient in terms of circuits per megahertz.

5. *Lower symbol rates mean less channel coding.* Another aspect of robustness is related to the signaling rate and the delay-spread issue. As discussed in Chapter 12, if we assume an average delay spread of, say, 2 μs in a particular environment, a system transmitting at, say, 340 kb/s (2 bits per symbol, or 170 kilosymbols per second) with a symbol time of about 5.9 μs would experience an average pulse overlap of at least 33%, generating significant intersymbol interference. On the other hand, a system with a 16 kilosymbol per second rate with a symbol time of 62.5 μs would experience average overlaps of about 3%, in the same environment.

High symbol rates will generate the requirement for intensive countermeasures that will impact both cost and capacity. For example, the

Ericsson proposal, which is typical of high symbol-rate proposals, necessitates not only significant channel coding, but also frequency hopping, to combat frequency-selective fading, and intensive adaptive equalization, to combat high levels of ISI. As a result, the ratio of actual information bits to transmitted bits is only 50% [10]. On the other hand, in the IMM system operating at much longer symbol rates (62.5 μs), the ratio is currently about 91%. In a cellular application, some additional channel coding might be applied, which could reduce this ratio to 75–80%. Nevertheless, the penalty of higher bit rates is obvious; the fixed delay spread becomes more and more of a problem as symbol times are reduced. In addition, the propagation delay, which is also fixed, becomes a larger proportion of the time-slot duration, which increases the percentage of the slot that must be devoted to guardbits. To maintain signal quality, a greater proportion of the bit stream must be set aside for error correction and other overhead. This will be a disadvantage for relatively wideband TDMA systems, including, quite possibly, the European GSM standards that are now emerging.

6. *Better cochannel-interference performance.* One of the most important advantages for digital transmission, in the eyes of many in the field, is the likelihood that digital systems will be able to operate under conditions of much higher cochannel interference, which will enable cellular designers to bring the reuse distance down and even reduce the reuse pattern from a 12-cell or 7-cell pattern to a 4-cell or even a 3-cell pattern. Some researchers believe that digital modulation, such as QPSK, with a C/I ratio of 12–13 dB, where the desired signal is 15–20 times stronger than the interfering signals, will be able to equal the audio quality levels of FM systems operating at 18 dB C/I, where the desired signal must be at least 63 times stronger than the interference.

This is a claim that needs to be evaluated with some caution. I believe that digital transmission will prove such an advantage, measured in terms of cochannel-interference performance; given the critiques now being developed of cellular geometries in general, however, one must be cautious about inferring reuse factors from raw C/I numbers. Nevertheless, it is clear that improved cochannel-interference performance is a good thing and will improve spectrum efficiency to some degree. What is not clear is exactly how much it will improve.

It is also not entirely clear whether there is a difference in cochannel-interference performance among the various alternatives discussed in the previous chapters. There is a lack of published empirical evidence on this point. Some calculations, however, suggest that TDMA architectures may be at an advantage over FDMA architectures for the following reason. In an FDMA system, with continuous transmission by all mobiles, cochannel-interference calculations must assume that all interfering mobiles are also transmitting continuously. By contrast, in a TDMA system, the likelihood is that at any given moment only a fraction of the mobiles (equal to the inverse of the number of time slots) are actually transmitting. The level of cochannel interference produced by a given population of mobiles should be reduced.

7. *Better adjacent-channel interference performance.* As noted in earlier chapters, today's cellular systems are designed such that at least first and second adjacent channels are restricted. In other words, if a particular frequency is occupied in a particular cell, the four nearest channels, two on either side, cannot be used, because the liberal emissions mask for cellular transmitters would result in excessive adjacent-channel interference. In effect, each cellular transmission today really takes up about three cellular channels. (This calculation may be subject to different interpretations; the figure of two restricted adjacent channels on either side is derived by reference to the much more stringent emissions mask for IMTS channels, which *are* designed to permit adjacent-channel operations [11].)

The effect of improved adjacent-channel interference performance upon spectrum efficiency is twofold. First, the ability to use adjacent channels will allow greater spectrum efficiency where dynamic channel allocations are desirable. For example, it may be that during the heart of the business day in New York City, most of the channels should actually be allocated to the cells serving Manhattan. In the evening, the channels should be allowed to "flow" outward with the traffic. Adjacent-channel restrictions make this very difficult to implement, if not impossible. Another case where poor adjacent-channel performance is very spectrum-inefficient arises with single-cell systems that might be set up in rural areas or along interstate corridors. A single-cell system that can only make use of every fourth channel and which cannot use the interstitial channels for some other purpose, even if not required for mobile traffic, is obviously a very inefficient system.

The other aspect of adjacent-channel interference performance that could be important is that very liberal emissions masks mean that adjacent-channel interference is probably complicating and intensifying the *co-channel-interference* problem noted above. For a mobile operating on a given frequency *x,* if an adjacent channel is occupied not in the same cell but in an adjacent cell, there may nevertheless be a significant amount of energy spilling over into the channel *x,* raising the level of cochannel-interfering energy that the mobile operating on channel *x* must contend with.

8. *Flexible bandwidth.* Finally, as noted in the discussions of the broader definitions of spectrum efficiency in Sections 15.1.5 and 15.1.6, the fixed circuit rate is inherently inefficient, or, more accurately, it leads to inefficient uses of the spectrum by not allowing users to adjust their bandwidth, and timing, to meet their actual communication needs. If we may be permitted an exotic metaphor, today's cellular systems are like a bar where patrons are permitted to buy beer only by the case; they must drink it on the spot, and cannot take anything with them. In such a situation, obviously a lot of beer would go to waste. Moreover, the cost of a drink would be quite high; somebody has to pay for the waste. By contrast, a digital cellular system that was capable of allocating bandwidth flexibly, as much or as little as the user actually needed, would be like the more familiar arrangement where it is possible to buy beer one bottle at a time.

Digital systems in principle are capable of implementing flexible bandwidth architectures with relative ease. TDMA systems are probably at an advantage here as well; the slot structure of a TDMA system is easier to reconfigure than an FDMA-SCPC system that is committed to fixed channels. This may seem to be a somewhat higher order solution to the spectrum-efficiency problem — and so it is. It is also likely to be the real answer to spectrum-efficiency concerns, as we look toward very widespread, extremely high capacity systems in the next century. It is difficult to see how we could serve, say, 30–40% of the population of the United States without abandoning the fixed-channel straitjacket of current SCPC systems.

15.1.8 Summary

To wrap up a very complex subject, I believe there are basically four questions that need to be posed for any prospective digital cellular architecture.

How many telephone circuits can be created per megahertz of spectrum in a single cell? This question subsumes the questions of information density, modulation efficiency, coder efficiency, and adjacent-channel performance.

How closely can two base stations operating on the same channel be located in terms of the ratio of the separation distance to the cell radius? This question subsumes the critical issue of cochannel-interference performance.

How flexible is the architecture in allocating channels to individual cells and to individual systems? Does the system allocate all cells the same amount of spectrum, regardless of differences in need? Does the architecture allocate the same number of channels to all systems, regardless of market size or traffic?

How flexible is the architecture in allocating bandwidth to match the actual communication needs of individual users? Does the system force the user to buy a case of beer, when all he wants is a bottle?

As to the still more general question of whether the overall allocation of spectrum among different services, different competing uses, such as television, or paging, or satellites, or mobile radio, I am afraid that this issue, if it is not to be thrown open to the forces of the free market, must remain one of the political imponderables with which regulators and industry lobbyists will continue to struggle.

15.2 COST

As described in Chapter 4, there are three components of cellular costs:

1. System costs: the costs of cell-site hardware, towers, real estate, the MTSO, and the intercell trunking facilities;
2. Mobile-unit costs: the cost of the cellular transceiver, control head, antenna and related hardware in the user's vehicle;
3. Operating costs: the usual administrative costs, marketing-selling costs, and the cost of leased facilities, *et cetera.*

15.2.1 System Cost Comparisons

An analysis of the economics of mobile telephony yields a fundamental conclusion: *There is a substantial inherent economic advantage for TDMA systems over FDMA (single channel per carrier) systems.* Indeed, the TDMA advantage overshadows *all* other factors in the economic evaluation of cellular-radio architectures.

As we saw in Chapter 4, the root of the high system costs, and high service charges, of today's analog cellular systems lies in their reliance upon uneconomic SCPC architecture. Each cell-site radio channel can support only one user at a time. The entire cost of the channel hardware — transmitter, receiver, amplifiers, *et cetera* — must be supported by that one user. By comparison, a TDMA system multiplexes several circuits on each carrier. In a 4-circuit per carrier system, e.g., the IMM system described previously, the cost of the radio hardware for each channel is borne by four users at any one time, instead of just one. To look at it another way, to serve a given population with a specified grade of service, the 4-circuit per carrier TDMA system needs only 25% of the radio-channel equipment at the cell-site. In the proposed European standard, each radio channel would support eight circuits per carrier, further spreading the cost burden. In practice, the savings from TDMA architecture in cost *per circuit* can be substantial, even assuming that TDMA channel hardware may be more costly *per channel* than FDMA-SCPC equipment.

The relationship between the number of circuits per carrier and the cost per circuit is dependent upon the proportions of fixed system costs, such as the cell-site real estate or the towers, and traffic-sensitive radio channel costs. Consider a hypothetical example, derived from the example developed in Section 4.3:

Case 1. Hypothetical Cell-Site Costs, FDMA-SCPC Architecture

Fixed Costs	$500,000	
plus		
Number of Channels	50	
Number of Circuits	50	
Hardware Cost	$10,000	(per channel)
Channel Costs	$500,000	
Total Costs	$1,000,000	
Loading	20:1	(typical mobile)
Total Customers	1,000	
Cost per Channel	$20,000	
Cost per Circuit	$20,000	
Cost per Customer	$1,000	

A TDMA architecture based on four circuits per carrier could serve the same population with fewer channels at lower costs per circuit, even with 20% higher channel costs:

Case 2. TDMA, Four Circuits per Carrier

Fixed Costs	$500,000	
plus		
Number of Channels	13	
Number of Circuits	52	
Hardware Cost	$12,000	(per channel)
Channel Costs	$156,000	
Total Costs	$656,000	
Loading	20:1	(typical mobile)
Total Customers	1,040	
Cost per Channel	$50,461	
Cost per Circuit	$12,615	
Cost per Customer	$631	

Certain factors can accentuate these differences. For example, the difference is widened in absolute dollars by the heavier traffic loadings characteristic of true telephone usage. At a concentration of 6:1, approximately equivalent to a telephone grade of service based on 0.1 Erlangs and a 0.5% blocking, the SCPC system outlined above would have a system cost per customer of some $3333, while the TDMA architecture would be about $2100 per customer.

The TDMA advantage also increases as the proportion of fixed costs falls. If the examples above are modified to reduce fixed cell-site costs to $125,000, a 20% proportion instead of 50%, the cost per customer for the FDMA system falls to $625 at 20:1 loading or to $2083 at 6:1 loading. The same costs for the TDMA architecture, however, fall to $270 or $900, respectively.

Systems with more circuits per carrier enjoy even lower per-circuit costs, although the economic gains appear to be greatest between SCPC and about 4–8 circuits per carrier. (See Figure 15.1.) Beyond eight circuits per carrier, two factors mitigate the impact of TDMA economics: first, the traffic-sensitive portion of system costs has been reduced to such a low percentage that fixed costs dominate; and, second, the cost of implementation of TDMA circuitry probably begins to increase on a per-channel basis — managing higher bit rates and more time slots per channel in the face of more serious delay-spread and propagation-delay effects brings about a situation of diminishing marginal returns.

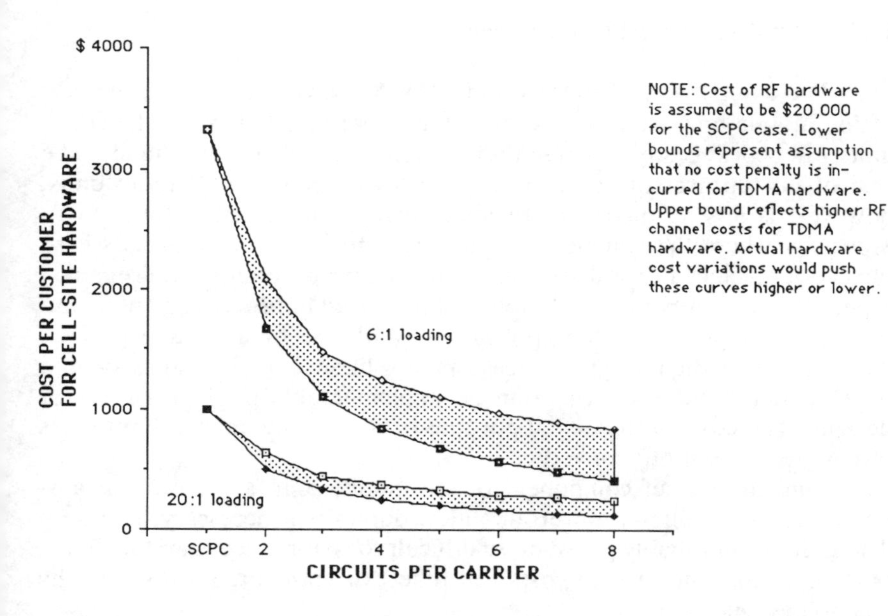

Figure 15-1 TDMA Economics: First Cost per User for Cell-Site Hardware (Variable Costs Only).

The advantage of time-division multiplexing is so significant, it is surprising that it has not received more attention from cellular architects to date. After all, the opportunity to implement TDM was the original purpose behind the development of T-carrier digital systems for wireline telephony: the designers of T-carrier were driven by the desire to share the costs of the channel facility, the wire pair, among a number of voice circuits. In this they were eminently successful: T-carrier has become the dominant multiplex format for interexchange trunks and has permeated dozens of other markets. Except in certain microwave niches, analog carrier (FDM) has probably been rendered very nearly economically obsolete in new installations throughout the voice network. Thus it should come as no surprise that TDM techniques promise tremendous economic advantages in digital mobile telephony as well. The FDMA approach, while it may gain additional capacity, does not change the economics of cost per circuit, and, as such, offers no relief to the cost dimension of the cellular crisis. TDMA systems gain capacity and *reduce the cost per circuit.* For this reason alone, I believe it is highly likely that the next generation of cellular radio will be based on digital TDMA architecture.

15.2.2 Mobile-Unit Cost Comparisons

The comparison of FDMA and TDMA designs in terms of the cost of the mobile unit are less clearcut. Many observers believe that FDMA units will be less costly, because they are felt to be inherently simpler. The cost of microprocessor-based systems such as digital mobile-radio units, however, is less a function of their perceived complexity of function — especially where the complexity resides in software or firmware, as is so typical of advanced digital systems — than of the processor requirements, especially the number of processors and the speed of processing and memory access required for the application. With the breathtaking pace of development in digital signal processors, it is likely that all designs — both FDMA and TDMA — will soon be implementable in single-processor designs. The costs of the digital portions of the mobile units in both types of designs will tend to converge.

The other chief component of the mobile unit is the radio; transmitter, receiver, filters, power amplifier, duplexer if necessary, *et cetera.* The radio will probably prove more difficult to cost-reduce than the digital portion of the unit. Radio costs will reflect at least three aspects of the technology selected:

1. FDMA systems inherently require a duplexer, since they must transmit and receive simultaneously; TDMA systems can usually be designed so that transmission and reception take place on different time slots, allowing a simple T/R switch to be implemented, which should reduce costs for TDMA units somewhat (perhaps 5–10%);
2. Constant-envelope modulation techniques do not require a linear amplifier and, consequently, should enjoy a modest cost advantage over linear-modulation techniques (10%?);
3. Modulation schemes which naturally produce very narrow emissions, narrow power spectra, may require less filtering, especially transmit filtering, compared to other approaches; the trend, however, will be toward the implementation of much filtering in software, so the impact of this factor will also be moderate and probably short-lived.

In short, the cost of the mobile units will be dominated more by manufacturing economies and product life-cycle considerations than by the broad architectural alternatives selected. Of course, costs will continue to constrain specific decisions: for example, many advanced voice coders may be held back from implementation until digital signal-processing chips reach certain cost-performance milestones. The likely advances in processor capabilities in the next five to ten years, however, will probably eliminate most sources of significant hardware cost differences for different algorithms. The more significant cost struggle will take place in the radio, and even there the differences will not be very large by comparison with the fundamental TDMA cost advantage at the system level.

15.2.3 Operating Cost Comparisons

It is difficult to separate out the effects of different architectures on operating costs *per se;* many of the effects are general and pervasive. For example, if TDMA can significantly lower capital requirements, facilitating pricing breakthroughs, then today's inordinately high marketing costs, which reflect the "hard sell" for a $150 a month end-user service and equipment charge, may be reduced. If better cochannel-interference performance permits an improved reuse factor, which in turn allows a given population to be served with fewer cells, this may also lower the system costs and help bring prices down.

An area where architecture may translate directly into lower operating costs involves the network-level decision of the degree of centralization in the cellular network. Today's centralized systems require extensive networks of trunks linking the cells to the central processor. The ability to implement more distributed architectures, with local switching and (wireline) network interface options will probably help reduce the backhauling costs as cellular systems grow. This effect, however, will probably be less pronounced for the early years of a cellular system than it may be later on, as densities go up and communities of interest begin to emerge within large urban systems.

15.3 OTHER CRITERIA: MODULARITY, DIGITAL COMPATIBILITY, OPENNESS TO NEW TECHNOLOGY

After cost and capacity, the chief criteria for evaluating alternative cellular architectures relate broadly to the question of flexibility and adaptability, to different operating environments, to changes in the public telecommunication networks, and to changes in the underlying digital transmission techniques.

15.3.1 Modularity and Geographical Flexibility

Designing a single architecture capable of efficiently serving the needs of massive metropolitan agglomerations like New York or Los Angeles while also meeting the quite different economic and operating requirements of smaller cities and rural areas is a challenge that cellular engineers have so far evaded. The implicit assumption that cellular telephony is an *urban* service pervades the current architecture. Yet the one-size-fits-all approach turns out to be inadequate for both the largest cities, which are still facing chronic spectrum shortages, and small towns, which are saddled with a top-heavy and economically infeasible architecture.

The issue of geographical flexibility has a great deal to do with the network-level design: more flexible cellular networks, with options adapted for smaller markets, will have to be developed. There is no particular magic in this; the wireline network has had to confront the same range of economic and operational variables. Indeed, the wireline infrastructure is quite different today in rural areas as compared to urban centers.

Narrowband systems are inherently more flexible in this respect than wideband systems. A wideband spread-spectrum system, in which the mobile unit "occupies" the entire spectrum during a single transmission, must be sized for the largest markets, assuming that we wish to avoid having different types of mobile unit for different size cities. Narrowband systems, on the other hand, can be built up modularly, on a channel-by-channel basis.

The requirement for geographical flexibility also accentuates the importance of adjacent-channel operations. The restriction on adjacent-channel utilization was a specific backward step (in the sense that precellular IMTS operators can and do operate on adjacent channels) that was taken by the original cellular architects *on the assumption* that cellular systems would *always* involve large numbers of cells, with subsets of channels allocated to individual cells. This is clearly based on an implicit metropolitan perspective. It is likely that many smaller markets will *never* require large multicellular systems; one or two cells will suffice. Indeed, if adjacent-channel operation were possible, a single cell operating with 333 frequencies would be capable of serving something on the order of 3000 to 6000 mobile units without any reuse at all. If digital techniques now on the horizon were applied, which could multiply this by a factor of three to eight, tens of thousands of mobile users might be served *without any cellular architecture at all!* This could well be adequate for all but the very largest markets.

15.3.2 Digital Network Compatibility

The slow and fitful evolution of the wireline world toward new digital network standards like ISDN has been viewed with skepticism by many in the short-fused world of cellular radio. When one is sprinting for break-even and trying to move revenues ahead of debt service payments as quickly as possible, there may be scant inclination to ponder the possible future of digital services ten or fifteen years hence. Even the next-generation thinkers have tended to dismiss ISDN on the assumption that any accommodation of cellular services to wireline standards must involve a radical compromise, that is, a reduction in bit rates for digital services by an order of magnitude.

Clearly, very narrowband FDMA systems, 15 kHz or less, will have a great deal of difficulty dealing with ISDN-type services, where the nominal end-user bit rate is 144 kb/s. On the other hand, TDMA proposals, including the evolving European standard, appear quite capable of handling near-ISDN rates. The IMM TDMA system described previously should be capable of providing nearly a full 2B + D rate in a standard 30-kHz channel [12]. Whether wireline ISDN compatibility is an important requirement, or whether ersatz reduced-rate ISDN services will be adequate, is one of the open questions for cellular planners.

15.3.3 Openness to New Technology

As we shall emphasize in the final chapter, the overriding fact about digital radio transmission technologies is that they are rapidly improving. A major concern for cellular planners is the degree of obsolescence that can be tolerated in a particular architecture. The inverse question is how open the architecture can be, to implement continuing technological improvements without forcing repeated, painful overhauls.

It is not possible to anticipate all technological advances that may occur in the next twenty or thirty years. It is possible, however, to identify some that are highly likely. For example, it is probable that coding techniques will move ahead more rapidly than modulation techniques. In particular, the likelihood is great that today's nominal 16 kb/s standard will be superseded by much lower bit rates in the next few years. How will cellular systems cope with, or take advantage of, such improvements?

The answer seems to indicate another clear advantage for TDMA architectures over FDMA-SCPC systems. In an FDMA-SCPC system built around 16 kb/s, for example, the only way to implement an 8-kb/s coder when it emerges will be to engage in another round of channel-splitting, which, especially if the adjacent channel issue has not yet been solved, will almost certainly produce a new upheaval in the channel allocations and the engineering of individual cells. Moreover, channel-splitting tends to become a trickier proposition as the bandwidth decreases; indeed, in the opinion of some observers, channel-splitting may become counterproductive due to cochannel-interference considerations [13]. Certainly, a whole set of system issues are raised concerning backwards compatibility of mobile units operating in the same system with different bandwidths and different power-spectra characteristics.

By contrast, the implementation of lower bit-rate coders in a TDMA system should involve a redefinition of the time-slot structure, but not of the channel itself. Transmission and cochannel-interference characteristics would probably remain stable; frequency plans and cell-site engineering would not be affected. Indeed, the IMM system described previously already operates in a dual mode, where either two or four time slots may be established in the channel, and different subscribers at different modulation levels can operate in the same system, even on the same channel [14].

The ability to implement flexible bandwidth transmissions under software control in a TDMA architecture is one of the chief advantages of TDMA systems.

15.4 SUMMARY

The next generation of cellular architecture will be based neither upon analog techniques, including SSB, which are inherently obsolescent, nor upon wideband spread-spectrum techniques, which are still apparently premature. Overall, the main dispute seems to be shaping up around the FDMA-SCPC *versus* TDMA decision. In my opinion, TDMA is a clear winner. The most significant advantages are (1) lower system costs due to multiplexing of base-station radio channels, which leads to lower equipment requirements, and (2) much greater flexibility in accommodating many of the technological advances anticipated in the coming years.

The preference for TDMA parallels the logic of TDM digital systems in the wireline network. The original motivation for T-carrier, for example, was the cost savings from sharing transmission facilities, the wire-pair, for interexchange trunking. One of the significant advantages of digital techniques in switching systems, in addition to cost savings, has been the flexibility of software-controlled system upgrades and enhancements. In the radio environment, the ability to reconfigure the TDMA channel through software, rather than hardware, is certainly a compelling advantage as well.

Finally, if, or when, ISDN does emerge as an important network capability, TDMA architecture should offer a much easier accommodation of ISDN-type services than the very narrowband systems suggested for FDMA-SCPC schemes.

As far as specific modulation techniques are concerned, it is a more difficult call. The early enthusiasm for constant-envelope techniques appears to have given way to a recognition of the spectral efficiency benefits of PSK-type linear modulations. The selection among the dozens of competing voice coders will be even more difficult. Indeed, it is in the area of coding technology that the search for a single standard may break down; the pace of coder development is certainly much faster than the rate of equipment change-out in the subscriber base. We may have to accept a variety of coding techniques operating, coexisting, within the overall cellular network.

REFERENCES

[1] John C. Bellamy, *Digital Telephony,* New York: Wiley, 1982, p. 273.
[2] D. Ridgely Bolgiano, "Spectrally Efficient Digital UHF Mobile System," *Proceedings of the 38th IEEE Vehicular Technology Conference,* Philadelphia, June 1988.
[3] Jan Uddenfelt and Bengt Persson, "A Narrowband TDMA System for a New Generation Cellular Radio," *Proceedings of the 37th IEEE Vehicular Technology Conference,* Tampa, June 1–3, 1987, pp. 286–292.
[4] Bolgiano, *op. cit.*
[5] D. R. Bolgiano, personal communication.
[6] Herschel Shosteck, *The Demand for Cellular Telephone: 1985-1995,* Silver Springs, Maryland: Herschel Shosteck Associates, September 1986, p. 11.
[7] Uddenfelt and Persson, *op. cit.*
[8] D. Ridgely Bolgiano, "Spectrally Efficient Digital UHF Mobile System," *Proceedings of the 38th IEEE Vehicular Technology Conference,* Philadelphia, June 1988.
[9] *Ibid.,* p. 287.
[10] *Ibid.,* p. 288.
[11] D. Ridgely Bolgiano, personal communication.
[12] Bolgiano, *op. cit.*
[13] William C. Y. Lee, Remarks at the FCC Seminar on "The Future of Cellular Radio," Washington, D. C., September 1987.
[14] Bolgiano, *op. cit.*

Part VI
MANAGING THE TRANSITION

Chapter 16
A LOOK AHEAD

It would be hard to find a historical parallel: an entire industry built upon a technology that does not yet exist! Yet that is, in a very real sense, what has happened to cellular radio. An industry of substantial size, serving more than one million customers in the United States, has been set upon an unsound technological foundation. Like builders working on soft soil, cellular operators are only becoming fully cognizant of the weaknesses of their situation as the edifice nears completion and its full weight begins to subside, like some technological Tower of Pisa. By that point the building is up, the mortgages have been contracted, and it is half full of tenants who are concerned about the cracks opening up in the walls.

The economics of analog cellular simply do not come together to ignite the classic chain reaction — low prices, which bring more customers to allow more efficient operations and higher volume manufacturing, which brings lower prices, *et cetera.* At the lowest feasible operating prices, the systems are still below the break-even point in subscriber numbers in many markets — there are only so many Cadillacs in Augusta, Georgia. In the few large markets where there are sufficient concentrations of upscale customers, cellular becomes, paradoxically, the victim of its "success," as the grade of service deteriorates back toward the blocking levels of pre-cellular mobile radio.

In fact, cellular radio does not appear to enjoy the classic leverage of most utility-like industries: steeply falling marginal costs per additional customer. In the wireline telephone network, per capita costs fall as more people can be served on the same cable route and the cost of deploying the cable circuits — which is usually the dominant cost factor in telephone service — can be spread across a larger base. Where population densities are high, cable routes can pick up hundreds of customers per route mile; in sparsely populated areas, there may be only a handful of customers per square mile. For this reason, wireline costs per customer are normally *much* lower in urban areas than in rural settings.

Unlike wireline telephony, electric power, broadcast radio or television, or cable TV, however, there are few *operating* economies of scale as a cellular system grows. If anything, the cost per circuit probably tends to *increase* as cell-splitting forces the construction of more expensive cell-sites and the trunking requirements rise as approximately the square of the number of cell-sites. The cellular operator never enjoys the marginal cost leverage of a cable route in wireline telephony or cable TV. The costs per customer probably do not differ significantly for a system designed for 10,000 customers as compared to one designed for 100,000.

The promise of cellular reuse — particularly the "breeder reactor" concept of cell-splitting to produce very large numbers of telephone circuits from a modest spectrum allocation — has also proved to be unsound. Cells smaller than about two miles in radius would appear to be unstable from a frequency planning standpoint [1]. The reality of this limit appears to be sinking in. At a recent meeting of concerned industry representatives (including Motorola, AT&T, Northern Telecom, General Electric, Ericsson, Novatel, among others), a flat statement was made: cell-splitting does not work. As an observer, I waited for someone to object, to quibble, to qualify. Around the room, heads nodded silently [2].

As Year 5 of the cellular era opens, there is indeed a gathering sense of disaster at such meetings. There is talk of a repetition of the CB radio collapse, of the imminent danger of "fundamental damage" to a young industry, if quick solutions are not forthcoming. There is a feeling of urgency, that normal technological and regulatory processes must somehow be speeded up, and new technology brought to the marketplace quickly before the Pisan Tower reaches its critical declination.

This foreboding may be exaggerated. I have tried to stress in Part II that the problems of mobile telephony, though serious, are developmental. In other words, they exist within, and because of, a specific technological framework, which can and will be superseded in a reasonable period of time. There is no doubt in my mind that mobile telephony will prosper, ultimately. Nevertheless, there is a palpable urgency that colors and, in some ways, distorts many discussions of the transition to the next generation of technology.

Most new technologies initially face a stiff battle against entrenched alternatives that have had years to hone reliability and reduce costs. When the first digital T-carrier systems were finally brought to market, they enjoyed hardly any cost advantage over existing analog carrier systems. There was considerable skepticism about the feasibility of digital techniques. In the words of the Bell historians, looking back on the difficulty of proving in new systems:

> A difficult cost comparison between an established system, well shaken down and in production, and a new one still in development occurred again and again from 1925 to 1975. It was not easy to prove in a coaxial system over carrier on pairs or early microwave radio over coaxial cable. . . . Management wisdom consisted less in accepting hard cost estimates on somewhat soft concepts than in seeing where a particular technology was headed, and which developments opened avenues unavailable to the existing systems [3].

Normally a new technology like digital cellular radio would enter into the market in an orderly way. It would be, initially, a "hard sell" to customers (cellular operators) who would have to be wooed from sunk investments and comfortable procedures. It would have to prove itself technically, and even then would be accepted only as its costs and capabilities began to show quantifiable advantages over existing techniques. An example of such a transition would be the introduction of digital optical discs into the recording industry, or the coming of high resolution television. But cellular radio is not "normal," as we have seen throughout. The "digital cellular" panacea is being conjured by people who in many cases know little about it, only that they need it, and fast. The danger exists, therefore, that the same mistakes will be made all over again. Instead of an orderly development cycle, we could have crash programs, with the emphasis on the word "crash." Instead of solutions we may have Band-Aids.

The cure for what ails cellular radio is not a single sword-in-the-stone, a technological gimmick or "appliqué" that will in one magic step release the industry from its difficulties. It will come instead from a more thoroughgoing change in industry mindset and "life-style": in particular, the attitude of the industry and its regulators toward technology will have to come in line with the realities of the digital revolution.

For forty years, that is for as long as the business career of anyone in the industry today, mobile telephony has operated in a world where there has been fundamental technological stability in the radio-transmission link. It was built on FM; there were no options to speak of. Competition was based on "features" and upon business competence (in manufacturing, reflected in costs, and in distribution, reflected in service), *not* upon fundamental technological innovation. In such a static technical environment, *uniform standards* were feasible. This produced a comfortable situation for customers, in the sense that they did not have to make difficult choices among competing standards. It made for a comfortable situation for the manufacturers, who could concentrate on honing their operations, rather than on risky R&D efforts. Also, it was comfortable for the regulators, who could write their rules around FM and go home feeling they had done their best.

That is all going to change now. The mobile-radio industry in the year 2000 will quite likely look a lot like the microcomputer business today. There will be a bewildering variety of technological alternatives, formats, standards *(de facto* and *de jure),* high-end and low-end products, products optimized for particular applications, splintering the mobile-communication market into dozens of niches that do not even exist as such today. There will be consumer frustrations about the impossible array of choices — and yet consumers will be making choices in increasing numbers. Some of the manufacturers will perhaps still cast a longing backward glance toward the simpler era of standards, but they will generally be too busy pushing the technological frontiers outward in search of competitive advantages to spend much time in nostalgia. Operators also will have learned to live with, and benefit from, the range of new options, just as computer users and wireline communication users have already learned to do. "Managing the transition," therefore, will mean not merely grafting a particular new technology onto existing cellular systems, but midwifing a whole new set of rules and a new way of thinking about technology and change in the mobile-radio business.

There are four broad aspects of the transition that we shall examine in this chapter:

1. The vexing question of standards;
2. The issue of backwards compatibility with existing analog systems;
3. The issue of forwards compatibility with future technological breakthroughs and enhancements;
4. The question of competition, in its many senses.

16.1 THE QUESTION OF STANDARDS

Many presentations on digital cellular radio begin with a call for a single, uniform standard. Much of the telecommunication business is built on standards; network planners habitually regard standardization as a prerequisite for orderly growth and often experience difficulty even conceiving of operating without standards. Wherever trouble has developed, there is a tendency to suspect inadequate standardization. Even though cellular radio has developed within a comprehensive framework of standards, some are convinced that the degree of standardization has not been thorough enough. The following excerpted dissertation, published in 1986 by Ian Ross, then the president of Bell Laboratories, is typical:

> As I examine the status and the future of cellular technology, I know how a parent feels as he tries to look objectively at a teenage child. It has now been 17 years since AT&T first filed with the Federal Communications Commission for permission to use the cellular technology we at AT&T Bell Laboratories invented. The story of our finally obtaining this permission contains a powerful lesson about how regulation can delay the introduction of new technology . . .

After reviewing some of the difficulties encountered by cellular radio, including the "ultimate, and major, limitation as to how small a cell can be designed," Ross continues to consider future technological solutions:

> What alternatives exist to increase the capacity of cellular systems . . . ? Practically speaking, I see only one viable alternative. For the 1990's, we clearly will need a better, more efficient and robust RF technology, able to accommodate a great number of customers using a limited radio-frequency spectrum. A number of new technologies have been proposed, ranging from the narrow channel on the one hand to spread spectrum and time-division multiple access on the other . . .
>
> Let us consider the major impediment to this progress — namely, the existence today of multiple cellular standards, which differ along various national lines. Today, four major standards are in use, for which equipment is being manufactured. These are the North American, the U.K., the Nordic and the Japan standards. In addition, there is the evolving Pan European standard. . . . What are the consequences of today's multiple standards? Cellular technology is well on its way to presenting users with an intolerable situation of limited connectivity . . .
>
> For the benefit of future generations of customers, we need uniform worldwide ISDN standards for cellular technology . . . We need to identify a common channel, a common transmission rate, and a common ISDN link format between mobile units and base stations . . . Ideally, there should also be a common frequency plan . . . [along with] protocols . . . for the format of digital information . . . [and] call set-up procedures . . . Even as we contemplate options for this next generation of technology, we have the lesson of the current generation and how it has grown and developed. The major "growing pain," of course, has been the proliferation and divergence of standards. We now have the opportunity to avoid this problem and to provide future generations of customers with universal connectivity — all by working together to agree on worldwide uniform cellular standards [4].

Other cellular manufacturers who have recently provided comments on the FCC's recent rule-making proceeding for the liberalization of cellular regulations have supported this view. Many benefits of existing and future standards have been adduced, including:

Enhanced availability of service. "In large part, the widespread availability of cellular service has been attributable to the standardization of technology by the Commission [5]."
Enhanced competition. "Competition will also be protected because if all manufacturers must comply with uniform standards no manufacturer will initially have a monopoly due to the fact that it is the first or only manufacturer using *de facto* standards that may later be adopted *de jure* [6]."
Lower costs. "If compatibility standards were adopted . . . the public would obtain the benefits of competition in the cellular industry without the increased costs resulting from parallel development of systems based on different standards [7]."
Interoperability. "Without standards for equipment and software, this [next-generation cellular system] would break down into a patchwork of small, balkanized networks separated from each other by a lack of interoperability, to the detriment of consumers [8]."

A very different note is sounded, however, by a key Japanese spokesman. Hiroshi Kojima, the director of the Telecommunications Systems Division for the Japanese MPT, the approximate equivalent of the FCC, where he has held chief responsibility for radio communication and broadcasting, takes a very different view of standards:

> Standardization was initially a concept of the 19th Century age of mass production. However, we are now entering an era of customer-oriented standards. To put it in a nutshell, as long as networks can be linked up, they can be utilized according to how individuals wish to use them. This may sound strange, but I think that standards may have to become "destandardized standards," that is to say, standardization will be based on user needs and shifted from single standards to sophisticated standards [9].

A strange idea, indeed! Yet, it points to a truism that is obscured beneath much of the rhetoric: the desire for a single standard is really driven by the manufacturer's goal of lowering production costs. The implicit bargain with the customer is that lower costs will translate into lower prices as competition forces the pass-through of manufacturing savings. (The other main driver for communication standards, namely the issue of interconnection and interoperability, is of more dubious validity. The definition of standard interfaces, which are much less all-encompassing than a total transmission standard, have allowed many sectors of the communication network to function perfectly well under conditions of "balkanization." We shall return to this below, in Section 16.5.)

From the customer's standpoint, other than the purported price break, standardization nearly always involves a sacrifice. "Off the rack" is cheaper than custom tailoring, but the suit never really fits as well. Sometimes the bargain is indeed severe, as when Henry Ford told us we could have any shade of black we wanted.

We no longer have only black Fords to choose from. The trend of product development in almost all modern consumer industries has been strongly *away* from excessive standardization, to win customers by customizing. If we examine the full product line of any large supplier of consumer items, whether automobiles or television sets or refrigerators, we find a bewildering proliferation of options, colors, features, and prices. We may almost become convinced that in many industries the trend toward destandardization may have gone too far. We understand, at least by presumption, that there is a cost that we all bear as a result of demanding that Pontiacs come in so many colors and styles. And yet, as consumers, we appreciate it, expect it, support it. Apparently we are increasingly willing to pay the price for customization.

The thrust of Kojima's comments is that communication networks up until now have been building black Model Ts; good, cheap, reliable, technologically solid, but unexciting. Kojima believes, as do many other observers, that communication users are reaching the point where they are willing to begin to pay for a greater degree of customization. We have enthusiastically welcomed the balkanization of the computer industry; for all that it complicates the decision-making process, computer users are eagerly customizing their own systems today in ways that were unthinkable twenty years ago. The difficulties of interfacing different formats, different hardware and software, are outweighed by the benefits of more precisely tailoring computer systems to users' needs.

The communication industry comes late to this kind of thinking. Even in the wireline network, there is still a strong tendency to look to central standards groups, populated by operators' and manufacturers' representatives, rather than to the end users, to determine communication standards. The idea of a "users group" — routine in the development and management of computer systems — is unheard-of in communication planning. In the current round of standards planning for digital cellular, the chief actors will be the FCC, the trade association of electronic manufacturers, and the association of cellular operators. Nowhere in the process are the actual users of cellular radio formally represented.

Another part of the price of standards is the potential for stunting technological progress. Standards are by definition anti-innovative. When the Model T runs well enough, this may be "acceptable" for a period of time, although eventually obsolescence overtook Henry Ford with a vengeance. When there is a fundamental inadequacy in the technological base, however, as is evidently the case with analog FM transmission in today's cellular radio, the standard acts as a serious impediment to the development of needed technological solutions.

Seizing upon this viewpoint, the FCC has proposed in its recent rulemaking proceeding to eliminate technical standards, at least as far as taking the Commission out of the role of a standard-setting body.

> Although we believe it is in the public interest to encourage the development of new, spectrum-efficient cellular technologies, inasmuch as many technologies are in various stages of development, we believe it would be inappropriate at this time to embark on a proceeding to select technical standards for future cellular systems. Such a course would be premature given the early stage of development of new cellular technologies and is likely to discourage technical innovations due to the uncertainties of the rule-making process. Instead, we believe that the appropriate means to accomplish this is to provide technical freedom in the rules in order to enable the introduction of new technologies. This would foster the development of competing technologies which could then be evaluated in the market. We wish to emphasize however that the development of new technologies need not lead to a lack of standardization over the longer term. We believe that a particular technology, or combination of technologies, may ultimately gain widespread marketplace acceptance and serve as a new standard for the next generation of cellular systems [10].

I believe that the FCC is correct: the imposition of a single technical standard has been, and would continue to be, an obstacle to the development of technological solutions to the problems with which the bulk of this book is concerned. I believe that most of the objections raised against the idea of operating without centrally mandated standards are spurious. The idea that cellular radio's problems are the result of too little standardization is simply incorrect. In fact, the United States has had a single standard for the current generation of cellular radio, covering all the aspects of the transmission system identified in Ross's analysis. The operational, technical, and economic problems of today's cellular systems have nothing to do with the fact that Sweden or Japan use slightly different formats. If the whole world adhered to the AMPS standard, it would not improve the technical or economic performance of the system in Los Angeles. Moreover, the argument made by some manufacturers that multiple standards will raise the manufacturers' costs unbearably is contradicted by their own participation in different markets, conforming to different standards.

Ericsson, the source of the comments quoted from [7] above, is probably the most international of all systems suppliers. As they indicated in their recent comments before the FCC:

> The Ericsson Group has supplied cellular systems using a variety of different standards. For example, the Ericsson Group's NMT cellular system is in operation in Denmark, Finland, Sweden, Norway, The Netherlands, Malaysia, Saudi Arabia, Thailand, and seven other countries. The NMT system has three different configurations. With the exception of the system in operation in The Netherlands, the NMT-450 system operates in the 420–480 MHz frequency band. It has 180 channels using 25 kHz spacing. In The Netherlands the NMT-450 system has 220 channels using 25 kHz spacing. There is also an NMT-900 system in Switzerland and the Nordic countries which operates in the 860–960 MHz band. It has 1999 channels and 25 kHz spacing. The Ericsson Group has supplied the United Kingdom, Ireland, and the Peoples Republic of China with the TACS cellular system [which] operates in the 872–950 MHz band. It has 1320 channels using 25 kHz spacing. [Ericsson] has supplied Australia, New Zealand, 29 U.S. markets, and 16 Canadian markets with the EIA or AMPS-type cellular system [which] operates in the 824–894 MHz frequency band. It has 832 channels using 30 kHz spacing. The Ericsson Group's breadth of experience as one of the largest suppliers of cellular systems in the world using a variety of different technical standards makes Ericsson uniquely qualified to comment in the proceeding [11].

Indeed, by the same logic one could argue for a single, uniform, worldwide standard for the design of telephone switches, a sinkhole of R&D funds far larger than cellular radio. Competitive research and development programs are a cost of doing business in a free-market economy.

Yet the question of whether a single standard for the next generation of cellular radio would discourage innovation or not is probably a moot point. It is *extremely* unlikely, in my opinion, that the United States will be able to develop such a standard as more than an interim measure, if at all. Can the industry agree on a standard? Could the FCC, or anyone else, enforce one? The question is not *should they?*, but, given the political system under which we operate and the tremendous diversity of digital technology, *will they?*

The key to the problem is the Three-to-Five-Year Paradox. The rate of generational technological change today in most of the relevant areas of digital technology, especially coding and modulation design, is about three to five years. Unlike FM transmission, which appeared both technologically stable and against which there were really no competing options in the early 1970s, digital transmission systems are permuted by the dozens. There are probably at least five or six strong contenders in the coding realm. The first half of the paradox holds that in five years time the performance characteristics of today's coders will have been superseded by a factor of two or more. The other half of the paradox is the observation that it would take at least from three to five years to develop, complete, evaluate, specify, and implement a standard for voice coding. The so-called standard would therefore be seriously obsolete by the time it reached the market. There are many other areas of digital signal processing where similar timing probably holds, including error correction and channel coding, adaptive equalization, diversity techniques, and modulation schemes.

Moreover, there is a great diversity of digital processing schemes. Each manufacturer has its own proposals to make for coding, equalization algorithms, forward error correction, *et cetera.* How are these to be evaluated? Manufacturer A, having invested \$10 million, or \$100 million, in a particular approach, is unlikely to acquiesce to an adverse decision, especially one that might not only fail to recoup any return on that development investment but which could also even involve the payment of royalties to a competitor? Where is the Supreme Court of Technology, especially given that the FCC has specifically, and properly, abdicated this responsibility as beyond its capabilities?

The value of *de jure* standards becomes even more dubious when it is recognized that all of the important alleged benefits of *technology standards* can be realized without setting any technical restrictions at all. The cellular market can remain open to new technology *if* certain *service standards* are imposed (see Section 16.5).

It is thus likely that the cellular industry will have to cope with a standardless world, whether it wants to or not! In fact, it will soon become apparent — once we have allowed ourselves to think the unthinkable thought — that it is precisely in the direction of "destandardized standards" that the salvation of the industry lies.

16.2 LOOKING BACKWARD: COMPATIBILITY OF THE NEXT GENERATION WITH THE INSTALLED ANALOG BASE

The standards question is also moot for an entirely different reason. Whatever new technology is implemented, it will almost certainly be incompatible with existing analog transmission standards. Assuming that service to existing customers will be maintained, the next generation of cellular systems will become *ipso facto* dual-standard systems. Under most scenarios, the system controller would have to distinguish between call requests from analog *versus* digital subscribers and handle the calls accordingly. In fact, once the system has been set up to handle and interface two incompatible groups of subscribers, the difficulty of handling three groups, or five, or eight, becomes much less forbidding in technical terms. It reduces to a question of economics — which is as it should be.

The plan for grafting multiple incompatible transmission technologies (at least two) into the same system raises serious questions for the operators, and for existing analog cellular customers. Both groups have a substantial unrecovered investment in a technology that is by definition obsolescent. Certainly the FCC, as the guardian of the public interest, has no desire to see the analog customer base abandoned or shortchanged in the rush to implement new digital technologies. Yet there is a possibility, even a likelihood, that this will happen.

There are essentially two approaches to marrying digital and analog generations: partitioned systems and nonpartitioned systems. In a partitioned system, a certain number of the total cellular channels that are currently allocated 100% to analog customers will be taken from the general pool and assigned for the exclusive use of the new digital customers. The partitioning means that in effect there are two distinct subscriber populations, which cannot share channels and which, therefore, exhibit two completely independent trunking efficiencies. Traffic calculations and grade of service for the two populations are independent. In a nonpartitioned system, all customers — digital and analog — would be equally able to access all channels; for traffic-engineering purposes, both digital and analog customers constitute a single population.

Most current proposals involve partitioned systems, probably because these are more straightforward and possibly somewhat less expensive for the operator, although this is questionable, as we shall discuss below. A typical plan, embodied in a recent paper from Bell Laboratories, calls for the gradual elimination of analog customers over a six-year period following the introduction of digital technology, in this case, a 10-kHz FDMA scheme. It is not entirely clear where the FM subscribers disappear to; nevertheless, the author points out that:

> The rate of decrease in 30 kHz radios is initially steep so that enough spectrum can be freed up to equip enough 10 kHz radios in the system to serve the new demand at very low blocking levels. . . . Blocking initially increases sharply for FM channel users while it is virtually 0 for new digital channel users [12].

In other words, the effect of introducing digital technology in a partitioned plan will be to substantially *worsen* the crowding of the existing subscribers by taking away some portion of the spectrum they already have. The operators will prosper because the cap on subscriber growth will be relieved. The new digital subscribers will enjoy excellent service on their freed-up channels. But the plight of today's analog customers — in whose name these changes are being sought — will intensify. There comes to mind an image of an East African waterhole in the dry season, crowded with hippopotamuses and catfish and other aquatic species, slowly shrinking under the sun, while the farmers drain even more water off to support their cash crops in nearby fields.

How many channels would be diverted for the new digital customers? Surprisingly, whereas the cellular manufacturers are unwilling as a rule to trust to the marketplace on issues of technology, they almost universally argue that the decision of how deeply to drain the analog pool should be left entirely to the operators' discretion, presumably so as not to cap their abilities to purchase digital equipment. Ericsson's comments to the Commission on this issue are typical:

> For example, if a cellular licensee determines that 20% of its traffic utilizes conventional analog service, *initially* it might want to allocate 20% of its capacity for conventional analog service. On the other hand, if a cellular licensee wanted to emphasize the provision of enhanced services to its subscribers at an earlier point in time, it might want to allocate more [!] of its capacity for advanced technology cellular service. This is clearly an area where the Commission should allow the marketplace to determine the speed with which such advanced technology cellular services should be introduced [13]. *(Emphasis added.)*

It may be that the waterhole will be very small indeed. The AT&T comments offer a rationale for service standards set by the marketplace:

> The [FCC] seeks guidance as to whether the rules should prescribe that a minimum percentage of cellular system capacity be set aside for use with conventional cellular mobile units. Here again, marketplace forces can best answer the question. The system provider has a strong incentive to provide quality service, to promote growth of his own service and to avoid erosion of market share to his competitor. If the service provider too quickly obsoletes his customers' equipment, they can migrate to his competitor . . . [14]

Let the market take care of it — the logic is fine, unless the profit potential of the new digital systems is different from that of today's analog systems. What if an operator can make more money serving four subscribers on a digital channel, even at lower usage charges, than he makes serving one customer with the same spectrum on an analog channel? What if the same economics hold for *both* operators in a given market? Will this invoke some sort of Gresham's Law, with profitable customers driving out the less profitable ones?

In my view, it is *very* likely that newer technologies will be substantially more profitable, will earn more per 30-kHz bandwidth, than today's analog systems. We must remember that in a partitioned plan each operator is effectively managing two independent businesses, each of which draws upon a common resource, the spectrum. Spectrum allocation to one group or the other becomes, like the allocation of capital, an investment decision. If analog customers are less profitable because analog costs are higher, we could find a distinct lack of interest on the part of the operators in service standards for current users. To make way for the digital generation, the analog customers might be dropped like so many bad pennies.

The superior alternative is to enforce nonpartitioned systems, where *all* subscribers, digital and analog, can access *all* channels at *all* times. To do so, the cell-site channel elements must be capable of operating in either mode, under system control, depending upon which type of customer happens to be the next in line. Trunking efficiencies are maintained, and there is virtually no discrimination in service standards between analog and digital customers. Analog customers are not automatically relegated to a second class of service. As the comments above indicate, however, this suggests that the FCC must be prepared to take an active role in promulgating and enforcing service standards, at least to the extent of nondiscriminatory or "equal" access in terms of grade of service. Note that this would also satisfy the concern about roaming that has been raised by others as a reason for seeking uniform transmission standards.

16.3 LOOKING FORWARD: OPENNESS TO FUTURE TECHNOLOGICAL ADVANCES

The Age of FM described in Part II was a long period of technological stability. Now the Digital Pleistocene has arrived, and the climate is changing. The technological stasis is dissolving. In 1974, for example, about the time that the cellular standards were being set, a standard text on mobile radio could marvel that voice-coding methods with "bit rates as low as 20 kb/s . . . have been reported that were still able to provide intelligible speech signals [15]." Today the threshhold of intelligibility *per se* is probably under 2 kb/s. Economical implementations of telephone-quality speech at 16 kb/s are becoming routine. Within two or three years, 8-kb/s telephone-quality speech in an economically implementable configuration is a virtual certainty. The same ferment is evident in all the other areas of digital signal processing, including modulation, equalization, error correction. These are all increasingly "algorithm-based" technologies, which benefit from the tremendous plasticity of software systems generally.

This is not going to end any time soon. Many in the mobile-telephone industry seem to harbor the vague hope that after a difficult period of "transition to digital," the technological stasis will be reestablished, and the old procedures, ways of thinking, business and regulatory strategies can be reinstated. The exact opposite will happen: the rate of technological change in the mobile-radio world is going to *accelerate* over the next ten to twenty years. There will be *more* options five years from now, not fewer: more competing coders, more interesting modulation schemes, more combinations optimizing different dimensions of mobile performance. This is going to make it very difficult for would-be standardizers. The problem of evaluating an increasing variety of alternatives is difficult; the problem of finding a standard that will not be obsolete by the time it is developed is almost impossible in today's environment.

We should be clear that technological obsolescence is not a term to be used lightly. All standards involve a certain amount of obsolescence; American television standards, established earlier than those in Europe, reflect a slightly inferior level of technical development. Yet the benefits of standardization are such that few in the United States have ever seriously considered changing over to the European format. With cellular radio and the projectable breakthroughs in digital radio technology in the next ten to fifteen years, however, we are dealing with a much more severe obsolescence. Consider the question of capacity alone. If we take today's system capacity as a baseline, it is likely that within from three to five years it will be possible to increase capacity by a factor of three to ten. In the *following* three to five years, the mid- to late-1990s, it is likely that system capacity could be more than doubled once again. In fact, based on current technology development efforts, I believe that *at least three or four doublings of system capacity lie beyond the immediate next generation* as a direct result of continuing improvements in digital signal processing. The doubling principle — "twice as much" or "half as much" every three to five years — will probably apply to all other relevant system parameters as well, such as cost. Even this statement may be conservative. It may be that system performance will continue to improve and advance along *all* dimensions at a rate such that the doubling rule will become a quasi-permanent aspect of the business.

Once again, some sense of this long-term technological turmoil may be gathered from a sideways glance at the computer industry. The evolution in computer hardware and software capabilities over the past twenty-five years displays a similar dual-trend of (1) accelerating rate of technical change and (2) increasing variety of systems and vendors. This has been unsettling for manufacturers, as even so dominant a market force as IBM has seen its ability to dominate emerging new niches undermined by the explosive pace of technical change. It is unsettling for customers who are often simply bewildered by the variety of apples and oranges and pomegranates that are on the market today and are always fearful that what they purchase today will be outdated by next year. It is unsettling for engineers who find that their own technical knowledge often seems to have a half-life of only a few years, and who are constantly being pushed by the younger generations of college graduates fresh from the cutting edge. And yet we have all learned to live with it; the computer industry is probably one of the healthiest and most innovative in the United States today.

The architecture for the next generation of mobile radio must be cognizant of these trends. In place of the traditional monolithic standards, we will have to develop a structure capable of accommodating a much more heterogeneous set of transmission technologies. For example, rather than asking which voice coder is "best," recognizing that "best" is both subjective-political and technically transitory, we should ask how can we develop an architecture that can accomodate many different voice coders, perhaps operating at different bit rates, representing different technical generations — all at the same time?

The broad question should be: What is the *minimum* that must be standardized? Carrier frequency? Channel bandwidth? Control channel identification and protocols for system access and call setup? Is it possible to do without a standard transmission link? Is it conceivable that many different modulation techniques and different voice coders could be accommodated within a single interoperable system?

There is reason for considerable optimism on this score. It derives from the very digital technology that has upset the applecart in the first place. There is a third general trend that characterizes the digital revolution: the decentralization of system control and the growing intelligence in all nodes, particularly at the periphery.

The era of FM is the era of "Dumb Radio": the individual mobile units are not adaptive. They are capable of operating only in a single fixed mode, with a fixed bandwidth and transmission format. They are not capable of adapting to the vagaries of the transmission channel, compensating for transmission errors. There is really not very much that an FM transmitter can do in the face of noise or interference other than to crank up the power. The decision-making in the system is almost entirely concentrated at the MTSO.

In keeping with the parallel drawn above between digital mobile-radio and digital computer systems, the trend toward distributed or decentralized processing will become much more marked. As late as 1979, in the official Bell system description of the AMPS architecture, the authors still harbored doubts about how much intelligence would have to be incorporated into the analog FM cellular mobile unit itself: "its design will *most likely* incorporate a microprocessor [16]." *(Emphasis added.)* By contrast, the next-generation subscriber unit will become essentially a special-purpose computer, with a variety of powerful processing elements to perform a wide range of signal-processing functions. We will have entered the era of "Smart Radio." One important result of the incorporation of this processing power directly into the mobile unit will be the opportunity for performing different format translations and even emulations right at the mobile itself. In other words, it will become possible for a mobile unit to operate in different modes — without wholesale duplication of hardware — under software control. And much as new capabilities are added to microcomputers today either by means of software upgrades loaded from ROMs or floppy discs, or through expansion boards that plug into preestablished communication slots, so the upgrade of future digital mobile subscriber units will enjoy a similar ease and economics.

16.4 COMPETITION

Unquestionably, the most significant nontechnological trend in the world of telecommunication in the past twenty years has been the astonishing victory of the idea of "competition" as a policy solution for a wide range of ills, from poor service and high costs to the slow rate of technological progress.

In the late 1960s, telecommunication in the United States was governed under a monopoly regime, subject to strict economic regulation by the state and federal authorities. Outside the United States, most other countries were even more committed to the public-utility concept of the communication industry; in most, the communication monopoly was government owned and operated, more or less along the lines of the Post Office.

During the 1970s, stimulated in part by the "successes" of deregulation in other industries from trucking to banking, telecommunication policymakers began to experiment with similar measures. They received added impetus from the long-standing anti-trust interests of the Justice Department, which — for the third time in sixty years — resulted in a major legal action against the Bell System. The procompetitive rulings of the courts, coupled with the incredible decision by AT&T to sever the Bell operating companies in its settlement with the Justice Department, effected a revolution in the institutional framework of the industry and gave a terrific boost to proponents of the new thinking.

By the late 1980s, the balm of "competition" was being lathered on all sorts of problems by regulators and policymakers. The enthusiasm had spread, incredibly, beyond the United States to telecommunication planners in many parts of the world. "Privatization," "deregulation," and "competition" were being proposed, discussed, and even implemented in places as diverse and unlikely as Britain, Sweden, Argentina, Japan, and Malaysia.

The radio communication world in one sense did not need to go through this revolution: competition had always been vigorous in most branches of mobile communication (until cellular radio). Once started, however, the juggernaut was hard to stop. The hazy principle of competition became a touchstone for a series of even more radical proposals. Some, such as the lottery method of licensing, which was designed to protect the competitive rights of the "little guy" to participate in the cellular bonanza, were actually implemented. Others, including such ideas as "flexible allocations" and even "spectrum farming," involved liberalizing the service restrictions on the use of the spectrum, even to the point of allowing licensees to decide fundamental channel and transmission parameters. Perhaps, in the future, there might not even be a distinction between "broadcast spectrum" and "mobile-radio spectrum"; blocks of electromagnetic real estate could be sold to the highest bidder, to do with what he will. For the conventional players in the affected industries, such notions are, to put it mildly, unsettling.

The strange thing about the romance with these new ideas is that there is still reluctance simply to *allow* "competition." Instead, it is felt to be necessary to *create* "competition" through all sorts of intricate procedures. Evidently there is still a concern that unbridled market forces may not actually constitute an adequate allocation mechanism for communication services. Also, since the definition of "competition" is still fuzzy, and the results expected from it are often unstated, the Rube Goldberg contrivances that are designed to "enhance competition," e.g., the cellular lottery, or the Byzantine trails of "equal-access" policy, often produce bizarre outcomes that may not clearly increase the level of "competition" in the industry. The cellular lottery, for example, which was designed to even out the odds against small competitors, has had the net end result of driving out many major nonwireline competitors and consolidating the hold of the wireline companies, especially the Bell operating companies, on the cellular industry, since the small players who won the licenses were generally most interested in selling the license to a major player.

How will the transition to a new generation of digital cellular technology affect, or be affected by, the trend toward competition? More specifically, what would be the impact of a single, uniform standard for the next-generation systems upon the level of competitiveness in the cellular industry?

Without pretending to a deep analysis of a classic issue in economic theory, we may observe that competition is generally viewed as having two beneficial effects:

1. Lower costs for equipment and services;
2. Increased technical innovation, as producers attempt to gain a competitive advantage based upon proprietary techniques.

The first result is readily visible in the trends in mobile-equipment prices in today's cellular markets. It has almost certainly been facilitated by the existence of uniform, or nearly uniform, technology standards. By making the original AMPS technology widely available, the standards process enabled many manufacturers to enter the mobile-equipment market without undergoing massive R&D programs. It also tended to concentrate the manufacturing of components in very large production runs that allowed a high degree of cost-reduction and production optimization. As a result, the cost of the typical low-end mobile unit has fallen approximately by a factor of three in the first four or five years.

The second desired outcome of competition, however, the stimulus it can provide to technological innovation, has been almost entirely lacking in the cellular world since the promulgation of the FM standards in the early 1970s. Here the existence of monolithic standards has acted as an obstacle to technical progress.

Standards tend to create a peculiar, debased form of competition that I have labeled *commoditization* (see Chapter 4). The process of commoditization has taken place in a number of telecommunication industries, including radio and television receivers, cordless telephones, CB radios, and, now, cellular radios. It is marked by a number of characteristics, including:

1. Virtually complete substitutability of different producers' equipment;
2. No proprietary technological base;
3. The preeminence of price competition, leading to migration of production to low-cost regions, especially the Far East, and to problems of dumping, inventory sell-offs, deep retail discounts;
4. The rise of pseudocompetition on the basis of minor cosmetic "features" that have no real functional significance;
5. The de-emphasis of new R&D, and the shift in overall competitive advantage from the technology-rich firms that tend to develop the

original innovations that create the industry to the technology-poor, low-cost manufacturers who can exploit lower foreign labor rates and the lack of development engineering overhead to capture market share.

Commoditization is a destroyer of industries. It undermines the incentive to innovate and places domestic producers at a disadvantage from which they can seldom recover (except occasionally by emulating their foreign competition and shipping all production offshore). It condemns the end user to obsolescence, while lulling him with low prices for technologically inferior products, analog cellular radio being a good example.

Overzealous standardization begins the process of erosion by taking away most of the incentive to continue technological investments. Some observers have viewed *all* proprietary technology — where one firm may develop qualitatively superior technology which it controls through patents — as anticompetitive. The patent effectively confers a monopoly upon its holder which allows that company to charge higher prices for its products. Therefore, it is reasoned, standards must be based on open technology, to reduce costs to the end user.

I believe this is very shortsighted. First, patents are very rarely absolutely controlling for a single company over long periods. The patent *per se* is only the tip of the iceberg: the unseen advantage of technology investment is the know-how and the lead time gained over competition in bringing new products to the market. It is the process of technology investment that must be maintained and fostered. Patents may create higher short-term prices, but those higher prices are what supports the technology investment process, and what gives corporate managers incentives to continue to search for breakthroughs. In the longer term — across technological generations — they result in lower costs as new R&D continues to receive support and the industry experiences repeated qualitative leaps forward.

Technology investment is the goose that lays the golden eggs. Again, we have only to consider the trends in costs and technical innovation in an industry such as the computer business, where standards have not stifled technology investment. Industries like computers or pharmaceuticals, where technology investment is paramount, have tended to remain strong, competitive, innovative, and have not been crushed by foreign competition. Even the automotive industry, where technology investment is a significant factor in market success, has maintained a reasonable balance of domestic and foreign production. By comparison, the cellular-radio business seems headed down the same path as the television-receiver business, with the same ravaging of domestic producers and the same loss of industry initiative to low-cost foreign manufacturers. Perhaps it is an exaggeration to attribute this trend entirely to the existence, or nonexistence, of uniform technical standards, but it is clear that standards play a very important part in creating the type of business environment where commoditization can take place.

16.5 A MODEST PROPOSAL

Finally, I would like to offer a brief proposal, aimed particularly at the regulatory community and the current system operators, for "managing the transition" to the next generation. The proposal is undoubtedly controversial. It goes against the grain of many traditional attitudes in the telecommunication industry. After sixteen chapters, however, I shall not attempt any further justifications. What follows is offered on a "For what it's worth, here's what one observer has concluded after a great deal of thought and analysis" basis.

1. Neither the FCC nor the industry should establish monolithic technology standards for the transmission methods to be used. In particular, the radio link — encompassing the voice-coding and modulation schemes — should be left as open as possible. I agree with the FCC's position stated previously as regards the role of standards in the market.
2. The industry, but not the FCC, may consider the establishment of *interface* standards, particularly at the cell-site interface to the switching network. This would facilitate the upgradability of cellular systems to accept new radio-link technologies and would allow different switches to operate with different transmission systems, supplied by different vendors, much as is done routinely today in the wireline network. (Ironically, this is one area where standardization does *not* exist in the cellular world today, which will greatly complicate the generational transition.)
3. The industry should consider whether to mandate the retention of (1) existing channelization schemes, or (2) compatible channelization schemes, as this will greatly facilitate interoperability. In particular, it would seem that the retention of existing control channels is probably desirable, to make sure that various generations of the cellular system can operate on a nonpartitioned basis.
4. The cellular systems should be capable of operating in a nonpartitioned mode; that is, all cell-site channels should be capable of operating in either FM or digital formats, and should be capable of servicing any cellular user.

5. The control architecture should be capable of recognizing many different user formats; the format translation should take place at the cell-site, or, conceivably, in some cases, at the mobile unit, rather than in the radio link.
6. The capability for analog FM operations should be preserved indefinitely as a backstop for roamers and for small system operators who may not upgrade to new technologies as rapidly as larger systems. This may be the best medium-term approach to retaining a nationwide roaming capability.
7. The FCC *should* regulate service standards. In particular, the FCC can utilize service standards to drive technological improvements, especially by reinforcing the move toward *telephony* service standards (in terms of blocking and traffic). The FCC must also use service standards to prevent discrimination against existing customers and to prevent the exercise of Gresham's Law–type logic from driving analog customers into a second-class of service. (The FCC should be prepared to allow for discriminatory pricing, however, so long as it reflects underlying cost-of-service differentials.)

REFERENCES

[1] Ian Ross, "Uniform Standards are Required to Ensure Global Cellular Growth," *Mobile Radio Technology,* May 1986, pp. 26–30.

[2] EIA Conference on Digital Cellular Standards, Washington, D. C., December 4, 1987.

[3] E. F. O'Neill, ed., *A History of Engineering and Science in the Bell System: Transmission Technology (1925–1975),* AT&T Bell Laboratories, 1985, p. 560.

[4] Ross, *op. cit.*

[5] Comments of General Electric Mobile Communications Business in the matter of FCC Docket 87–390, January 15, 1988, p. 4.

[6] Comments of Ericsson North America, Inc., in the matter of FCC Docket 87–390, January 15, 1988, p. 11.

[7] *Ibid.,* p. 11.

[8] Comments of General Electric, *op. cit.,* pp. 5–6.

[9] Hiroshi Kojima, " 'De-Standardization' Proposed for 21st Century," *The Telecom Tribune,* August 1987, p. 2.

[10] Notice of Proposed Rule-Making, FCC Docket 87–390, 1987, Paragraph 11.

[11] Comments of Ericsson, *op. cit.,* p. 3.

[12] Samuel W. Halpern, "Introduction of Digital Narrowband Channel Technology into the Existing Cellular Spectrum in the United States," *Proceedings of the 37th IEEE Vehicular Technology Conference,* Tampa, June 1–3, 1987, p. 149.
[13] Comments of Ericsson, *op. cit.,* p. 17.
[14] Comments of the American Telephone & Telegraph Company in the matter of FCC Docket 87–390, January 15, 1988, p. 7.
[15] William C. Jakes, ed., *Microwave Mobile Communications,* New York: John Wiley and Sons, 1974, p. 219.
[16] Z. C. Fluhr and P. T. Porter, "Advanced Mobile Phone Service: Control Architecture," *Bell System Technical Journal,* Vol. 58, No. 1, January 1979, p. 45.

Index